Progress in Colloid and Polymer Science · Volume 119 · 2002

W0055392

Springer-Verlag Berlin Heidelberg GmbH

Progress in Colloid and Polymer Science

Editors: F. Kremer, Leipzig and G. Lagaly, Kiel

Volume 119 · 2002

Analytical Ultracentrifugation VI

Volume Editors:

W. Borchard and A. Straatmann

Springer

The series Progress in Colloid and Polymer Science is also available electronically (ISSN 1437-8027)

- Access to tables of contents and abstracts is *free* for everybody.
- Scientists affiliated with departments/institutes subscribing to Progress in Colloid and Polymer Science
  as a whole also have full access to all papers in PDF form. Point your librarian to the LINK access registration form
  at http://link.springer.de/series/pcps/reg-form.htm

ISSN 0340-255X
ISBN 978-3-662-14641-5
ISBN 978-3-540-44672-9 (eBook)
DOI 10.1007/978-3-540-44672-9

Springer-Verlag a member of BertelsmannSpringer Science + Business Media GmbH
http://www.springer.de
© Springer-Verlag
Berlin Heidelberg 2002
Originally published by Springer-Verlag
Berlin Heidelberg New York in 2002
Softcover reprint of the hardcover 1st edition 2002

Typesetting: SPS, Madras, India

Cover Design: Estudio Calamar,
F. Steinen-Broo, Pau/Girona, Spain
Cover Production: design & production, D-69121 Heidelberg

SPIN: 10835 839

Printed on acid-free paper

Progr Colloid Polym Sci (2002) 119: V
© Springer-Verlag 2002

PREFACE

Since 1978 symposia on Analytical Ultracentrifugation have been performed in Germany, where now the 12th meeting has taken place in Duisburg from March 1st to 2nd 2001. As in the years before various fields of ultracentrifugation were covered concerning general theory, research problems in biochemistry, biophysical chemistry, macromolecular chemistry, size distributions of particles interacting systems, hydrodynamics and applications to crosslinked swollen systems. In addition a new evaluating software was presented. The development of the ultracentrifugal technique in the last years is mainly characterized by an enhanced treatment of online available data from optical detection systems and also the development of a new analytical ultracentrifuge of Beckman equipped with maintenance-free drives.

Twenty contributions of the recent conference have been selected for publication. They are arranged as can be taken from the contents corresponding mainly the groupings of posters and lectures during the meeting.

The 12th international symposium on Analytical Ultracentrifugation was generously sponsored by BASF (Ludwigshafen), Bayer AG (Leverkusen), Degussa AG Coatings and Colorants (Marl) and Beckman Coulter GmbH (München). Thus it was possible to support colleagues from Eastern countries of Europe to cover a small part of their travel expenses. We express our thanks to the companies. Especially we reward the numerous unnamed reviewers for helping to improve the contributions of this special volume of Progress in Colloid and Polymer Science.

W. Borchard
A. Straatmann

Progr Colloid Polym Sci (2002) 119: VI–VII
© Springer-Verlag 2002

CONTENTS

Progr Colloid Polym Sci (2002) 119:1–10
© Springer-Verlag 2002

W. Mächtle
M. D. Lechner

Evaluation of equilibrium and nonequilibrium density gradients in an analytical ultracentrifuge by calibration with marker particles

W. Mächtle (✉)
Kunststofflaboratorium,
Polymer Physics, BASF AG
67056 Ludwigshafen, Germany
e-mail: walter.maechtle@basf-ag.de
Tel.: +49-621-6048176
Fax: +49-621-6092281

M. D. Lechner
Physical Chemistry
University Osnabrueck, Barbarastrasse 7
49069 Osnabrueck, Germany

Abstract Density gradient measurements inside an analytical ultracentrifuge (AUC) are an excellent tool for characterizing nanoparticles in the 10–1000-nm diameter range. Because of its very high resolution (i.e. its fractionation power according to the particle density and its high precision) it is possible to analyze the chemical nature of nanoparticles, especially of complex colloidal mixtures. This means that AUC density gradient measurements are a kind of particle density spectroscopy. Usually, the relation between the radial position and particle density, $\rho(r)$, inside an AUC density gradient is calculated by using the well-known (barometrical) equilibrium equation in the formulation of Hermans and Ende (1963); however, this equation for an ideal bimodal density gradient mixture fails in some cases. The higher the content of the heavy component, and the bigger the difference between the actual density gradient being formed and the equilibrium gradient, the bigger the failure or the deviation from ideality. We report a systematic study of these deviations using a marker nanoparticle system of 11 precisely characterized ethylhexyl acrylate/methyl acrylate copolymer latices with known nearly equidistant particle densities. During this study we also learned to use this 11-marker system as a pragmatic and simple calibration system for aqueous density gradients, thereby reducing considerably the error of the measurements in absolute nanoparticle densities. Some application examples are presented. One advantage of the new calibration technique is that higher particle densities are now accessible. Another advantage is the reduction in the measuring time. We no longer have to wait till equilibrium is reached (sometimes up to 90 h!); instead already after 9 h we get reasonable results. This means our static density gradient now approaches a "dynamic" one.

Key words Analytical ultracentrifuge · Density gradient · Particle density · Colloidal nanoparticles · Polymer characterization

Introduction

Density gradient measurements inside an analytical ultracentrifuge (AUC) are an excellent tool for characterizing nanoparticles in the 10–1000-nm diameter range [1–8]. Because of its very high resolution (i.e. its fractionation power according to the particle density and its high precision) it is possible to analyze the chemical nature of nanoparticles, especially of complex colloidal mixtures, and to study reactions on particle surfaces.

In principle, a density gradient inside an AUC measuring cell is built up by spinning at high rotor speed a mixture of a light main component (mostly water) and a heavy dissolved component (often a sugar, like metriza-mide) till equilibrium is reached. Small amounts of nanoparticles dispersed in this mixture will sediment or float during this buildup time to a specific radial position inside the cell, where the particle density and the local gradient density are identical. Mixtures of chemically different nanoparticles are fractionated in this way and appear at different radius/density gradient positions. This is a kind of particle density spectroscopy. This statement is illustrated in Fig. 1. In the upper part of Fig. 1 the most common industrially important homopolymer latices are arranged along a density ρ-axis, from 0.9 to 1.5 g/cm^3, according to their particle density, ρ, from polybutadiene to poly(ethylhexyl acrylate) (PEHA), polystyrene (PS), poly(methyl acrylate) (PMA), poly(vinyl chloride) to poly(acrylic acid). By copolymerization, a huge number of copolymer latices with particle densities in-between are producible.

Fractionation of a mixture of such latices, at high resolution, according to their particle density, could be described as "particle density spectroscopy". This is exactly what is possible with AUC density gradients. It is demonstrated in the lower part of Fig. 1, that a ρ range of 0.85–1.4 g/cm^3 is really accessible by AUC water/metrizamide density gradients, not with just one gradient, but with a set of nine gradients with different metrizamide contents from 5 to 25 mass%. For low particle densities one adds methanol to water in order to lower the density of the mixture; for high densities one replaces normal water by D$_2$O.

Usually, the relation between the radial position and the particle density, $\rho(r)$, inside an AUC density gradient is calculated by using the well-known (barometrical) equilibrium equation in the formulation of Hermans and Ende [9]; however, this equation for an ideal bimodal density gradient mixture fails in some cases. The higher the content of the heavy component, and the bigger the difference between the actual density gradient being formed and the equilibrium gradient, the bigger the failure or the deviation from ideality. Deviations also result if one changes from a bimodal to a trimodal density gradient mixture such as water/methanol/metrizamide.

We report here a systematic study of these deviations using a marker nanoparticle system of 11 precisely characterized ethylhexyl acrylate (EHA)/methyl acrylate (MA) copolymer latices with known nearly equidistant particle densities of $\rho = 0.980/1.000/1.021/1.043/1.066/1.089/1.114/1.140/1.167/1.196/1.225$ g/cm^3. During this study we also learned to use this 11-marker system as a pragmatic and simple calibration system for aqueous density gradients, thereby reducing considerably the error of the measurements in absolute nanoparticle densities and the measuring time. We also describe this new marker calibration technique and present some application examples.

Theory of AUC density gradients

The literature exhibits several possibilities for the determination of the density gradient in an AUC, i.e. the

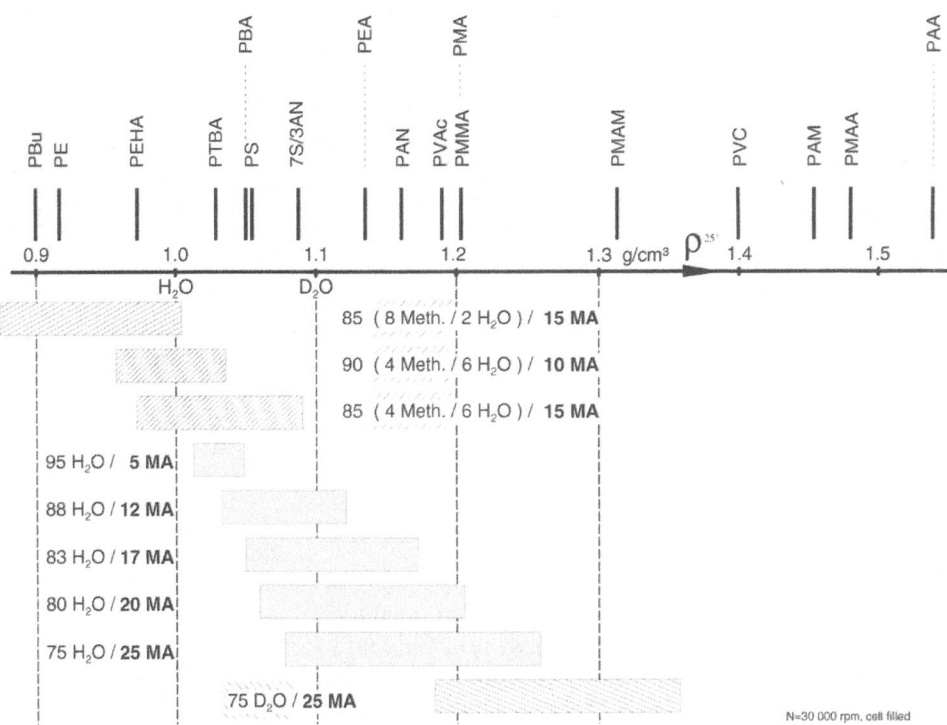

Fig. 1 Particle densities of the most common homopolymer latices arranged along a density ρ-axis (*upper part*) and accessible density ranges of nine differently composed static equilibrium water/metrizamide density gradients in an analytical ultracentrifuge (*AUC*) (*lower part*)

radial dependence of the solution density inside the AUC measuring cell (ρ–r relation). One of the oldest methods is the theoretical calculation of the equilibrium density profile by Hermans and Ende [9]; however, the Hermans–Ende equation/theory is restricted to ideal solutions where both the mixing enthalpy and the mixing volume are zero. The method/theory of Hearst and Vinograd [10] requires detailed knowledge of the chemical potential of the components in a solution and has been, up to now, restricted to aqueous solutions of several salts.

The procedure of Munk [11] gives an empirical relationship with several constants which may be determined with marker polymers of known density and several equilibrium runs at different rotor speeds. The method of Lechner and coworkers [12–14] calculates the density gradient for real solutions. As the resulting equation is an implicit function one has to calculate the density gradient by regression procedures which can cause complications in some cases. The direct refractometric ρ–r determination of the equilibrium and non-equilibrium density gradient requires a precise baseline [1, 2, 15]. This may be well established in the case of time-dependent dynamic density gradients [2]; however, in the case of equilibrium or near-to-equilibrium density gradients one has to use double-sector cells for the establishment of the precise baseline [15].

Owing to difficulties with respect to the procedures described earlier we describe in the following a new marker calibration method for the evaluation of equilibrium and nonequilibrium density gradients in an AUC. Since the Hermans–Ende equation is the mathematical basis of the new procedure as well, we give a brief outline of the Hermans–Ende theory.

Considering an ideal binary mixture of two components with indices 0 and 1 the ratio of the volume fractions of components 1 and 0, φ_1 and φ_0, as a function of the rotor distance, r, was given by Hermans and Ende [9] as follows:

$$\varphi_1/\varphi_0 = \alpha \exp(\beta r^2) \ , \tag{1}$$

with

$$\varphi_k = \left(n_k V_k^0\right) \bigg/ \left(\sum_k n_k V_k^0\right) = \left(m_k v_k^0\right) \bigg/ \left(\sum_k m_k v_k^0\right); \quad k = 0, 1 \tag{2}$$

and

$$\varphi_0 + \varphi_1 = 1 \ , \tag{3}$$

where n_k, m_k, V_k^0 and v_k^0 are the amount of matter, the mass, the molar volume and the specific volume of the component k. α is an integration constant for sector-shaped cells,

$$\alpha = \frac{\exp\left[\beta \varphi_1^{in}\left(r_b^2 - r_m^2\right)\right] - 1}{\exp\left(\beta r_b^2\right) - \exp\left(\beta \phi_1^{in} r_b^2 + \beta \phi_0^{in} r_m^2\right)} \ , \tag{4}$$

where φ_k^{in} is the initial volume fraction of component k, r_m is the rotor distance at the meniscus of the cell, r_b is the rotor distance at the bottom of the cell and

$$\beta = \omega^2 \left(M_1/\rho_1^0\right)\left(\rho_1^0 - \rho_0^0\right)/(2RT) \ , \tag{5}$$

where ω is the angular velocity, and M_1, ρ_0^0, and ρ_1^0 are the molar mass of component 1 and the densities of components 0 and 1, respectively. R is the gas constant and T the absolute temperature.

The density of an ideal solution, ρ, is given by

$$\rho = \varphi_0 \rho_0^0 + \varphi_1 \rho_1^0 \ . \tag{6}$$

Combining Eqs. (1), (3) and (6) results in the Hermans–Ende equation in the ρ–r from:

$$\rho = \frac{\rho_0^0 + \rho_1^0 \alpha \exp(\beta r^2)}{1 + \alpha \exp(\beta r^2)} \ . \tag{7}$$

Under the assumption $\alpha \exp(\beta r^2) \ll 1$ the following equation holds:

$$\rho = \rho_0^0 + \rho_1^0 \alpha \exp(\beta r^2) \ . \tag{8}$$

The following typical example demonstrates that in many cases $\alpha \exp(\beta r^2) \ll 1$. Light component water, with 1 mass% alkyl sulfonate added to stabilize the latex particles, heavy component metrizamide, $\rho_0^0 = 0.9987$ g/cm^3, $V_0^0 = 18.02$ cm^3/g, $\rho_1^0 = 2.1552$ g/cm^3, $V_1^0 = 370.3$ cm^3/mol, $M_1 = 789.1$ g/mol, rotor speed $N = 30.000$ min^{-1}, rotor distances $r_m = 5.9$ cm and $r_b = 7.2$ cm, $T = 298$ K, $\beta = 8.434 \cdot 10^{-2}$ cm^{-2}, $\alpha = 1.955 \times 10^{-3}$, $\alpha \exp(\beta r^2) = 3.382 \times 10^{-3}$ (with r≈6,5 cm = middle of the cell). The value of $\alpha \exp(\beta r^2)$ is much smaller than 1.

Equation (8) holds for ideal mixtures. For real mixtures one has to correct the parameters ρ_0^0, ρ_1^0, α and β with correction factors or, in other words, to replace them by three adjustable parameters a, b and c; this gives

$$\rho = a + b \exp\left(cr^2\right) \ . \tag{9}$$

Experimental

The following materials were used: double-distilled water, alkyl sulfonate C$_{12}$–C$_{20}$ (BASF, Ludwigshafen, Germany), metrizamide (Nyegaard, Oslo, Norway) and PS latex (BASF). The physical constants of water/alkyl sulfonate (1 mass%) and metrizamide are given in the theoretical part of this work. All the values and measurements refer to 25 °C.

All the polymer latices used were polymerized with a standard emulsion polymerization technique in our BASF research laboratories using BASF monomers. The properties of the PS latices show a particle density of $\rho_{PM} = 1.055 \pm 0.002$ g/cm^3 and a mass-average diameter of $D = 160$ and 210 nm. Our 11 marker particles are EHA/MA copolymer latices which were all polymerized separately. All 11 latices had nearly the same particle diameter of 200 nm, but the particle densities were different because of the 11 different copolymer compositions [$w_{MA} = 0/10/20/30/40/50/60/70/80/90/100$ mass%]. These particle densities (calculated theoretically from the homopolymer densities and verified using a Paar densitometer within $\Delta\rho = \pm 0.003$ g/cm^3) were $\rho_{PM} = 0.980$ (PEHA)/1.000/1.021/1.043/1.066/1.089/1.114/1.167/1.196/1.225 (PMA) g/cm^3. After the separate characterization we mixed a part of these 11 latices in equal parts (weight/weight). This 11-marker-calibration mixture was added to differently composed water/metrizamide density gradient solutions. The total latex concentration was 1 g/l. The "contaminated" latex Acronal XY originated from a BASF production plant.

The measurements were conducted with the following equipment: Kratky-type vibrating tube densitometer with a glass capillary (A. Paar, Graz, Austria) and a modified OPTIMA XL AUC (Beckman-Coulter, Palo Alto, Calif.). This AUC was equipped with a Beckman eight-hole rotor and home-made Schlieren optics with a multiplexer which we installed in this machine [7]. We had to use Schlieren optics for our steep density gradients (i.e. for metrizamide contents higher than 15 mass%) because the interference optics fails in this case [7]. The evaluation of the Schlieren photographs was done with a graphics tablet and a computer. The data were transposed into an ASCII file which was accessible to a personal computer. In all cases we had the same running conditions: 3-mm single-sector cell, –2° wedge window, 30,000 rpm, cell nearly completely filled (5.9–7.2 cm), running time 22 and 44 h. The Schlieren photographs are taken on 70-mm Ilford film (100 ASA, black and white) using an exposure time of 3 s.

Results and discussion

The principle of the Schlieren optics in our AUC [7] is shown in Fig. 2. The rotor bears two cells: a measuring cell and a counterbalance cell with radius reference marks at radial positions of 5.7 and 7.3 cm. The elements of the Schlieren optical light path are a flash lamp with a green filter (546 nm) in front of it, a Schlieren slit, a collimating lens, a deflecting mirror, a phase plate, a camera lens, a cylindrical lens and a photographic plate (or film or digital television camera) at the end.

In the top-right corner of Fig. 2, a superimposed top view of the measuring and counterbalance cells is shown. Inside the sector-shaped measuring cell we see sedimenting latex particles. These are visualized on the three Schlieren photographs below as sedimenting turbidity bands, taken at different running times of 5, 15 and 40 min.

These three Schlieren photos are shown again on the left-hand side of Fig. 3. They are the result of a sedimentation run, which yields, by means of a turbidity detector and the Stokes equation, the complete particle size distribution of this PS latex [3, 16].

From this sedimentation run we now come to the density gradient run on the right-hand side of Fig. 3, by adding to the "light" water 11 mass% of the "heavy" sugar, metrizamide, $\rho = 2.1$ g/cm^3. We then had to wait 16 h to reach a radial equilibrium density gradient from $\rho = 1.03$ to 1.10 g/cm^3. Within this time all the PS particles gathered within a narrow turbidity band. The radial band position, r, yields the particle density $\rho = 1.055$ g/cm^3 for the PS latex particles measured.

The transformation of the linear r-axis into the exponential ρ-axis (both below the 16-h Schlieren photographs in Fig. 3) is usually done by means of the Hermans–Ende equation (Eq. 7).

This transformation/calculation for the 16-h measurement example of our PS latex in Fig. 3 is shown in Fig. 4. The ρ–r diagram according to the Hermans–Ende equation in Fig. 4 for this 89 H$_2$O/11 metrizamide density gradient reaches from $\rho = 1.03$ to 1.10 g/cm^3

Fig. 2 Schlieren optical setup in an AUC with a photographic plate at the end of the optical light path and resulting Schlieren photographs of sedimenting polystyrene (*PS*) latex particles inside an AUC measuring cell

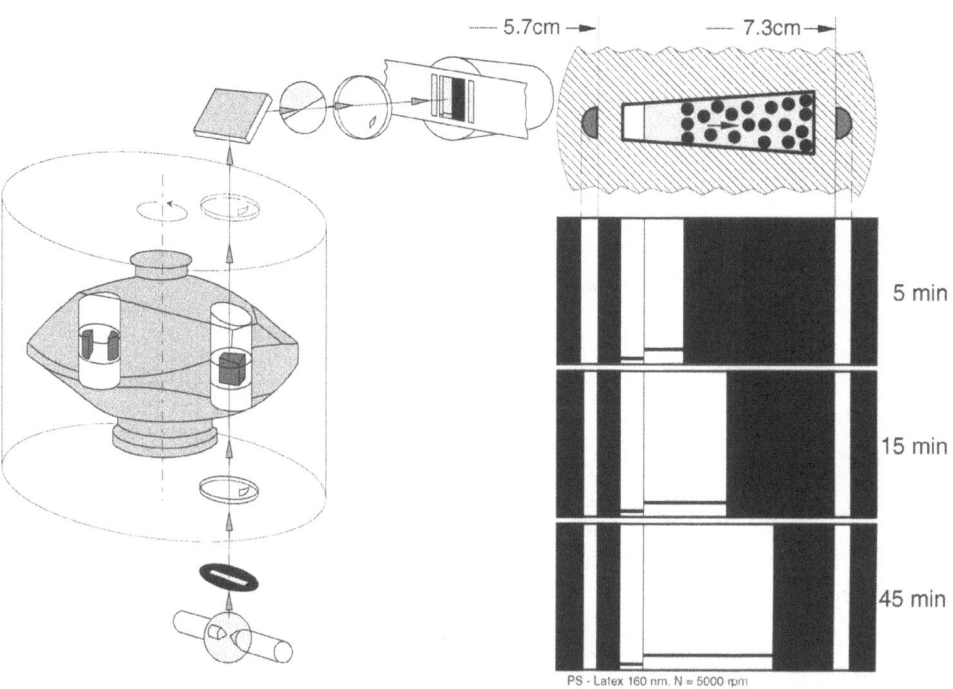

5 min

15 min

45 min

PS - Latex 160 nm, N = 5000 rpm

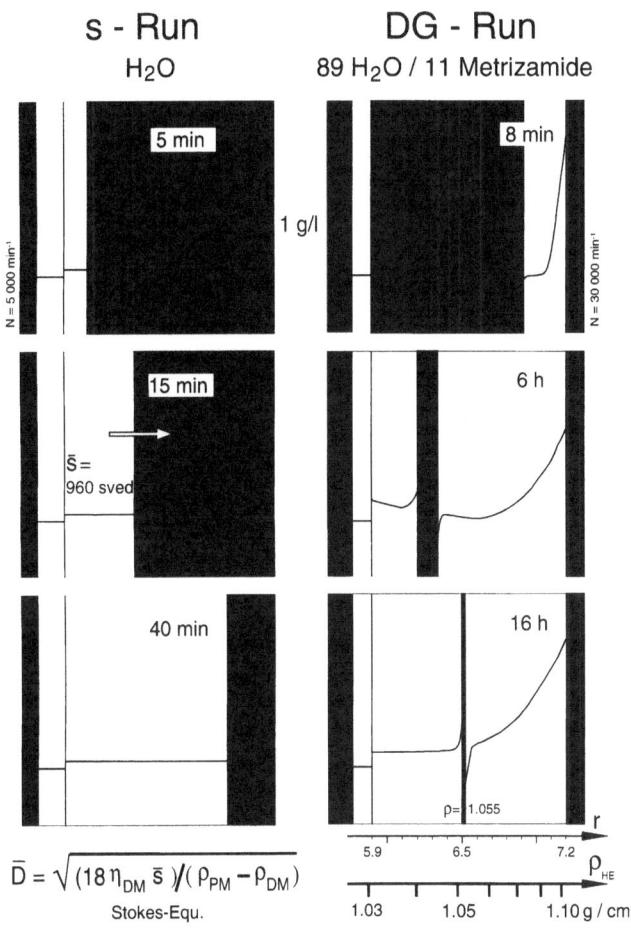

s - Run
H₂O

DG - Run
89 H₂O / 11 Metrizamide

$$\bar{D} = \sqrt{(18\,\eta_{DM}\,\bar{s}\,)/(\rho_{PM} - \rho_{DM})}$$

Stokes-Equ.

Fig. 3 AUC Schlieren photographs of PS latex particles during a sedimentation run (*left*) and a static equilibrium density gradient run (*right*). *D* is the diameter, *s* is the sedimentation coefficient and ρ_{PM} is the density of the latex particle; η_{DM} and ρ_{DM} are the viscosity and density of the dispersion medium (mostly water)

for a nearly complete cell filling height of $r = 5.9$–7.2 cm. The equations in Fig. 4 are Eqs. (4), (5), (7) and (8) already discussed in the theoretical part.

In some cases the Hermans–Ende equation (Eq. 7) fails:

1. If we do not really reach equilibrium.
2. If the mass percentage of metrizamide is too high.
3. If we change from the bimodal water/metrizamide density gradient system to a trimodal system (for example by adding methanol).

A demonstration of that systematic gradual failure is shown in Fig. 5. For this purpose we put the same PS latex into 4 H₂O/metrizamide density gradients with different metrizamide contents of 6, 8, 12 and 15 mass%. On the left-hand side in Fig. 5 we see the four resulting Schlieren photographs and on the right-hand side, schematically, the evaluation according to the Hermans–Ende equation. As indicated, the calculated Hermans–Ende values, $\rho_{HE} = 1.049$, 1.054, 1.064 and

1.070 g/cm³, are not identical. The discrepancy between expectation and calculation in Fig. 5 increases with increasing content of metrizamide.

We became aware of this fact by knowing precisely the density of the measured PS latex particles, $\rho_{PS} = 1.055 \pm 0.002$ g/cm³. This fact gave us the idea to replace the Hermans–Ende evaluation method with a better one, namely with a pragmatic calibration method, using marker particles with precisely known particle densities. Because such markers were not available on the market, we synthesized our own 11 EHA/MA copolymer latex markers described in the Experimental section. The main advantages of these 200-nm marker particles for our experiment are a uniform, precisely known particle density and a nearly equidistant graduation of the different particle densities.

We put this mixture of 11 markers into seven H₂O/ metrizamide density gradients with different metrizamide contents. To two of them we added methanol. The corresponding seven Schlieren photographs after a running time of 22 h are shown on the left-hand side of Fig. 6. Indeed we see all single 11 marker particles in the form of narrow turbidity bands, distributed over the seven different gradients. For example, in the upper gradient we see markers 1, 2 and 3, in the 88 H₂O/ 12 metrizamide gradient, in the middle we see markers 4–8 and in the lower gradient we see markers 6–11.

Again we transform the common linear radius *r*-axis (from 5.9 to 7.2 cm) of these seven 22-h Schlieren photographs into a particle density ρ-axis by means of the Hermans–Ende equation. The upper part of Fig. 7 shows the result of this transformation in the usual Hermans-Ende scheme. The vertical lines across the whole diagram indicate the known particle densities of our 11 markers. For better visualization and comparison the position line of marker 6 is dashed. Inside the seven different density gradient boxes the turbidity bands of the different 11 markers are indicated as black bars at the ρ positions calculated with the Hermans–Ende equation.

There are differences between calculation and expectation. In the upper methanol gradient the differences are small. In the 88 H₂O/12 metrizamide gradient in the middle the differences are bigger. In the lower gradients, with high metrizamide contents of 17, 20 and 25 mass%, the differences for the high-density markers 9, 10 and especially 11 are further increased. Surprisingly small are the differences for marker 6, which is visible in five gradients.

One possible reason for these differences between calculation and expectation, i.e. the gradual failure of the Hermans–Ende equation, is that after 22-h running time we have not really reached equilibrium. We were able to prove this by increasing the running time from 22 to 44 h. The right-hand side of Fig. 6 shows the result of this 44-h run, again in the form of seven Schlieren photographs.

Fig. 4 Schematic representation of the Hermans–Ende equation (Eq. 7) to calculate the ρ–r relationship for the static equilibrium AUC density gradient presented in Fig. 3 at 16-h running time. (All the *symbols* in the equation are explained in the theoretical part of this work)

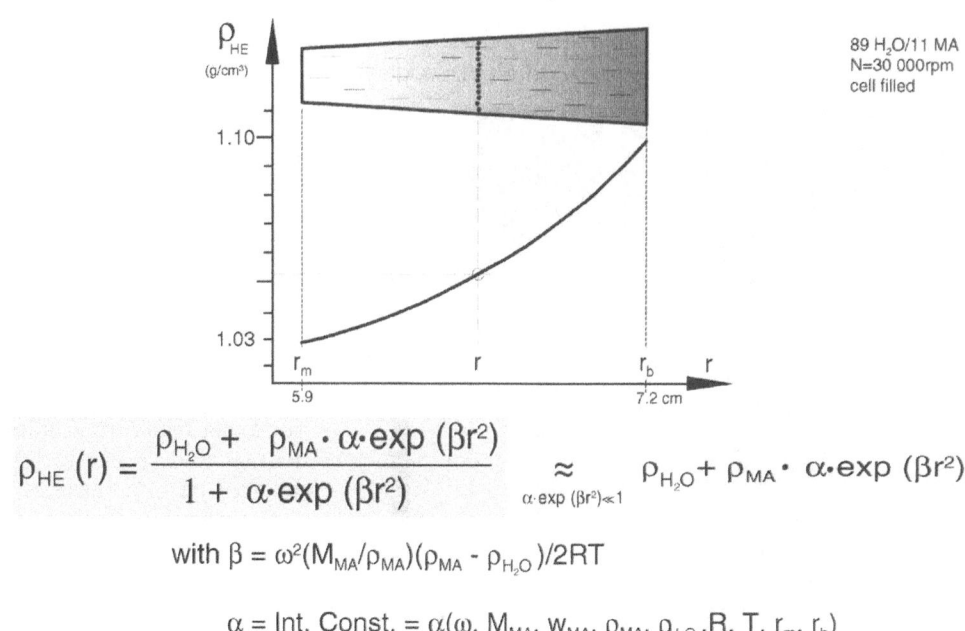

$$\rho_{HE}(r) = \frac{\rho_{H_2O} + \rho_{MA} \cdot \alpha \cdot exp(\beta r^2)}{1 + \alpha \cdot exp(\beta r^2)} \underset{\alpha \cdot exp(\beta r^2) \ll 1}{\approx} \rho_{H_2O} + \rho_{MA} \cdot \alpha \cdot exp(\beta r^2)$$

$$\text{with } \beta = \omega^2(M_{MA}/\rho_{MA})(\rho_{MA} - \rho_{H_2O})/2RT$$

$$\alpha = \text{Int. Const.} = \alpha(\omega, M_{MA}, w_{MA}, \rho_{MA}, \rho_{H_2O}, R, T, r_m, r_b)$$

Fig. 5 Four static AUC density gradients of a 210-nm PS latex (with $\rho_{PS} = 1.055$ g/cm³) in water/metrizamide mixtures of different composition. The resulting Schlieren photographs at 44 h are on the *left*. The corresponding evaluation scheme according to Hermans–Ende is on the *right*

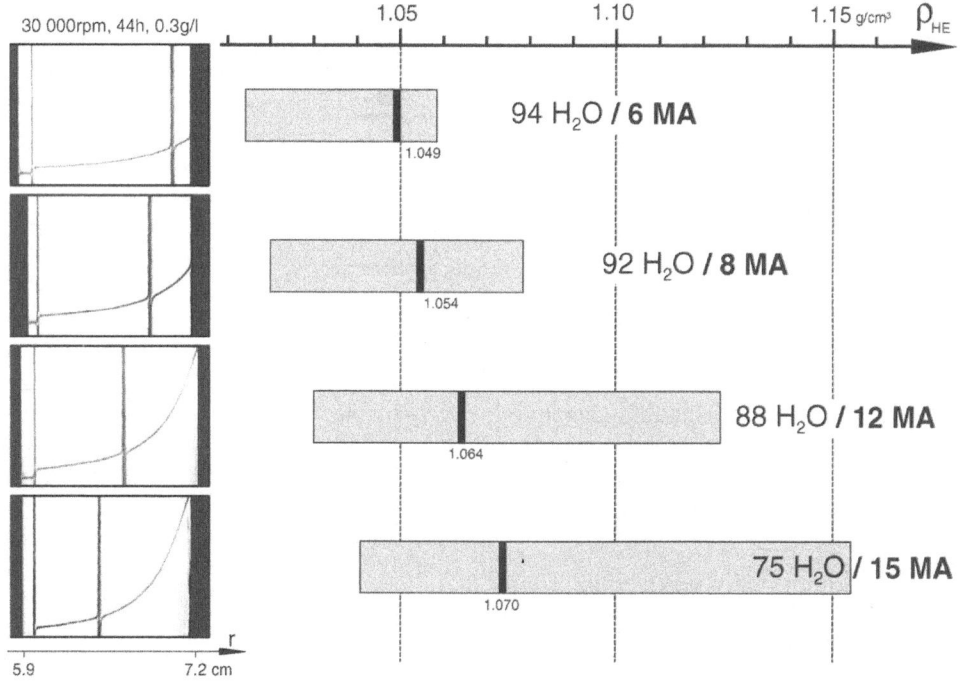

Indeed, we had not reached equilibrium at 22 h. This can be seen from the differences between the Schlieren photographs at 22 and at 44 h. For example, in the upper gradient in Fig. 6 we see the three markers 1–3 at 22 h, but at 44 h we additionally see marker 4 near the cell bottom. Also in the 83 H₂O/17 metrizamide gradient we see additionally marker 4, but now near the cell meniscus. The same is valid for marker 5 in the 75 H₂O/25

metrizamide gradient at 44 h. This means that the 44-h gradients are steeper than the 22-h gradients.

The central part of Fig. 7 shows the Hermans–Ende scheme of these 44-h gradients (right-hand side of Fig. 6). Now, surprisingly, although we are nearer to the equilibrium, the differences between calculation and expectation are bigger than after 22-h running time. This is clearly seen at marker 6. Also the differences for the

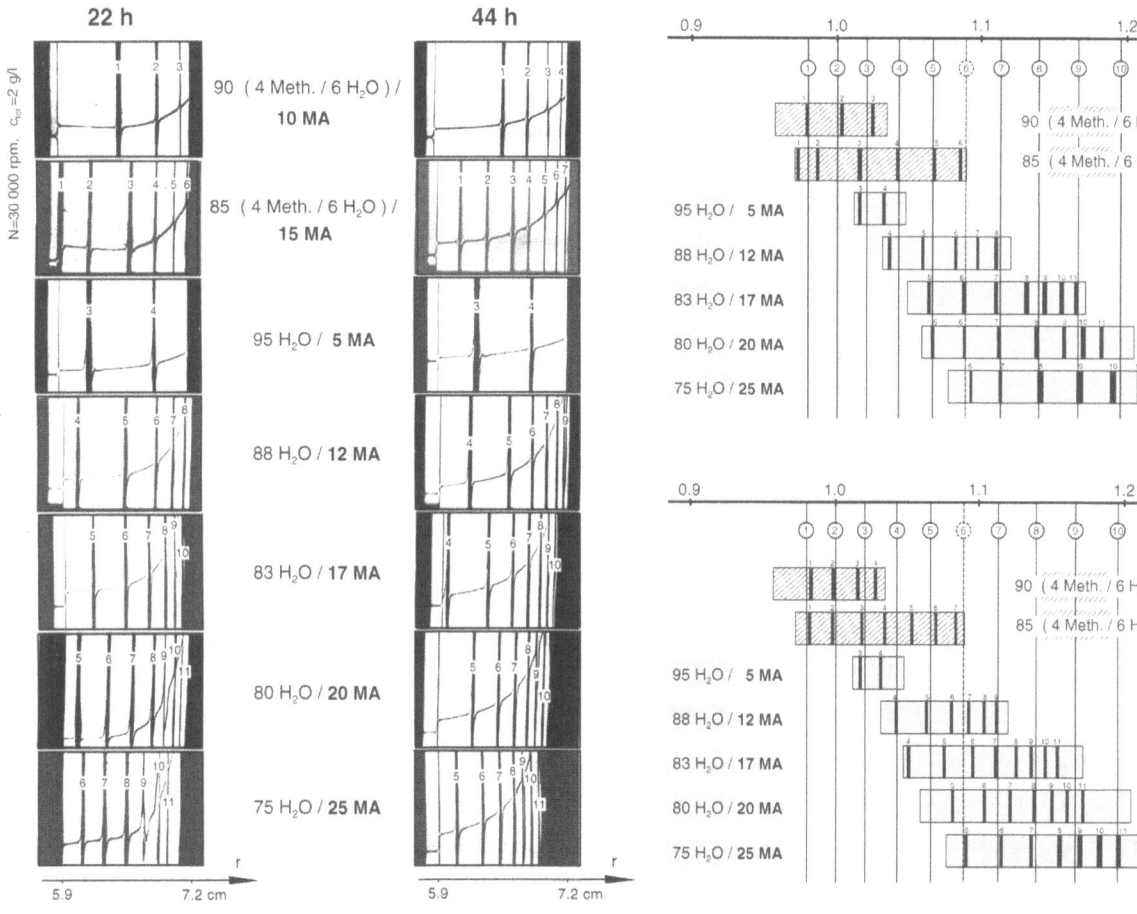

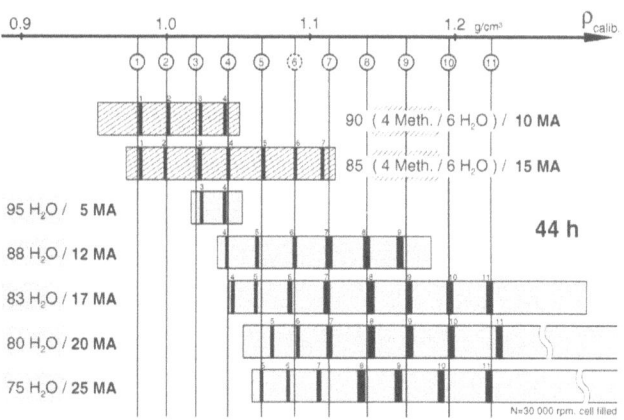

Fig. 6 Schlieren photographs of seven static water/metrizamide density gradients of different composition (taken at 22 and 44-h running time) in which our 11 ethylhexyl acrylate/methyl acrylate (*MA*) marker mixture was dispersed. The different markers appear as narrow turbidity bands at different radial positions

high-density marker 11 are increased in comparison to 22-h (upper part of Fig. 7).

This increasing failure of the Hermans–Ende equation with increasing metrizamide content and especially in the high-density area of the AUC density gradients near the cell bottom is a principle. We will not discuss this here (a discussion can be found in Refs. [12, 14]). Instead we describe in the following our pragmatic solution of the problem by replacing the Hermans–Ende method by our new calibration method for the evaluation of AUC density gradients, using such marker particles with known particle densities.

For this purpose Fig. 8 shows again, only for the 88 H_2O/12 metrizamide gradient, the Schlieren photographs at 22 and at 44 h and the corresponding Hermans–Ende scheme. The radial positions of the visible six markers 4–9 along the linear *r*-axis are precisely known by measurement of the photographic plates on a digital tablet. Also the particle densities of

Fig. 7 Evaluation scheme of the seven static 11 marker density gradients presented in Fig. 6. The *long vertical lines* represent the expected density positions of the known particle densities of the 11 markers. The *short black bars* represent the calculated ρ positions from the *r* positions within the Schlieren photographs in Fig. 6. *Upper part*: calculation according to Hermans–Ende using the 22-h photographs. *Middle part*: calculation according to Hermans–Ende using the 44-h photographs. *Lower part*: calculation using our new marker calibration procedure (optimal exponential ρ–r calibration curves, see Fig. 10)

these markers are precisely known. We use both together to construct pragmatic ρ–r calibration curves. These are

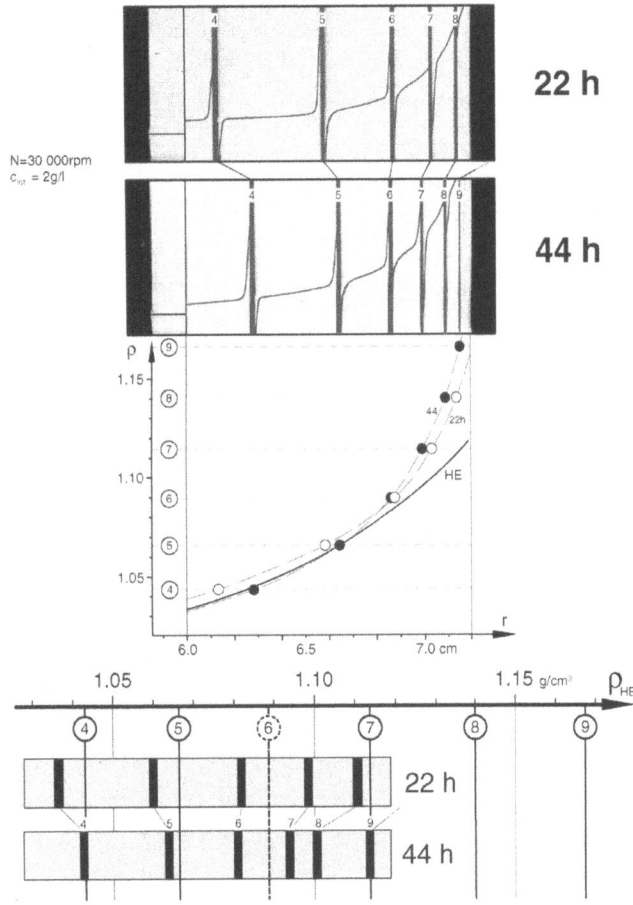

Fig. 8 AUC Schlieren photographs and Hermans–Ende evaluation scheme of static 88 water/12 metrizamide density gradients at 22 and 44 h. The ρ–r diagram shows the different ρ–r marker points. The *curves* through these points are the pragmatic calibration curves. For comparison the calculated Hermans–Ende curve is also shown

shown in the middle of Fig. 8: one curve for 22 h and a different, steeper one for 44 h. For comparison the Hermans–Ende (equilibrium) curve, which is "valid" for 22 and 44 h, is also introduced into this ρ–r diagram. There is a considerable discrepancy between the experimental marker points and the Hermans–Ende theory at high particle densities near the cell bottom.

The ρ–r-diagram of Fig. 8 is shown again in Fig. 9, but now only at 22 h. There was the question: what is the optimal mathematical approximation curve through these five single marker points? First we tried it with the polynomial approximation indicated, with three adjustable parameters a, b and c; however, this was not satisfactory. So we tried the exponential approximation indicated, again with three adjustable parameters a, b and c, which is similar to the Hermans–Ende equation in the case $\alpha \exp(\beta r^2) \ll 1$, i.e. to Eq. (8) or Eq. (9). This exponential approximation was satisfactory. So we also used it for all other gradients.

All the exponential approximation curves are shown in a ρ–r summary diagram for our seven different density gradients at 22 h in Fig. 10. This diagram also shows all single marker points of our 11 markers: the base of the constructed seven ρ–r calibration curves. This set of seven calibration curves is stored in the form of seven parameter triples a, b and c in our computer (and not in the form of all single marker points). So we are able to calculate continuously the corresponding ρ value for every radial position inside an AUC measuring cell.

The same procedure as for the 22-h gradients in Figs. 9 and 10 was carried out for our seven 44-h gradients. The result (not shown in the form of a diagram) is similar to Fig. 10. but all seven calibration curves are steeper at high densities near the cell bottom.

Figure 7, lower part, shows a first application of this 44-h calibration. The same seven 44-h density gradients

Fig. 9 Calculated optimal ρ–r calibration curves through the marker points. We used a polynomial and an exponential approximation curve through the 22-h marker points of Fig. 8, both with three adjustable parameters a, b and c. For comparison the calculated Hermans–Ende curve is also shown

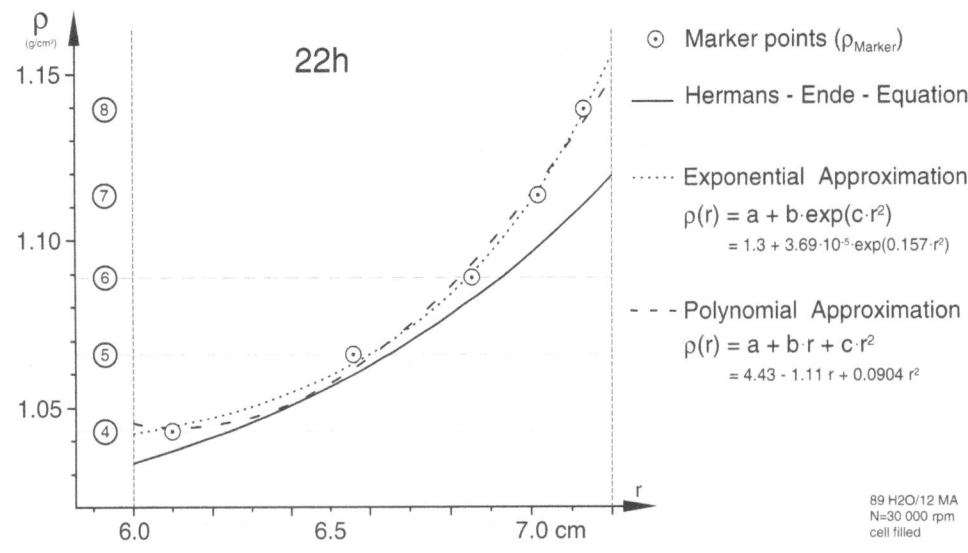

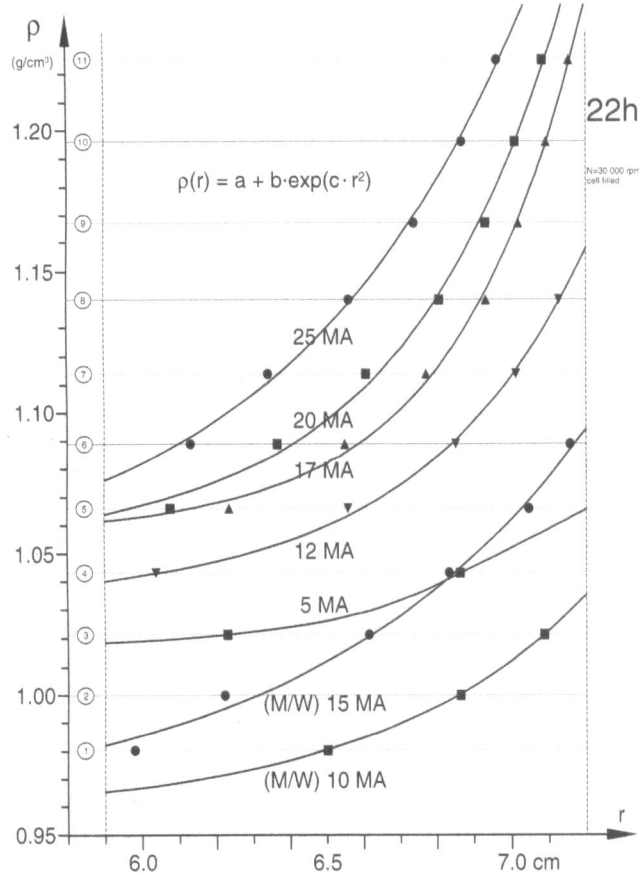

$$\rho(r) = a + b \cdot \exp(c \cdot r^2)$$

22h

N=30 000 rpm
cell filled

25 MA

20 MA

17 MA

12 MA

5 MA

(M/W) 15 MA

(M/W) 10 MA

Fig. 10 Calculated optimal exponential ρ–r calibration curves through the corresponding 22-h marker points of our set of seven differently composed static water/metrizamide AUC density gradients, shown in Fig. 6

are seen in the middle part of Fig. 7 evaluated with the Hermans–Ende method, and we find there, as mentioned before, big differences between calculation and expectation, especially for markers 6 and 11. Now, evaluated with the new calibration method, these differences are drastically reduced as shown in the lower part of Fig. 7. Surely, this agreement between calculation and expectation could still be better, look at marker 7, but, nevertheless, we considerably improved the precision of absolute particle density measurements in AUC density gradients to about $\Delta\rho = \pm 0.003$ g/cm^3.

Another advantage of our new calibration method is that we do not have to reach equilibrium. Reasonable results are also obtained at 20 or at only 8 h, if the corresponding calibration constants are used. This is a drastic reduction in the measuring time.

A second, real industrial application example of the new calibration method is shown in Fig. 11: the analysis of a contaminated BASF latex, called Acronal XY, which showed a wrong application performance. This latex is a 50 butyl acrylate/50 methyl methacrylate copolymer, so we expect a uniform particle density of $\rho_{PM} = 1.126$ g/cm^3. In addition, in the first (upper) 85 H$_2$O/15 metrizamide gradient in Fig. 11, we see the expected uniform Acronal XY particles at a density position of $\rho = 1.120$ g/cm^3; however, surprisingly we see other unexpected components with lower densities, one at $\rho = 1.072$ g/cm^3 and especially one fixed at the meniscus.

So, additionally we put Acronal XY into a second 85 (4 methanol/6 H$_2$O)/15 metrizamide gradient in a lower-density range and we saw two further components, at $\rho = 0.996$ and 1.038 g/cm^3. We identified these three

Fig. 11 Particle density spectroscopy of a contaminated 50 butyl acrylate /50 methyl methacrylate copolymer latex (Acronal XY, $\rho_{expected} = 1.126$ g/cm^3), by means of two differently composed static AUC density gradients. The new marker calibration evaluation scheme is also shown

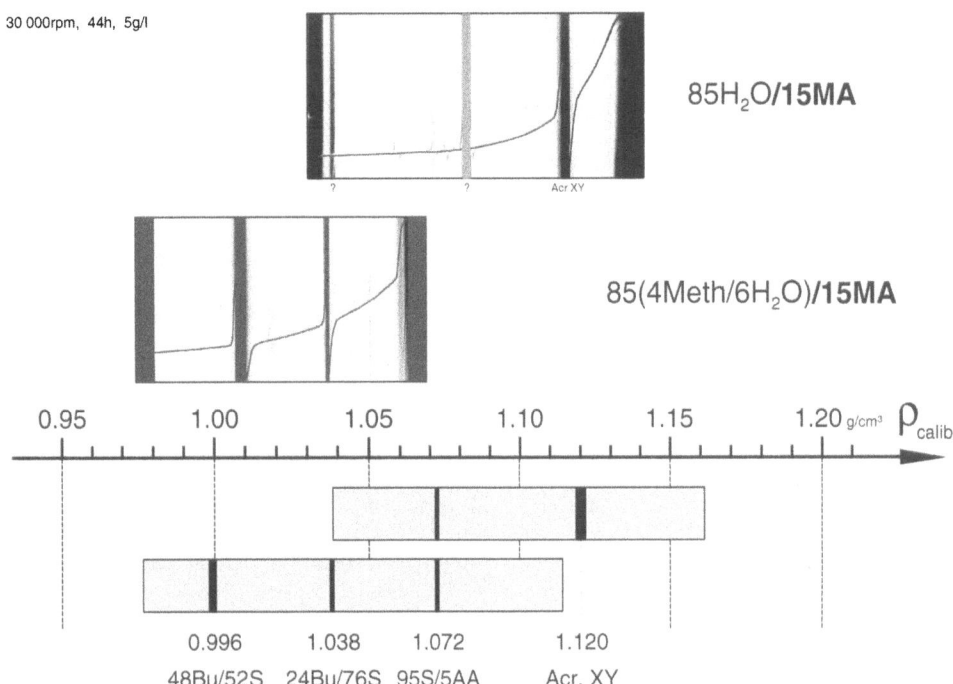

30 000rpm, 44h, 5g/l

85H$_2$O/**15MA**

85(4Meth·6H$_2$O)/**15MA**

0.996   1.038   1.072   1.120

48Bu/52S   24Bu/76S   95S/5AA   Acr. XY

unexpected components as 48/52 and 24/76 budadiene/styrene and as 95/5 mass% styrene/acrylic acid copolymer latices.

So, the explanation of the problem was rather simple: A distribution valve was defective and different latices were, unintentionally, transferred into the same storage vessel.

Conclusion

The second application example in Fig. 11 is a convincing demonstration that it is really possible to practise a kind of particle density spectroscopy within AUC density gradients. The high-resolution power according to the particle density is inherent in this fractionation method. By decreasing the rotor speed it is possible to create flat density gradients, so very small density differences down to $\Delta\rho = \pm 0.0003$ g/cm^3 between two particle components are detectable.

Our replacement of the Hermans–Ende evaluation method for AUC density gradients is a new pragmatic calibration method, using markers of precisely known particle densities, does not improve the high-resolution power of AUC density gradients, but the absolute precision of particle density measurements. This is important for practicing particle density spectroscopy. Using our present 11 EHA/MA markers the absolute error of measurement is $\Delta\rho = \pm 0.003$ g/cm^3. We are looking for more and better markers, especially in the density ranges $\rho = 0.85$–1.00 and 1.2–1.6 g/cm^3 in order to further reduce this absolute error. Especially the high density area, with high metrizamide contents of 20–50 mass%, where the Hermans–Ende method fails completely, becomes more and more important because of new types of nanoparticles, for example, organic/inorganic nanocomposites.

An important advantage of our new calibration method is the reduction in the measuring time. Density gradients with very high metrizamide contents need up to 150 h to really reach equilibrium; however, we do not need equilibrium. Our method works as well at 20 or even at 8 h. This means that our static density gradient is now a "dynamic" one [2]. For precise results a new density gradient run has to be done in exactly the same manner as the marker calibration run: same metrizamide content, cell filling height, rotor speed and running time. For new conditions we have to perform a new calibration run. This is easily done by running in the eight-cell rotor additionally one cell filled with our calibration mixture. The present 11 EHA/MA marker mixture was stable and showed no change within the last 2 years. We always got the same results as presented in Figs. 6 and 7.

Because we chose large diameters of about 200 nm for our marker particles, their migration velocity is also high enough to follow immediately the slow buildup of the H_2O/metrizamide density gradient. Within minutes the markers sweep to their corresponding radial density position within the measuring cell. Thus, measurements of the radial marker position at a lot of different running times, like in Fig. 8, are an excellent tool to study the time dependence of the buildup of AUC density gradients. Perhaps theorists will use these kinds of measurements to develop fast dynamical density gradients (like in Refs. [1, 2]) with precisely calculable $\rho(r, t)$ functions within the measuring cell. Another possibility is to use such measurements to improve the old Hermans–Ende equation theoretically by taking into account higher H_2O/metrizamide interactions, i.e. higher virial coefficients. The first attempts were reported in Refs. [12–14].

Acknowledgements The authors thank M. Stadler for the further mechanical and optical improvement of the OPTIMA XL-SO setup and H. Roth and M. Kaiser for their doing all the AUC measurements.

References

1. Lange H (1980) Colloid Polym Sci 258:1077–1085
2. Mächtle W (1984) Colloid Polym Sci 262:270–282
3. Mächtle W (1992) In: Harding SE, et al (ed) AUC in biochemistry and polymer science. Royal Society of Chemistry, Cambridge, pp 147–175
4. Mächtle W, Ley G, Rieger J (1995) Colloid Polym Sci 273:708–716
5. Mächtle W, Ley G, Steib J (1995) Colloid Polym Sci 99:144–153
6. Kirsch S, Doerk A, Bartsch E, Sillescu H, Landfester K, Spieß W, Mächtle W (1999) Macromolecules 32:4508–4518
7. Mächtle W (1999) Prog Colloid Polym Sci 113:1–9
8. Leyrer RJ, Mächtle W (2000) Macromol Chem Phys 201:1235–1243
9. Hermans JJ, Ende HA (1963) J Polym Sci Part C Polym Symp 1:161–177
10. (a) Hearst JE, Vinograd J (1961) Proc Natl Acad Sci USA 47:999–1004; (b) Hearst JE, Vinograd J (1961) Proc Natl Acad Sci USA 47:1005–1014
11. Munk P (1982) Macromolecules 15:500–505
12. Lechner MD (1997), Macromol Rapid Commun 18:781–786
13. Lechner MD, Mächtle W, Sedlack U (1997) Prog Colloid Polym Sci 107:148–153
14. Lechner MD, Borchard W (1999) Eur Polym J 35:371–376
15. Lechner MD, Mächtle W, Sedlack U (1997) Prog Colloid Polym Sci 107:154–158
16. Mächtle W (1999) Biophys J 76:1080–1091

Progr Colloid Polym Sci (2002) 119:11–18
© Springer-Verlag 2002

Gordon Lucas
Lars Börger
Helmut Cölfen

Solubility equilibrium gradients in the analytical ultracentrifuge: an approach towards the isolation of critical crystal nuclei in solution

G. Lucas · L. Börger · H. Cölfen (✉)
Max Planck Institute of Colloids
and Interfaces, Colloid Chemistry
Research Campus Golm, 14424 Potsdam
Germany
e-mail: coelfen@mpikg-golm.mpg.de
Tel.: +49-331-5679513
Fax.: +49-331-5679502

Abstract Sedimentation equilibrium methods which are able to establish a solubility gradient for inorganic species in an ultracentrifugal field are presented. These methods are generally based on the concepts of density gradient and sedimentation equilibrium ultracentrifugation. However, in addition to the formation of a density or an equilibrium gradient, physicochemical parameters such as the pH or the solvent quality are varied throughout the solution column if appropriate high-density or high-molar-mass solutes are chosen or a mixture of solvents with different density is applied. This is demonstrated for two types of solubility gradients including for the case of a pH gradient; parameters for the adjustment of the overall pH are discussed. pH gradients were formed up to 3 pH units and are sensitive to addition of electrolytes, so they can only be applied to sparingly soluble salts. The gradual variation of the solubility of inorganic particles leads to the dissolution of the particles upon sedimentation when the dissolution point is reached. The ionic species formed show increased diffusion compared to the sedimenting particles, so they can diffuse back to regions of lower solubility and thus form a crystal again. This finally leads to an equilibrium situation for the critical crystal nucleus. In the case of a pH gradient with CdS, it is demonstrated that a transition from particles to dissolved ions indeed takes place and can be monitored in the analytical ultracentrifuge. For $BaCrO_4$, the transition to the more soluble $BaCr_2O_7$ with decreasing pH can readily be monitored via the associated spectral changes, clearly demonstrating the possibility to perform chemical reactions in a pH gradient.

Key words Sedimentation equilibrium · Sedimentation velocity · Density gradient ultracentrifugation · Critical crystal nucleus

Introduction

The transition state between soluble and crystalline species, termed the "critical crystal nucleus", has been intensively investigated and has been a matter of controversial discussion throughout the last decades. So far, conditions and sizes of critical crystal nuclei were mainly a matter of theoretical predictions applying various nucleation theories without experimental proof [1–5]. Although several successful attempts for the controlled nucleation of colloidal crystals from monodisperse micron-sized latices were reported, these systems cannot be treated as a test system for general nucleation theories. Most crystals nucleate from ionic or low-molar-mass species showing much higher diffusion. Additionally, electrostatic contributions may have a significant influence. Only very recently, experimental investigations into the size domain of only a few nanometres were reported. The first experimental observation of critical crystal nuclei was of apoferritin crystal building units of about 12-nm size by atomic force microscopy [6], but this finding is already heavily debated [7] and is still above the size scale of regular crystallization events. Although critical crystal nuclei have already been observed by analytical ultracentrifugation in rare cases and for very dense colloids [8], they were never isolated in solution despite their fundamental importance for nucleation theories as well as crystal growth processes. Gas-phase reactions, on the other hand, which produce small clusters of defined sizes, are not representative of solution reactions despite their use for the investigation of cluster formation. Therefore, it is a major goal for fundamental as well as for applied colloid science to isolate preparatively critical crystal nuclei in a stable dynamic equilibrium between dissolution and crystallization in solution. This goal can, in principle, be achieved by solubility gradient centrifugation where the crystal solubility increases towards the cell bottom. This is outlined schematically in Fig. 1.

When a particle sediments towards the bottom of an ultracentrifuge cell in a solubility gradient, it dissolves into a region of ions where diffusion is a much stronger force than sedimentation; hence, the ions will diffuse back into regions of lower solubility where crystal formation is favoured. This should then lead to a dynamic equilibrium between crystallization and dissolution, in analogy to the well-established sedimentation equilibrium. This concept is investigated in the present study, with the focus on the various possibilities of solubility gradient generation.

Experimental

A Beckman XL-I analytical ultracentrifuge (Beckman Coulter, Palo Alto, Calif.) was used by simultaneously applying Rayleigh interference and UV–vis absorption optics. For the preparative

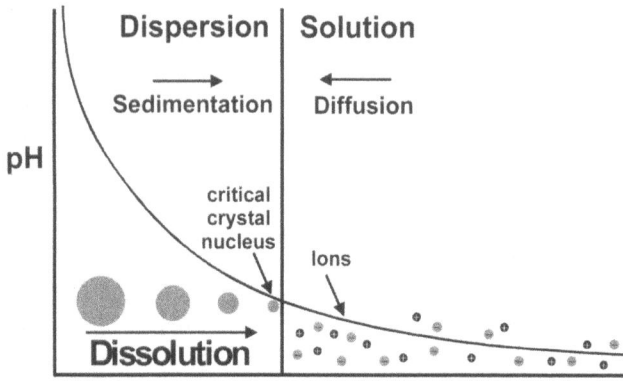

Fig. 1 Dynamic equilibrium between crystallization and dissolution in a solubility gradient for the example of a pH gradient

runs, a Beckman Optima L-70 ultracentrifuge (Beckman Coulter, Palo Alto, Calif.) was used with a Ti 70.1 rotor [poly(sodium 4-styrene sulfonate), PSS, gradients] or a Ti 70 rotor (HI gradients) using polyalloymer tubes. UV–vis spectra were recorded using a UVICON 931 spectrophotometer (Kontron, Nev Fohrn) and dn/dc was measured with an NFT Scanref (Göttingen) at 632.8 nm. pH measurements were done with a Metrohm model 716 DMS Titrino and a Metrohm model 691.

All the chemicals were purchased from Aldrich unless otherwise stated: PSS $M_w \approx 70,000$ g/mol, PSS standard $M_w \approx 140,000$ g/mol, $M_w/M_n = 1.05$ (in house), $BaCrO_4$ (98%), tetrahydrofuran (THF) analytical grade, copper sulfate pentahydrate (99.999%). CdS was freshly prepared by mixing 0.1 M Na_2S and 0.1 M $CdCl_2$ followed by exhaustive washing of the precipitate with distilled water.

Results and discussion

Two different possibilities for the generation of a solubility gradient in an ultracentrifuge cell were investigated in this study. One possibility is the variation of the solvent composition of a mixture of fully miscible solvents with different density. These solvents form a density gradient upon the action of the ultracentrifugal field. A second possibility is the generation of a pH gradient by the sedimentation of an acid or base with a very dense or polymeric counterion.

Solubility gradient

A solubility gradient between THF ($\rho = 0.888$ g/ml at 25 °C) and water ($\rho = 0.997$ g/ml at 25 °C) can be established in an ultracentrifugal field. Wavelength scans of a 30:70 vol% THF/water mixture are shown in Fig. 2 at various radial positions along an ultracentrifuge cell, after running at 60,000 rpm for 14 h,

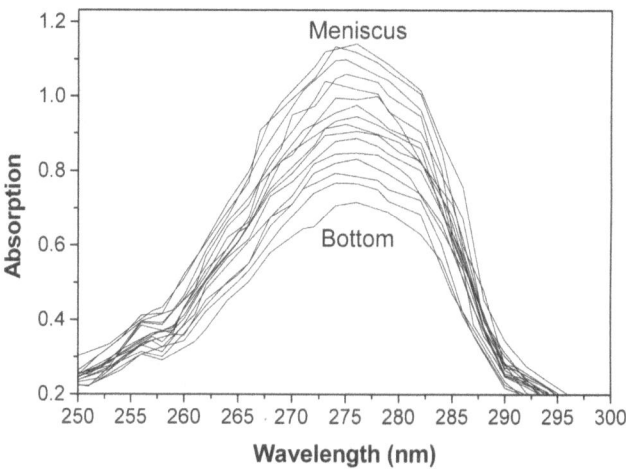

Fig. 2 Equilibrium concentration profile of a 30% tetrahydrofuran (*THF*), 70% water mixture at 60,000 rpm and 10 °C measured vs. water. The distance from axis of rotation of the individual scans are 6.30, 6.35, 6.40, 6.45, 6.50, 6.55, 6.60, 6.65, 6.70, 6.75, 6.80, 6.85, 6.90, 6.95, 7.00 and 7.10 cm

demonstrating the radial change of the THF concentration and thus of the solvent composition. Radial scans at 275 nm allow the calculation of the local THF concentrations after determination of the THF extinction coefficient (Fig. 3).

It can be seen that the solvent composition varies from 23.6 to 36.1 vol% THF throughout the ultracentrifuge cell. An experiment was performed using this solubility gradient on a saturated copper sulfate pentahydrate solution of 30 vol% THF in water. Copper sulfate pentahydrate is reported to be very soluble in water (20.4 g/100 g H_2O at 20 °C), but is found to be insoluble in THF. This was confirmed by a density measurement of the supernatant of a saturated copper sulfate solution in THF, which was identical to the density of THF to at least three significant figures. UV–vis measurements also showed the absence of copper sulfate in the THF supernatant of dispersed crystals.

Absorption scans following the copper sulfate pentahydrate showed that the concentration increased towards the bottom of the cell, where the greater fraction of water was present. To date, however, no evidence of critical crystal nuclei of copper sulfate forming at a critical THF fraction expressed in a discontinuous slope of the concentration gradient has been observed in the analytical ultracentrifuge. Nevertheless, a solubility gradient would be the cleanest way for the generation of critical crystal nuclei as it works without an additive as a gradient-forming material, which has the potential to interact with the crystals. Therefore, future investigations in this field will be more directed towards ultracentrifugation using solubility gradients with different model systems which preferably exhibit quantum size effects.

pH gradient

The first question addressed in this study was whether the generation of a real pH gradient is possible in an ultracentrifugal field. It is known that H^+ ions are the most rapidly diffusing species present, which would establish a constant pH throughout the ultracentrifuge cell unless drastic conditions are applied that lead to the accumulation of H^+ ions. One possibility towards local H^+ accumulation is the gradient formation of a dense or high-molar-mass (poly)anion such as PSS or I^-, so that H^+ ions accumulate by electrostatic interaction towards the bottom of the ultracentrifuge cell. Indeed, this leads to a pH gradient as shown in Fig. 4 for the case of PSS visualized by the indicator bromocresol purple, which changes its colour from purple to yellow between pH 6.8 and 5.2. The same can be observed for HI, where the coloured iodine is indicative for the position of I^- and thus the accumulation of H^+ (Fig. 4).

The pH change of various gradient materials was investigated in more detail in a preparative ultracentrifuge to elucidate the maximum possible pH gradient and to investigate the potentiality of pH adjustment. The results are summarized in Table 1.

It can be elucidated that the largest pH gradients are formed with a polyacid as the sedimenting material. Low-molecular-weight acids, like acetic acid derivatives, do not generate a pH gradient owing to the lack of sedimentation; however, such low-molecular-weight acids can be used to increase the pH gradient as well as to adjust the pH to a desired range, which is demonstrated by the PSS/bromosuccinic acid mixture. Also, it is possible to produce gradients with radially increasing pH as demonstrated for CsAc. It is thus possible to produce pH gradients over a wide range of conditions.

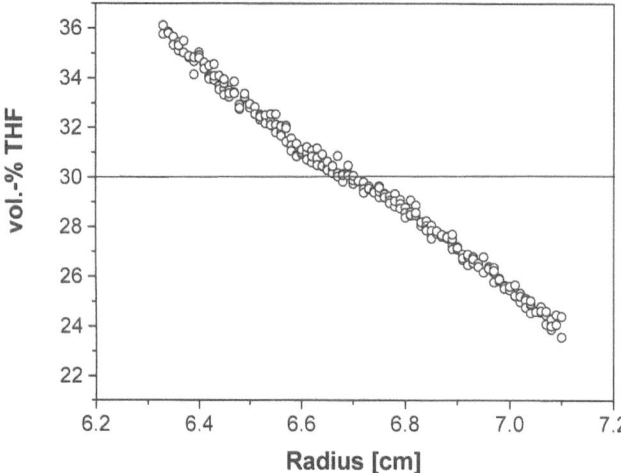

Fig. 3 THF concentration profile along the axis of rotation

Fig. 4 Demonstration of pH gradient formation by bromocresol purple as a pH indicator in a poly(sodium 4-styrene sulfonate) (*PSS*) (140,000 g/mol) gradient (*left*) centrifuged for 2 days at 60,000 rpm at 25 °C and 1.5 wt% HI centrifuged at 60,000 rpm and 25 °C for 3 days (*right*)

Table 1 Generation of pH gradients with various materials: poly(sodium 4-styrene sulfonate) (*PSS*)

Substance	pH				Force (*g*)
	Top	Middle	Bottom	ΔpH	
PSS 70,000 g/mol, 5 g/l	6.35	6.02	5.67	0.68	266,000
PSS 70,000 g/mol 5 g/l + 0.01 M bromosuccinic acid	6.66	5.57	5.14	1.52	118,000
Poly(acrylic acid) 750,000 g/mol, 4.2 g/l	4.42	3.93	3.27	1.15	266,000
Bromosuccinic acid (0.1 M)		1.83	1.63	0.2	184,000
meso-2,3-Dibromosuccinic acid (0.1 M)	1.65	1.52	1.46	0.19	184,000
2-Bromopropionic acid (0.1 M)	2.05	2	1.86	0.19	184,000
Trifluoroacetic acid (1 M)	0.34	0.32	0.25	0.07	104,000
Trichloroacetic acid (0.1 M)	1.14	1.13	1.12	0.02	104,000
Trichloroacetic acid (0.01 M)	1.59	1.59	1.56	0.03	266,000
Cesium acetate (0.5 M)	7.21	7.39	7.66	−0.45	184,000

Table 2 The effect of salt on the formation of a pH gradient

Substance	pH				Force (*g*)
	Top	Middle	Bottom	ΔpH	
PSS 70,000 g/mol, 5 g/l	6.35	6.02	5.67	0.68	266,000
PSS + 0.01 M NaCl	6.09	5.91	5.48	0.61	266,000
PSS + 0.1 M NaCl	5.69	5.63	5.52	0.17	266,000
PSS + 1 M NaCl	5.32	5.31	5.26	0.06	266,000

In addition, the possibility of adjusting the pH of polyacids/bases by neutralization was investigated; however, the accompanying salt formation reduced or even destroyed the pH gradient. This is demonstrated in Table 2 for the case of NaCl added to PSS pH gradients.

This finding shows that the pH gradients rely on the electrostatic interaction of the polyacid and the H^+ ion; therefore, the pH gradients can only be used for the dissolution of sparingly soluble salts to keep the additional counterion concentration as low as possible.

Figure 5 shows that pH gradients are real equilibrium gradients, which can be demonstrated by the path

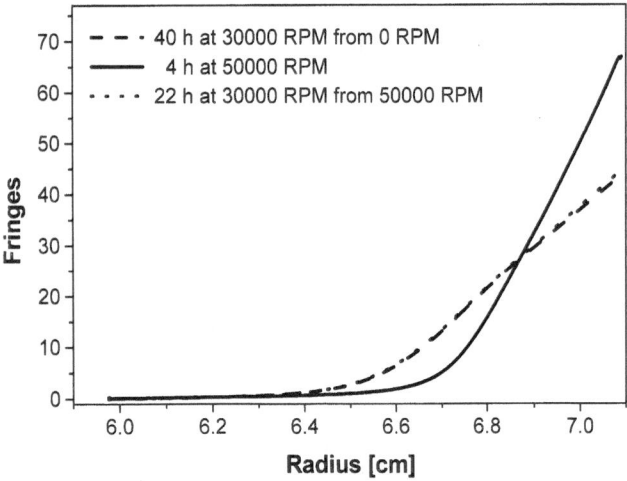

Fig. 5 Proof of PSS (140,000 g/mol) pH gradient equilibrium at 25 °C

independence of reaching an equilibrium. Furthermore, it can be seen that the pH gradient range can be adjusted through the variation of the polyacid gradient by changes in the rotational speed. Additionally, the strong deviation of the concentration gradient from the exponential shape near the cell bottom suggests high nonideality, which should be expected for such a condition.

It is possible to convert the detected fringe shifts to a PSS concentration and thus to local pH values in the ultracentrifuge cell. Using $\Delta n = J\lambda/a$, where n is the refractive index, J is the fringe shift, λ is the wavelength of light and a is the optical pathlength, the PSS concentration can be determined using the measured refractive index increment dn/dc of 0.19 ml/g (PSS 140,000 g/mol) neglecting a change in dn/dc by counterion condensation. The concentration dependence of the solution pH of PSS (140,000 g/mol) in the range 0–50 g/l could be well fitted by a first-order exponential decay function, thus allowing the calculation between pH and fringe shift:

$$pH = 3.6 + 3.1\exp(-0.05J) \ . \tag{1}$$

To determine whether inorganic crystals completely dissolve and reform in such pH gradients as outlined in Fig. 1, CdS and $BaCrO_4$ were chosen as the sparingly soluble salts which dissolve in acidic media. When CdS sediments in an HI gradient, a normal equilibrium gradient of CdS is established for initial HI concentrations up to 2 wt%. Wavelength scans show that the particle size increases from 0.8 to 1.2 nm as revealed from the band-gap positions in the UV–vis spectra, which shift from 292 to 301 nm at the cell bottom [9]. This implies that a pH threshold value must exist for the start of the CdS dissolution (which did not take place in this experiment) so that the particle size increases towards the cell bottom. This is in contrast to the expected particle size decrease for a gradual dissolution with pH according to Fig. 1. The situation changes for the more acidic gradient using 5 wt% HI. Again, an equilibrium CdS gradient is observed up to a certain point near the cell bottom, where a steep decrease in the UV absorption is detected (Fig. 6a).

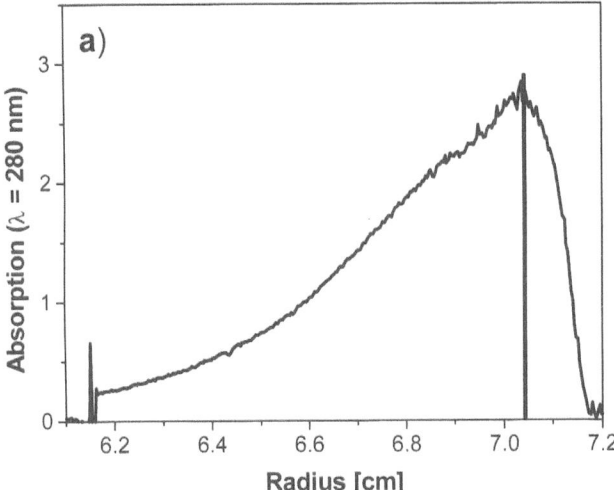

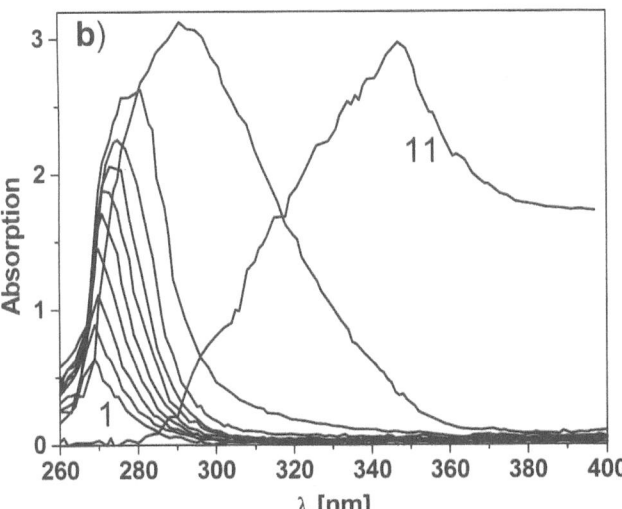

Fig. 6 **a** CdS in a HI pH gradient detected at 280 nm and 60,000 rpm (c_{CdS} = 1.4 g/l in 5 wt% H_2S saturated HI, reference 5 wt% H_2S saturated HI). **b** Wavelength scans for different radii in **a**: scan 1 = 6.2 cm, scan 11 = 7.2 cm, scan interval = 0.1 cm. Note that below 270 nm, HI absorbs, so no reliable signal could be detected for λ < 270 nm

This indicates dissolution of CdS near the cell bottom where the pH is lowest, so critical crystal nuclei could be expected at the point where the absorption just starts to drop (Fig. 6a). UV–vis spectra taken at various radial positions indicate agreement with the experiment performed at lower initial HI concentrations: the CdS size increases towards the cell bottom in the equilibrium gradient from 0.9–1.2 nm as elucidated from the bandgap positions which shift from 294 to 301 nm at r > 7.05 cm; however, the UV–vis spectra clearly differ from all those acquired at r < 7.05 cm. The spectra are now indicative of a soluble complex. Indeed, it is known that Cd halogenides form complexes at high concentrations according to [10]

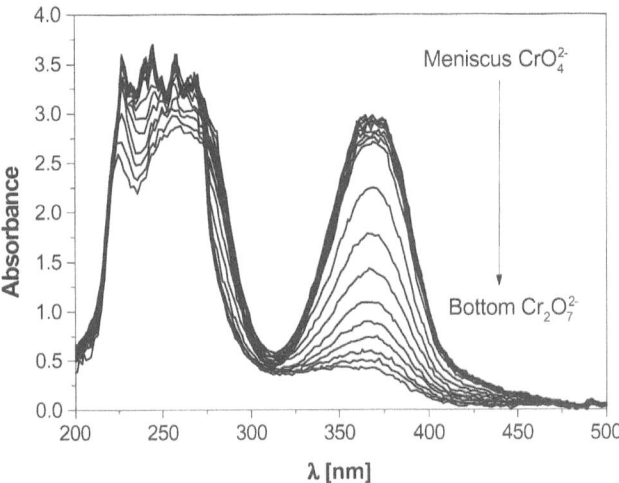

Fig. 7 UV–vis absorption spectra of $BaCrO_4$ from an initially saturated solution taken at different radial positions in a pH gradient of 5 g/l PSS (140,000 g/mol) at 50,000 rpm and 25 °C

$$2CdI_2 \rightarrow Cd[CdI_4] \ . \tag{2}$$

As a further system, we looked at the dissolution of $BaCrO_4$ in a PSS pH gradient. $BaCrO_4$ ($K_p^{25\,°C} = 1.17 \times 10^{-10}$) reversibly transforms into the more soluble $BaCr_2O_7$ upon a decrease in pH according to

$$2CrO_4^{2-} + 2H_3O^+ \leftrightarrow Cr_2O_7^{2-} . + 3H_2O \tag{3}$$

This transformation is also expressed as a colour change from yellow (CrO_4^{2-}) to orange ($Cr_2O_7^{2-}$) and can thus be followed via UV–vis absorption spectra, where $Cr_2O_7^{2-}$ shows a typical peak at 375 nm, whereas CrO_4^{2-} shows a shoulder starting at 350 nm.

Figure 7 confirms the transition of chromate to dichromate with decreasing pH in a PSS gradient. $BaCr_2O_7$, on the other hand, has a much higher solubility compared to $BaCrO_4$, determined as 0.11 mol/l at 25 °C via the absorption at 375 nm at pH 1.

Three regions can be detected in a pH gradient (Fig. 8a). In the first region, near the meniscus, which is essentially depleted of PSS as indicated by the fringe shift of 0, the pH has a constant value of 6.7 and the $BaCrO_4$ remains unchanged as indicated by the constant absorption at 375 nm in this range. The absorption also stays constant in region 2, where the pH drops to 6.5 owing to a slightly increasing PSS concentration. In region 3, however, the PSS concentration steeply increases, which leads to a pH drop to pH 4 at the cell bottom (see also Fig. 8b). The pH drop at r > 6.4 cm leads to $BaCrO_4$ transformation into the soluble $BaCr_2O_7$, so the occurrence of critical crystal nuclei can be expected in that region; however, the radial drop in the absorption at 375 nm is broad which may be the result of the nonnegligible UV absorption of dichromate at that

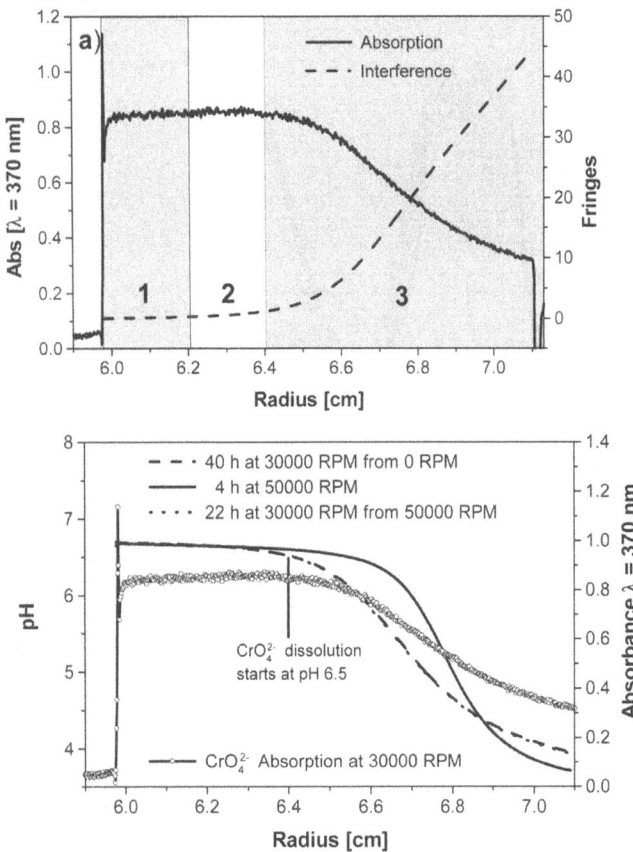

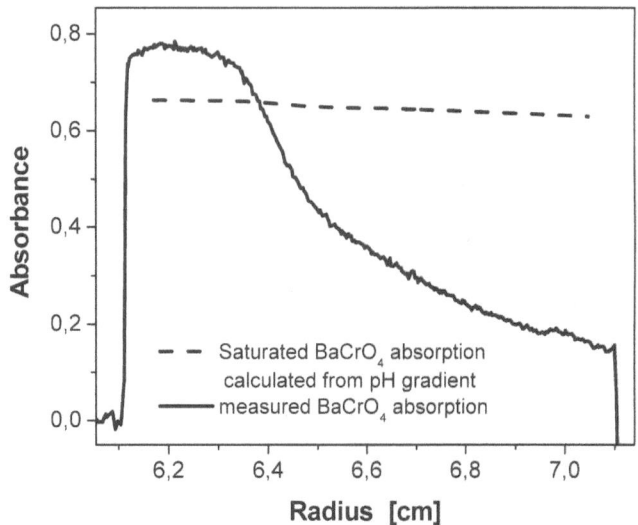

Fig. 9 Absorption of barium chromate at 375 nm, 30,000 rpm and 25 °C in 70,000 g/mol PSS (2 g/mol,) + BaCrO$_4$(saturated) compared to the absorption of barium chromate calculated via the local pH along the cell

Fig. 8 a Different regions in a pH gradient as monitored by UV absorption at 375 nm to follow the BaCrO$_4$ concentration and Rayleigh interference optics to determine the PSS (140,000 g/mol) concentration. **b** Local pH in the ultracentrifuge cell and the response from the BaCrO$_4$ monitored by the UV absorption at 370 nm

wavelength. The CrO$_4^{2-}$ concentration gradient detected at 375 nm has the same shape as the radial pH profile, showing that the mineral system directly responds to pH changes.

In order to determine the likely position of critical crystal nuclei in the ultracentrifuge cell, a graph of barium chromate absorption in relation to pH at different positions in the cell was produced. By following the PSS concentration in the analytical ultracentrifuge using the interference optics, the corresponding chromate absorption gradient can be calculated using a calibration plot. An example of the actual barium chromate absorption compared to the expected absorption of saturated barium chromate calculated from the pH at the same position in the cell is shown in Fig. 9. In this graph, a region of increased chromate concentration compared to the expected value can be identified. Thus, the solution is supersaturated and particles can be expected in that region. Near the intersection between the expected and the measured concentration profile, critical crystal nuclei should be present even if the previously described treatment neglects the dichromate absorption.

The experiment was repeated in a preparative ultracentrifuge. The fractions that were suspected to contain critical crystal nuclei were isolated in an attempt to prove the existence of critical crystal nuclei by an independent method. Transmission electron microscopy, however, only revealed spherical particle aggregates with $d \approx 50$ nm consisting of $d \approx 5$ nm primary particles. Five nanometres is a rather large size for a critical crystal nucleus, so it is likely that drying upon sample preparation led to particle growth and aggregation. Wide-angle X-ray scattering was also applied on the solutions, but the concentration was far too low to detect any crystalline species; therefore, it could not be proved by an independent experiment that critical crystal nuclei are formed by pH gradient ultracentrifugation.

It is known that analytical ultracentrifugation can yield highly resolved particle size distributions with almost atomic resolution down to sizes below 1 nm [8, 11–13]; therefore, a sedimentation velocity experiment was performed in order to determine the likelihood that particle size distributions can be measured despite the high concentration of PSS. The results are presented in Fig. 10; PSS was detected via Rayleigh interference optics and the chromate concentration was followed by UV–vis absorption at 375 nm.

It becomes evident, that the chromate–dichromate transition rapidly responds to pH changes induced by the PSS gradient. From Fig. 10a, it can be seen that steep changes in the PSS gradient composition are accompanied by steep changes in the chromate concentration. After 400 min the gradient approaches equilibrium, indicating that even at the highest centrifugal field it is impossible to fully sediment all compounds. This is

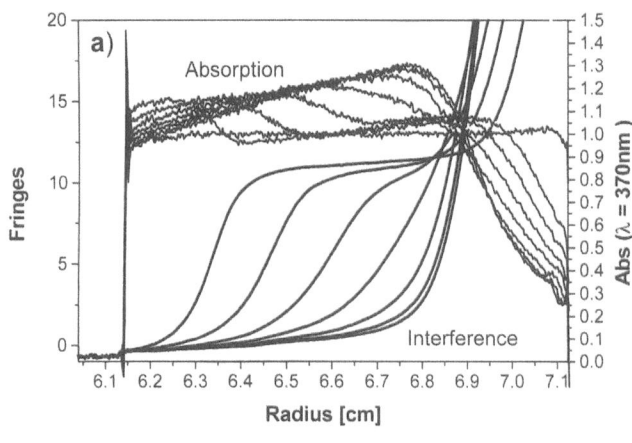

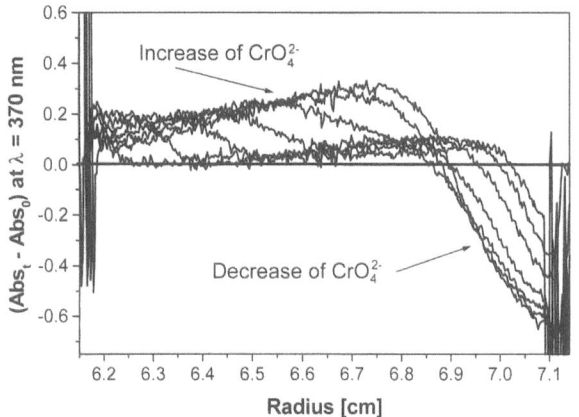

Fig. 10 a Sedimentation velocity experiment with PSS (140,000 g/mol, 5 g/l) + BaCrO₄(saturated) at 60,000 rpm and 25 °C, scan interval = 50 min. **b** Absorption scans for the same experiment after subtraction of the absorption for $t = 0$

reasonable, considering the charges in the polyelectrolyte and its high concentration.

If the chromate velocity scans are corrected for the initial chromate concentration at $t = 0$ by subtracting the initial scan from all the scans taken at later times throughout the experiment, an estimate can be obtained about regions of chromate formation and chromate dissolution (Fig. 10b). The chromate formation results from a pH increase compared to the initial solution pH, whereas the chromate dissolution responds to a pH drop. These results are in qualitative agreement with the finding from Fig. 9 which indicates a region of a supersaturated solution with particles. In Fig. 10, especially towards the end of the experiment, the absorption increases radially up to the point of the comparatively steep decrease due to dissolution. This shows a radially increasing chromate concentration up to the dissolution point, in agreement with the CdS results. Besides these methodological results, this experiment clearly demonstrates that sedimentation velocity experiments cannot be used to characterize the particles isolated from a preparative pH gradient, as pH gradients form very

quickly, thus disturbing the original equilibrium conditions; therefore, other independent methods have to be sought for the characterization of these particles.

Conclusion

Solubility gradients for inorganic particles inside the field of an analytical ultracentrifuge can be established by various methods. Whereas a solvent composition gradient only offers limited flexibility owing to the relatively small changes in the solvent composition, pH gradients can cover much higher ranges of the solubility variation of an inorganic mineral. This can be attributed to the pH gradient material, whose sedimentation equilibrium concentration can be varied widely by the polymer molar mass, rotational speed, etc. Nevertheless, the involvement of a polymeric species endangers many potential variations of the crystal growth as shown in Ref. [14]. Also, it was shown that the formation of a pH gradient relies on strong interactions of H^+ to the polyanion, and salt addition leads to the destruction of the pH gradient; therefore, pH gradients are only suitable for the dissolution of sparingly soluble salts. The ideal solubility gradient would be produced by a solvent mixture, if a high range of different solvent compositions could be reached. For the case of a pH gradient, it could be shown that true equilibrium gradients are established; therefore, a major requirement for the preparative isolation of slices containing critical crystal nuclei is fulfilled. However, the independent observation of these nuclei has not yet been realized. The characteristics of crystalline species in solubility gradients are radially exponentially increasing concentrations, as expected for a normal sedimentation equilibrium concentration gradient, followed by a concentration decrease at a defined point, indicative of particle dissolution. In the case of CdS, UV–vis spectra were observed which differ from those of crystalline CdS and are likely to represent soluble Cd complexes. For BaCrO₄, regions of supersaturation as well as BaCrO₄ depletion were identified. With all these indications of a border region between sedimenting crystalline particles and dissociated ions at the cell bottom, solubility gradient ultracentrifugation is a very promising tool for the preparative isolation of critical crystal nuclei in solution. The main advantage of this method is that these nuclei can be isolated in a stable dynamic equilibrium condition and are thus suitable for any kind of further characterization, which does not disturb the equilibrium.

Acknowledgements We acknowledge financial support from the Max Planck Society and the DFG (SFB 448). H.C. furthermore acknowledges financial support from the Dr. Hermann Schnell foundation.

References

1. Gibbs JW (1961) The collected works of JW Gibbs. Yale University Press, New Haven
2. Kashchiev D (1982) J Chem Phys 76:5098
3. Oxtoby DW, Kashchiev D (1994) J Chem Phys 100:7665
4. Oxtoby DW (1998) Acc Chem Res 31:91
5. Mutaftschiev B (1993) In: Hurle DTJ (ed) Handbook of crystal growth. Elsevier, Amsterdam, p 189
6. Yau ST, Vekilov PG (2000) Nature 406:494
7. Oxtoby DW (2000) Nature 406:464
8. Cölfen H, Pauck T (1997) Colloid Polym Sci 275:175
9. Vossmeyer T, Katsikas L, Giersig M, Popovic IG, Diesner K, Chemseddine A, Eychmüller A, Weller H (1994) J Phys Chem 98:7665
10. Hollemann AE, Wiberg N (1985) Lehrbuch der anorganischen Chemie, 91st edn. de Gruyter, New York, p 1042
11. Cölfen H, Pauck T, Antonietti M (1997) Progr Colloid Polym Sci 107:136
12. Börger L, Cölfen H (1999) Progr Colloid Polym Sci 113:23
13. Börger L, Cölfen H, Antonietti M (2000) Colloids Surf A 163:29
14. Qi L, Cölfen H, Antonietti M (2000) Angew Chem Int Ed Engl 39:604

Progr Colloid Polym Sci (2002) 119:19–23
© Springer-Verlag 2002

Peter Lavrenko
Kirill Yevlampiev
Darya Volokhova

Hybrid polymer heterogeneity and boundary profile in a low-speed sedimentation experiment

P. Lavrenko (✉) · D. Volokhova
Institute of Macromolecular Compounds
Russian Academy of Sciences
199004 St. Petersburg, Russia
e-mail: lavrenko@mail.macro.ru

K. Yevlampiev
St. Petersburg State University
Physical Department, Petergoff
198904 St. Petersburg, Russia

Abstract Solution properties of a new hybrid polymer (fullerene-containing polymer, FP) with a regular structure of macromolecules have been studied. In the starlike polymer molecule, six equal polystyrene chains and six poly(*tert*-butyl methacrylate) ones are attached as the arms to the fullerene C_{60} molecule used as a core. Hydrodynamic characteristics of the FP sample were compared with parameters of the six-arm star FP. The boundary profile in a low-speed sedimentation experiment obtained in several solvents with different refractive indices was analyzed to determine the inhomogeneity of the hybrid polymer sample.

Key words Fullerene-containing polymer · Low-speed experiment · Boundary profile · Hide-and-seek technique · Diffusion

Introduction

Analytical ultracentrifugation is a powerful method for precise analysis of polydisperse systems of macromolecules – particularly for the characterization of such newly engineered macromolecules such as fullerene-containing star chains [1, 2]. The present work deals with an analysis of the solvent–solution boundary spreading in a low-speed regime which provides the diffusion coefficients and the heterogeneity parameters. In this regime, the boundary was formed in a synthetic-boundary cell. The method has been applied to a hybrid star polymer using the "hide-and-seek" expedient to display information on the sample heterogeneity. Direct treating of the W-like boundary profiles was also performed.

Experimental

Sample and solutions

The fullerene-containing polymer (FP) with a regular starlike structure of the macromolecules was synthesized as described earlier [1]. Each of the macromolecules consists of the fullerene C_{60} core and twelve arms: six polystyrene (PS) arms and six poly(*tert*-butyl methacrylate) (P*t*-BMA) ones. The masses of the PS and the P*t*-BMA arm chains were 4 and 5 kDa, respectively; hence, the molecular weight of the FP sample was expected to be close to 55 kDa.

The solutions were prepared in benzene, bromoform, and *o*-dichlorobenzene (*o*-DCB) and the viscosity, density, and refractive index values measured at 26 °C were as follows:.

	$\eta_0 \times 10^2$ (g cm^{-1}s)	ρ_0 (g/ml)	n_D
Benzene	0.590	0.8740	1.4980
Bromoform	1.850	2.8912	1.5956
o-Dichlorobenzene	1.380	1.2973	1.5454

Methods

Analytical ultracentrifugation was performed in a 3180 analytical ultracentrifuge (AUC) from the Hungarian Optical Works MOM equipped with interference optics as described in Ref. [3]. The rotor speed was 8,000 rpm with an accuracy of ± 20 rpm. The temperature of the rotor (26.0 °C) was maintained with an absolute accuracy of ± 0.1°C and with a constancy of regulation of ± 0.05 °C. A polyamide synthetic boundary double-sector cell with a 12-mm light path and quartz windows was used.

In long-time experiments, the free diffusion of the polymer molecules in solution was investigated using a gradient method in which a sharp boundary was formed between the solvent and the solution in a Teflon cell [4] with $h = 2.0$-cm light path. A Tsvetkov interferometric diffusometer with spar twinning, a, (0.10 cm) and a compensator fringe interval, b, (0.15 cm) was used [5].

All the data were evaluated via nonweighted linear regression analysis and refined by least squares.

Results and discussion

The type and the form of the boundary profile, $\delta(x)$, depends on the value and the sign of the difference between the refractive indices of the solvent and the polymer. Remember

$$\delta(x)\partial n/\partial x = (dn/dc)\partial c/\partial x \; ,$$

where c is the solute concentration (in grams per liter) and dn/dc is the specific refractive increment of the polymer–solvent system. Thus, different profiles can be expected in experiments with solvents of different optical density. Particularly, if $dn/dc = 0$ then $\delta(x) = 0$. In other words, polymer molecules are invisible in a solvent with the same refractive index.

Boundary profile in FP solution in benzene and bromoform

A contour of the typical experimental fringe, $\delta(x)$, and its change with time is shown in Fig. 1 as observed in a solution of FP in a low-speed experiment. Positive (in benzene) and negative (in bromoform) peaks were observed. This is due to the fact that the refractive index of the polymer, n_{FP}, is higher than that of benzene, n_{benz}, but lower than that of n_{brom}; therefore, dn/dc is positive (in sign) in the FP–benzene system and negative in the FP–bromoform one.

The dn/dc value was estimated from the interference fringe area, Q, by

$$dn/dc = (Q/c)(\lambda/abh) \; ,$$

where λ is the wavelength of light. The average values of $(dn/dc)_{546}$ thus obtained are 0.036 ml/g in benzene and −0.08 ml/g in bromoform, which agree well with the content of the FP sample and the refractive indices of the solvents.

At low frequency of the AUC rotor rotation, spreading of the concentration solvent–solution boundary with time is mainly due to the diffusion phenomenon. Let us now analyze the translational mobility of the macromolecules in the solvents used.

Diffusion of macromolecules in benzene and bromoform

The curves in Fig. 1 are symmetric in form, and the experimental time is sufficiently long for a significant change in the curve shape to calculate the diffusion coefficient, D. The interference curves were treated with the "height–area" method [3]. The second central moment (dispersion), $\overline{\sigma^2}$, of the $\partial c/\partial x(x)$ distribution was evaluated by

$$\overline{\sigma^2} = (a^2/8)[\text{argerf}(aH/Q]^{-2} \; ,$$

where H is the maximum ordinate and argerf means an argument of the probability integral. The results are represented in Fig. 2, where the $\overline{\sigma^2}$ values are plotted against the time/viscosity ratio to eliminate the solvent viscosity effect. The experimental points 1 and 2 form a linear function (open and full circles, respectively). D was evaluated from the slope of the $\overline{\sigma^2}(t)$ dependence by $D = (1/2)\partial\overline{\sigma^2}/\partial t$, where t represents time. D was extrapolated to vanishing solute concentration. The centrifugal field effect was estimated to be negligible in the experiments.

It is important to note that points 1 and 2 in Fig. 2 form a common function. This means that reduced

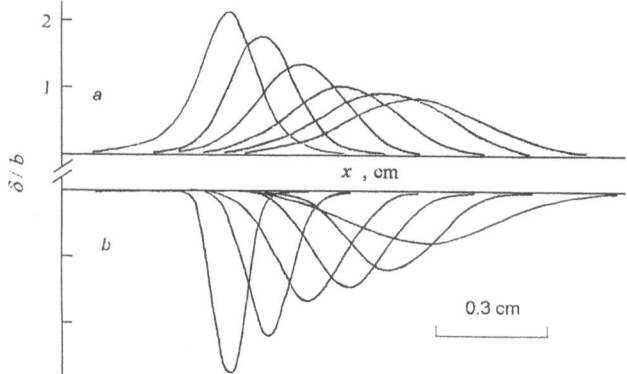

Fig. 1 Solvent (*left*)–solution (*right*) boundary profile and its change with time as observed in a solution of fullerene-containing polymer (*FP*) in benzene (*a*) and bromoform (*b*) at 26 °C at (from *left* to *right*) a 0.25, 1, 2, 3, 4, and 5 h and at b 0.7, 2, 4, 5, 7, and 18 h, respectively, after the start. The solute concentration was 3.00 g/l (*a*) and 0.80 g/l (*b*). For convenience, every second curve is shifted to the right by 0.10 cm. Here and in Fig. 3, δ is the fringe-shift number and b the fringes interval

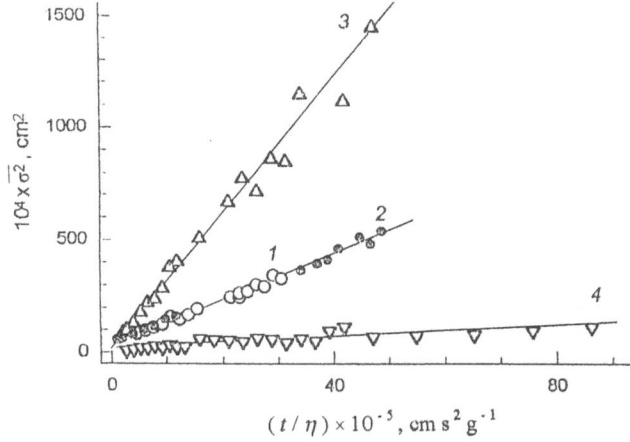

Fig. 2 Dispersion, $\overline{\sigma^2}$ versus the time/viscosity ratio as calculated from the $\partial c/\partial x$ curves presented in Fig. 1 and treated by the "height–area" method for FP in benzene (*1*), bromoform (*2*), and *o*-dichlorobenzene (*3, 4*) at 26 °C. The solute concentration was 2.06 (*1*), 0.80 (*2*), and 30.0 g/l (*3, 4*)

translational mobility of the particles responsible for the FP mass transport in benzene solution coincides with that in bromoform. Remember again that the refractive index of benzene coincides practically with that of Pt-BMA, $n_{benz} \approx n_{Pt\text{-}BMA}$, and, hence, in an experiment with FP–benzene solution we "do not see" the distribution of the Pt-BMA chains. In turn, $n_{brom} \approx n_{PS}$. Hence, in FP–bromoform solution we "do not see" the distribution of the PS chains. We can conclude, therefore, that the particles responsible for the diffusion phenomenon in a solution of FP are composed of both PS and Pt-BMA chains bound together. This result obtained with the "hide-and-seek" expedient produces substantial evidence for the FP molecule structure.

As a result, we have the $D_{FP} \times \eta_0$ value (4.8×10^{-9} gcm) which is lower than that obtained in the same solvent earlier for the six-arm fullerene-containing star PS [2]. We now combine the diffusion–viscometric data to form an expression, $T^3/(D_{FP} \times \eta_0)^3[\eta]$, which is sensitive to molecular weight only [6]. For FP, its numerical value is twice that of the six-arm star PS, and, hence, the mass of the FP molecule is twice that of the six-arm star PS. This is again in accord with the expected FP molecule structure.

Boundary profile in FP solution in o-DCB

The refractive index of o-DCB, n_{DCB}, lies in the middle of the interval from n_{benz} to n_{brom}, and, therefore, the FP molecules are practically invisible in this solvent: $n_{FP} \approx (n_{benz} + n_{Brom})/2$; however, a small amount of the

two other components in the sample was detected when the solute concentration was increased many times. The principal pictures obtained in this way are represented by curves 1–3 in Fig. 3. The W-like boundary profile is formed here as a sum of two peaks: a positive ($+$) and a negative one ($-$), the positive peak being wider and faster spreading in time. The areas of the corresponding curves are indicated in the following as Q^+ and Q^-, respectively.

To characterize the components, these peaks were graphically resolved and treated separately in the Gaussian approximation as described earlier. The results are represented by points 3 and 4 in Fig. 2. One can see that these components (impurities) differ significantly from the main substance in optical and hydrodynamic properties. For a low-molecular-weight component with dn/d$c > 0$, we obtained $(D\eta_0)^+ = 13 \times 10^{-9}$ gcm. This value is very close to that of the separate PS arm chains [2], and, hence, the positive peak in curves 1–3 in Fig. 3 can be referred to the initial PS which was not involved in the synthesis. Part of the PS component does not exceed 7% (by weight) as estimated from the positive part, Q^+, of the experimental curve area.

For a high-molecular-weight component with dn/d$c < 0$, we obtained $(D\eta_0)^- = 0.54 \times 10^{-9}$ gcm. The negative dn/dc value is expected for the Pt-BMA–o-DCB system. The part of this component in the FP sample does not exceed 3% (by weight) as estimated from the negative curve area, Q^-, with molecular weight [7] $M > 1 \times 10^6$. If we take into account insolubility of Pt-BMA in o-DCB, we have to assume that the high-molecular-weight component consists mainly of Pt-BMA bound with a small amount of a substance which improves its solubility.

Inhomogeneity parameters from the low-speed sedimentation experiment

The W-like boundary profile was also analyzed without independent treatment of the resolved positive and negative peaks using the simple method derived later for the profiles in the Gaussian approximation (ideal diffusion).

First, we analyze the experimental curves 1–3 in Fig. 3. The area of the summary curve is proportional to the difference between the refractive indices of solvent and solution, respectively, $\Delta n = Q\lambda/abh$, and is 7.1×10^{-5}. In a solvent with $n = n_{DCB} + \Delta-/2$ (such as o-DCB + bromoform, 0.072wt%), we have $Q^+ = Q^-$, and the polymer substance is not visible, $\delta(x) = 0$, at the start at any x. Later, if $D^+ \neq D^-$, the x-distribution of the macromolecules looks like curves 4–6 in Fig. 3. The distances x_0, X_0, y, and Y are obviously the principal parameters of the W curve. Correlation between these parameters reflects clearly the difference between the

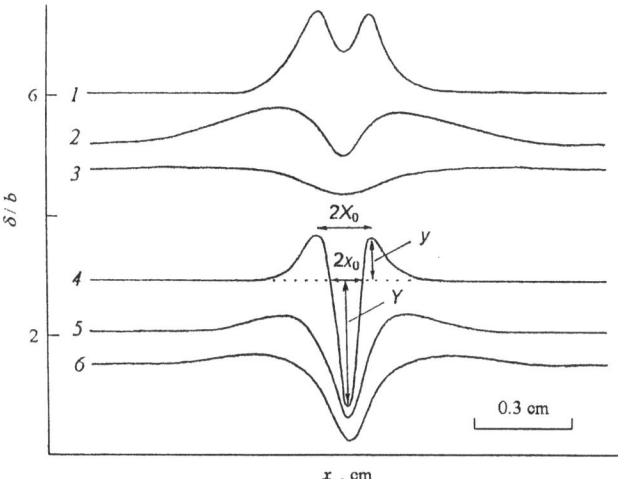

Fig. 3 Principal solvent (*left*)–solution (*right*) boundary profile and its change with time as observed in an experiment with a solution of FP in o-dichlorobenzene at 26 °C and the solute concentration 30.0 g/l (curves *1–3*), and corresponding normalized distributions $(1/c_0)\partial c/\partial x$ (*4–6*) at times 1 (*1, 4*), 7.5 (*2, 5*), 55 (*3*), and 18 h (*6*), respectively, after the start. See the text for more details

diffusion coefficients of the components and can be used for the D^- and D^+ determination.

The boundary profile can be approximated by the Gaussian function

$$\delta(x) = (1/c_0)\partial c/\partial x = \left[(2\pi)^{1/2}\sigma\right]^{-1}\exp\left(-x^2/2\overline{\sigma^2}\right) ,$$

where σ is the standard deviation, $\sigma = \left(\overline{\sigma^2}\right)^{1/2}$. The dispersion, $\overline{\sigma^2}$, of the x spectrum increases with time during the low-speed run as $\overline{\sigma^2} = 2Dt$ (we use here the D symbol instead of the D_s quantity usually obtained from a sedimentation experiment). For each of the components, $i = 1$ and 2, we write

$$\delta_i(x) = (\pi D_i t)^{-1/2}\exp(-x^2/4D_i t) . \tag{1}$$

Let $D_2 \geq D_1$, $(dn/dc)_1 < 0$, and $(dn/dc)_2 > 0$. The summation leads to

$$\delta(x) = \delta_2(x) + \delta_1(x) .$$

At the start, $\delta(x) = 0$ at any x. Latter, if $D_1 \neq D_2$, it looks like curves 4–6.

In accordance with the scheme given in Fig. 3, $\delta(x) = 0$ if $x = x_0$. For this abscissa point, we readily obtain

$$x_0^2 = \frac{2\ln(D_2/D_1)}{1/D_1 - 1/D_2}t . \tag{2}$$

In turn, $\partial\delta/\partial x = 0$ if $x = X_0$. For this abscissa point, we obtain

$$X_0^2 = \frac{6\ln(D_2/D_1)}{1/D_1 - 1/D_2}t . \tag{3}$$

Let A denote the numerical multiplier in Eq. (3):

$$A \equiv \frac{6\ln(D_2/D_1)}{1/D_1 - 1/D_2} . \tag{4}$$

The A value thus obtained is represented in Fig. 4 as a function of D_1 and D_2.

On the other hand, if $x = 0$, then

$$\delta(x = 0) \equiv Y = (\pi t)^{-1/2}(D_1^{-1/2} - D_z^{-1/2})$$

and we write the $\delta(x = 0)/\delta(x = X_0) \equiv Y/y$ ratio as

$$Y/y = \frac{D_1^{-1/2} - D_2^{-1/2}}{D_1^{-1/2}\exp\left(-\frac{3}{2}\frac{\ln D_2/D_1}{1-D_1/D_2}\right) - D_2^{-1/2}\exp\left(-\frac{3}{2}\frac{\ln D_2/D_1}{D_2/D_1-1}\right)} . \tag{5}$$

The ratio Y/y thus evaluated is represented in Fig. 5 as a function of D_1 and D_2.

The system of transcendental Eqs. (3) and (5) was resolved graphically when the approach was applied to treating of the experimental W-like boundary profiles.

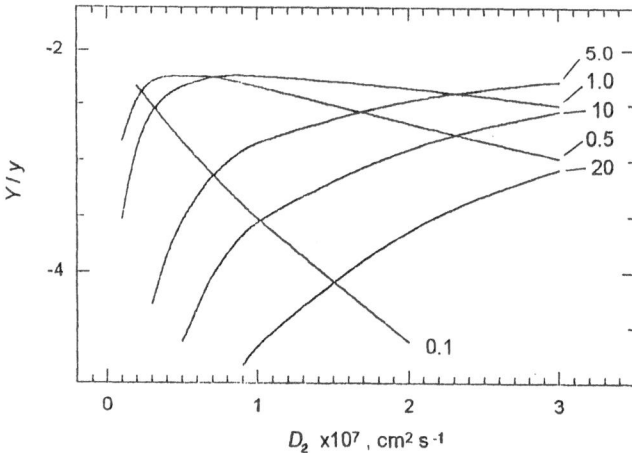

Fig. 5 Plot of Y/y versus D_2 as calculated by Eq. (5) with D_1 ($\times 10^7$) from 0.1 to 20 cm^2/s

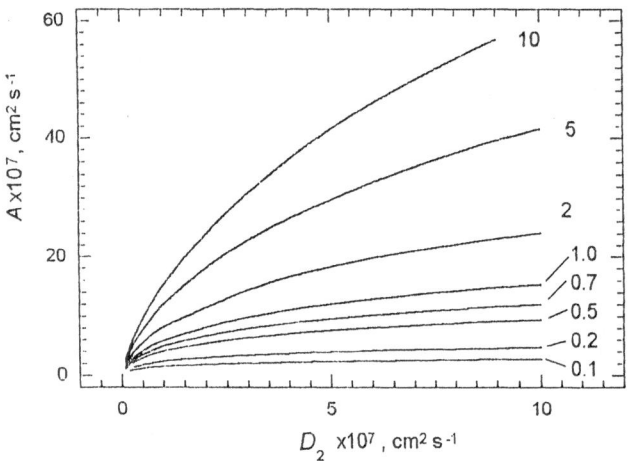

Fig. 4 Plot of A versus D_2 as calculated by Eq. (4) with D_1 ($\times 10^7$) from 0.1 to 10 cm^2/s

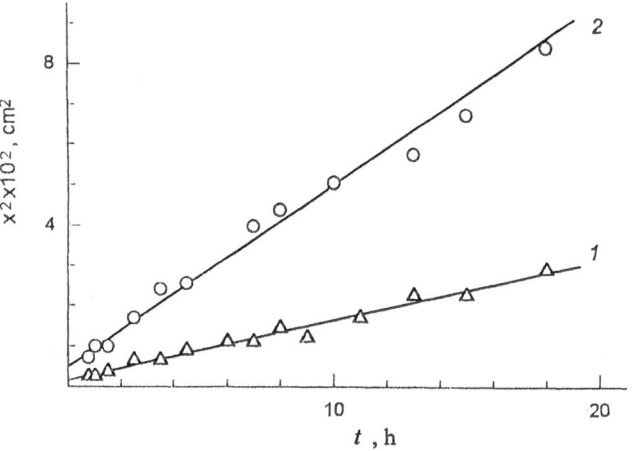

Fig. 6 Quantities x_0^2 (*1*) and X_0^2 (*2*) versus time for a normalized distribution $(1/c_0)\partial c/\partial x$ in the experiment with FP solution in *o*-dichlorobenzene at 26 °C and a solute concentration of 30.0 g/l

Application to the experimental data

We apply the method to the experimental data obtained for FP in solution in o-DCB and represented in Fig. 3. The values of X_0^2 and x_0^2 are plotted in Fig. 6 against time as obtained from curves 4–6 in Fig. 3. The experimental points are well approximated by a linear function, and, within experimental error, the slope of curve 2 is three times that of curve 1. This is in accord with Eqs. (2) and (3). The slope yields a value of 10.3×10^{-7} cm^2/s for the A parameter. This value is available if D_1 is correlated with D_2. The correlation is represented by curve 1 in Fig. 7 as derived from Fig. 4.

For the same profile curves, the ratio Y/y was evaluated to be 4.0 ± 0.5, which leads to the correlation between D_1 and D_2 represented by the shaded area between curves 2 and 3 in Fig. 7 as derived from Fig. 5. The common part of the curves corresponds to the expected D_1 and D_2 values which are in a good agreement with the D^- and D^+ quantities as evaluated previously by the independent method. These results confirm the validity of the method and the approximations used.

Conclusions

The method was derived for treating the W-like boundary profiles obtained in the low-speed regime of analytical ultracentrifugation. The method yields characteristics of the heterogeneous polymer components: weight part and difference in the diffusion coefficients. The validity of the method was confirmed by direct comparison with the independently obtained free-diffu-

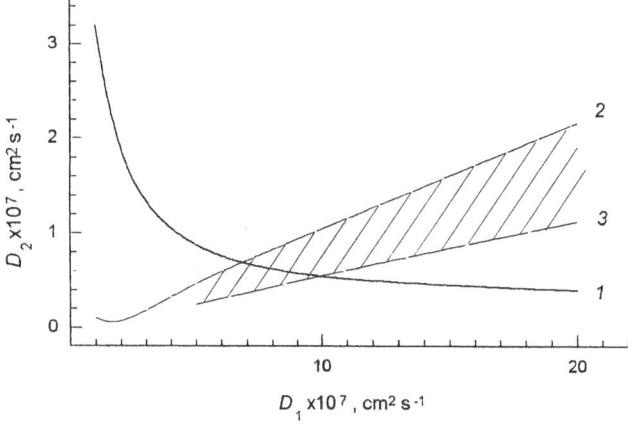

Fig. 7 Correlation between D_2 and D_1 available for FP in o-dichlorobenzene at 26 °C as obtained from the experimental A value (curve 1) and from the Y/y ratio of 4 ± 0.5 (*shaded area between curves 2 and 3*)

sion data. This technique is a formal analogy with the differential sedimentation method [8] combined with the "hide-and-seek" expedient [9]. It may be useful in addition to other methods [10–20] which were developed earlier for characterization of the polymer inhomogeneity using either the stationary diffusion data or the results of a low-speed sedimentation experiment.

Acknowledgements The authors express their gratitude to L. Vinogradova for providing the polymer sample. The work was supported in part by the Russian Foundation for Fundamental Researches through grant no. 00-03-33083a.

References

1. Vinogradova LV, Lavrenko PN, Amsharov KY, Zgonnik VN (2001) Vysokomol Soedin (in press)
2. Lavrenko PN, Vinogradova LV (2000) Polym Sci Ser A 42:726
3. Lavrenko P, Lavrenko V, Tsvetkov V (1999) Prog Colloid Polym Sci 113:14
4. Lavrenko PN, Okatova OV, Khohlov KS (1977) Instrum Exp Tech USSR 20:1477
5. Tsvetkov VN (1989) Rigid-chain polymers. Consultants Bureau, New York
6. Tsvetkov VN, Lavrenko PN, Bushin SV (1984) J Polym Sci Polym Chem Ed 22:3447
7. Kozhokaryu M, Skazka VS, Berdnikova KG (1966) Vysokomol Soedin 8:1063
8. Kirschner MW, Schachman HK (1971) Biochemistry 10:1900
9. (a) Frenkel SY (1960) Vysokomol Soedin 2:731; (b) Frenkel SY (1966) Vysokomol Soedin 8:1657
10. Gralen H (1941) Kolloid-Z 95:188
11. Charlwood PA (1953) J Phys Chem 57:125
12. Daune M, Freund L (1957) J Polym Sci 23:115
13. Sundelof L-O (1966) Arkiv Kemi 25:1
14. (a) Cherkasov AN, Klenin SI, Eizner YE (1965) Vysokomol Soedin 7:902; (b) Cherkasov AN, Klenin SI, Eizner YE (1968) Vysokomol Soedin Ser A 10:1971
15. Cussler EL (1965) J Phys Chem 69:1144
16. Fomin GA (1973) Vysokomol Soedin Ser A 15:1917
17. Brown JC, Pusey PN (1974) J Phys D 7:131
18. Porsch B, Kubin M (1974) Collect Czech Chem Commun 39:3494
19. Okabe M, Matsuda H (1983) J Appl Polym Sci 28:2325
20. Bushin SV, Astapenko EP, Smirnov KP (1993) Vysokomol Soedin Ser A 35:573

Progr Colloid Polym Sci (2002) 119:24–30
© Springer-Verlag 2002

Studies on the partial specific volume of a poly(ethylene glycol) derivative in different solvent systems

Christos Tziatzios
Andrei A. Precup
Christian H. Weidl
Ulrich S. Schubert
Peter Schuck
Helmut Durchschlag
Walter Mächtle
Jacomina A. van den Broek
Dieter Schubert

C. Tziatzios (⊠) · J.A. van den Broek
D. Schubert
Institut für Biophysik, Johann Wolfgang
Goethe-Universität, 60590 Frankfurt am
Main, Germany
e-mail: tziatzio@biophysik.uni-frankfurt.
de
Tel.: +49-69-63015835
Fax: +49-69-63015838

A. A. Precup · C. H. Weidl
U. S. Schubert
Laboratory of Macromolecular
and Organic Chemistry,
Eindhoven University of Technology,
5600 MD Eindhoven, The Netherlands

P. Schuck
Division of Bioengineering and Physical
Science, ORS, National Institutes of
Health, Bethesda, MD 20892-5766, USA

H. Durchschlag
Institut für Biophysik und Physikalische
Biochemie, Universität Regensburg, 93040
Regensburg, Germany

W. Mächtle
Kunststofflaboratorium, BASF AG, 67056
Ludwigshafen, Germany

Abstract The specific volume of charged supramolecular compounds dissolved in organic solvents varies considerably with the solvent system applied; in addition, it is influenced by the presence of salt. In this study we determined the specific volume of an uncharged molecule from the same molar mass range in order to find out whether it shows the same dependencies. To allow application of solvents of widely differing polarity, including water, a poly(ethylene glycol) derivative of molar mass 3,650 g/mol was used as a model system. The primary method applied for determining the specific volume was the buoyant density method, in which sedimentation equilibrium experiments using solvent mixtures of different density are performed and the specific volume is obtained as the reciprocal of that solvent density for which the compound is neutrally buoyant. A second method applied for determination of the specific volume was digital densimetry. We found that the strong influence of the solvent on the specific volume observed with charged compounds is also shown by the uncharged poly(ethylene glycol) derivative, the differences in the specific volume between different solvent systems amounting up to 15%; however, no significant dependence on the presence of salt was observed. We also found that, with the compound studied, a simple rule relating the specific volume and solvent polarity apparently does not exist.

Key words Partial specific volume · Sedimentation equilibrium analysis · Organic solvents · Uncharged compounds

Introduction

Supramolecular compounds are important objects of study in current chemistry owing to their potential usefulness in nanotechnology and material sciences [1]. There are, however, serious obstacles to be overcome before these compounds can be successfully applied. One of them is the control of their association behaviour in solution [2], which, in one case, was already shown to be crucial for the formation of ordered surface films [3]. To address this problem, our group has demonstrated that sedimentation analysis in the analytical ultracentrifuge is a highly useful method for the characterization of the state of association of solubilized supramolecular compounds [4, 5]. First studies applying the technique have been reported [6–10].

In all work applying sedimentation analysis and in many other techniques including small-angle X-ray and neutron scattering, the partial specific volume, $\bar{v}$, or related volume quantities of the macrosolutes under analysis, represent essential parameters. They are required for the evaluation of molar masses, interaction parameters (preferential binding and exclusion of principal solvent and/or third components), and other molecular characteristics [11–14].

In multicomponent solutions, volume determinations may be performed at constant molality or constant chemical potential of added solvent components, thereby yielding partial specific or isopotential specific volumes. While the value for the partial specific volume does not change significantly with the experimental conditions, the isopotential specific volume may exhibit drastic changes as a response to changes in the environment, for example, the presence of high concentrations of (charged) cosolvents [13–15]. Consequently, pronounced differences between these two quantities may occur, even at vanishing concentration of additional solvent components (Fig. 1 in Ref. [15]). This has been demonstrated in detail for biological polyelectrolytes such as proteins and nucleic acids [13, 14]. Interaction parameters may be calculated only if both volume quantities are known. Preferably, volumes at vanishing solute concentration should be compared to avoid unexpected concentration interferences. The specific volume of small molecules, in particular salts or other electrolytes, shows a marked dependence on the concentration [13, 14].

For proteins or nucleic acids in aqueous solutions, the determination of specific volumes does not present a serious problem, and the variation in the specific volume following changes in buffer composition in most cases is small and can be easily corrected for [13, 14] (but see Refs. [16, 17]). In contrast, with charged supramolecular compounds in organic solvents the situation is more complex: changing the (organic) solvent or adding a supporting electrolyte (necessary to suppress the primary charge effect [18, 19]) leads to large and up to now unpredictable changes in the specific volume [7, 20]. In addition, significant dependencies of the specific volume on the compound concentration were observed. This makes the application of the standard technique for determining the specific volume, digital densimetry, questionable, owing to the much higher compound concentrations required compared to those used in sedimentation analysis [13, 14]. Thus, determining the specific volume seems to be the vulnerable spot in sedimentation analysis of supramolecular compounds, in particular of charged ones.

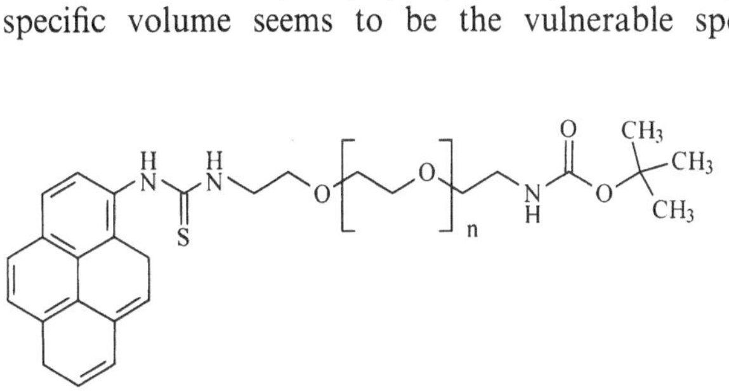

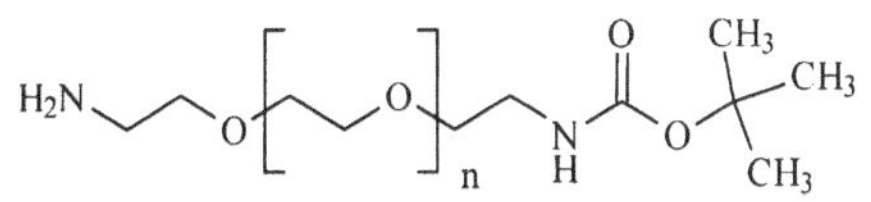

To examine whether the observations described are restricted to electrically charged compounds, we studied the dependency of the specific volume on solvent for an uncharged molecule from the same molar mass range as used for the earlier studies [4, 7, 20]. To allow application of solvents of widely differing polarity, including water, we used a poly(ethylene glycol) (PEG) derivative as a model system. Most of our experiments made use of the buoyant density method, in which the "effective" specific volume, $\bar{v}_{eff}$, is determined, by sedimentation equilibrium experiments performed in solvent mixtures of varying density, as the reciprocal of that solvent density for which the compound does not sediment. In addition, classical digital densimetry was applied; the partial specific volulme obtained will be termed "$\bar{v}$". We show that the large variability of the specific volume described previously is also found with the uncharged compound studied now. We also show that the results of the buoyant density method can be highly misleading with respect to the specific volumes in the pure solvents, if inadequate solvent combinations have been used.

Materials and methods

The PEG derivative used as a model system (Fig. 1a) was obtained by modifying a commercial precursor (Fig. 1b; nominal $M_w = 3,400$ g/mol; Shearwater Polymers): To a solution of 50 mg precursor in 3 ml dimethylformamide (DMF), a solution of 5.3 mg 1-pyrene isothiocyanate (Molecular Probes) in 2 ml DMF was added, and the mixture was stirred for 48 h at 40 °C. After evaporation of the solvent, the compound was redissolved in dichlormethane and purified by size-exclusion chromatography on Bio Beads SX1 (Bio Rad; 70 × 5 cm) in the same solvent. According to matrix-assisted laser desorption ionization time of flight mass spectroscopy using a Voyager DE Pro apparatus (Perceptive Biosystems) with α-cyano cinnamic acid/tetrahydrofuran (THF) as a matrix, the purified compound was characterized by $M_n = 3,637$ g/mol and $M_w = 3,679$ g/mol.

The organic solvents and salts used were purchased from Merck or Sigma/Aldrich and were of analytical grade or for spectroscopy. D_2O (99.7%) was a gift from the Karlsruhe Nuclear Research Centre, and $D_2^{18}O$ (98%) was purchased from Johnson Matthey.

Sedimentation equilibrium experiments on the PEG derivative were performed using a Beckman Optima XL-A ultracentrifuge, using an An-50Ti rotor. The cells were equipped with titanium double-sector centrepieces (BASF) and polyethylene gaskets. The rotor temperature was 20 °C and the rotor speed 40,000 rpm. The sample volume was 200 μl and the sample concentration approximately 0.3 g/l. Sedimentation equilibrium was established within approximately 25 h. The absorbance versus radius distributions were recorded between 340 and 350 nm (i.e., in a broad absorbance maximum of the dye-labelled PEG), depending on the solvent system used. They were evaluated as described in Refs. [4, 5, 7] using the computer program DISCREEQ [21, 22]. Effective molar masses, $M_{eff} = M(1 - \bar{v}_{eff}\rho_0)$, were determined from one-component fits with zero baseline position. If these fits were of insufficient quality, multicomponent fits were applied, the effective weight-average molar mass, $M_{w,eff}$, replacing M_{eff}. The latter fits assumed the presence of several oligomers characterized by integral multi-

Fig. 1 The poly(ethylene glycol) (*PEG*) derivative **A** used as a model compound and **B** its precursor

ples of the effective protomer molar mass, $M_{1\text{eff}}$ (as calculated from the known M_1, the actual solvent density, and the $\bar{v}_{\text{eff}}$ value determined from the buoyant density of the compound in the same solvent mixture). $M_{\text{w,eff}}$ was obtained from the relative contributions, $\bar{c}_i$, of the different oligomers to total compound concentration, averaged over the sample volume (i is the number of protomers per oligomer) [21, 22]:

$$M_{\text{w,eff}} = \frac{\sum_{i=1}^{n} \bar{c}_i i M_{1\text{eff}}}{\sum_{i=1}^{n} \bar{c}_i}.$$

The calculated $M_{\text{w,eff}}$ was insensitive against variations in the oligomer set or the $\bar{v}_{\text{eff}}$ value applied as long as the fit was of good quality. $M_{\text{w,eff}}$ or M_{eff} is, of course, sensitive against variations in the baseline position (except for very small M_{eff} values). We have found, however, that with samples which have undergone gel permeation chromatography (or recrystallization) the baseline position is close to zero [4, 7, 20]. In the present study this is confirmed by the virtual agreement of the $\bar{v}_{\text{eff}}$ values determined either from data points around $M_{\text{eff}} = 0$ or from the slope of a straight line fitted to a major range of the data (see later).

An Optima XL-I ultracentrifuge, interference optics, and Epon centrepieces were used for sedimentation velocity runs, in phosphate-buffered saline, on the starting material for the compound studied (Fig. 1b). The rotor speed was 50,000 rpm and the sample volume 400 μl. The fringe displacement profiles obtained were modelled using Lamm equation solutions, applying the program SEDFIT, as described in Refs. [23–25].

A Paar DMA 02 density meter was used for the measurements of $\bar{v}$ and the solvent density by digital densimetry at 20 °C [26]; the solute concentrations were 5–10 g/l. The calculations for predicting the partial specific volume of the dye-labelled PEG derivative in aqueous solution made use of an algorithm based on volume increments and certain corrections [27, 28]. For conductivity measurements a WTW LF521 device (Wissenschaftlich-Technische Werkstätten, Weilheim), which was calibrated by 0.1% NaCl in water, was applied.

Results

The specific volume of charged supramolecular compounds dissolved in organic solvents depends on the nature of the solvent used; in addition, it is influenced by the presence of salts [7, 20]. The experiments on the PEG derivative shown in Fig. 1 were aimed at finding out whether or not similar behaviour is shown by this uncharged compound. In addition, they were aimed at comparing the results of the two methods applied for determining the specific volume: the buoyant density method and digital densimetry. The solvent systems used were acetonitrile/propylene carbonate, THF/propylene carbonate, THF/chloroform, and $H_2O/D_2O/D_2^{18}O$ with the buoyant density method; acetonitrile, propylene carbonate, THF, chloroform, H_2O, and a THF/chloroform mixture (23:77 v/v) with digital densimetry.

Determination of $\bar{v}_{\text{eff}}$ by the buoyant density method

The goal of these experiments was the determination of the $\bar{v}_{\text{eff}}$ values of the PEG derivative in the solvents

acetonitrile, THF, chloroform, and water. For this purpose, sedimentation equilibrium runs were performed in four different series of binary solvent mixtures, and the effective molar mass, $M_{\text{eff}} = M(1 - \bar{v}_{\text{eff}}\rho_0)$, of the solute or its multioligomer equivalent, $M_{\text{w,eff}}$, was determined as a function of the solvent density, ρ_0. The dependency of the evaluated M_{eff} or $M_{\text{w,eff}}$ on the solvent density for each solvent system, in the absence and, with two of the mixtures, also in the presence of salt (tetrabutylammonium hexafluorophosphate), is shown in Fig. 2a and b. According to the Fig. 2 the transition from sedimentation to flotation ($M_{\text{eff}} = 0$) is well-defined by the data and thus allows determination of $\bar{v}_{\text{eff}}$ with good accuracy (either on the basis of data points around $M_{\text{eff}} = 0$ or by fitting a straight line to a major range of the data).

As shown by Fig. 2a, the solvent pairs acetonitrile/propylene carbonate and THF/propylene carbonate yield $\bar{v}_{\text{eff}}$-values quite close to each other. On the other hand, the $\bar{v}_{\text{eff}}$-values in $H_2O/D_2O/D_2^{18}O$ and THF/chloroform are distinctly different both from the

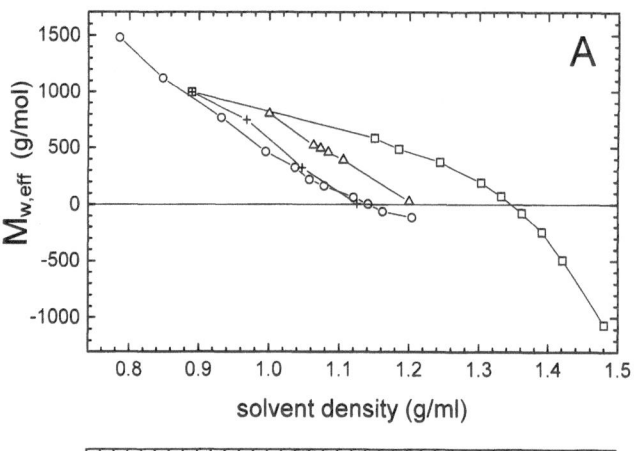

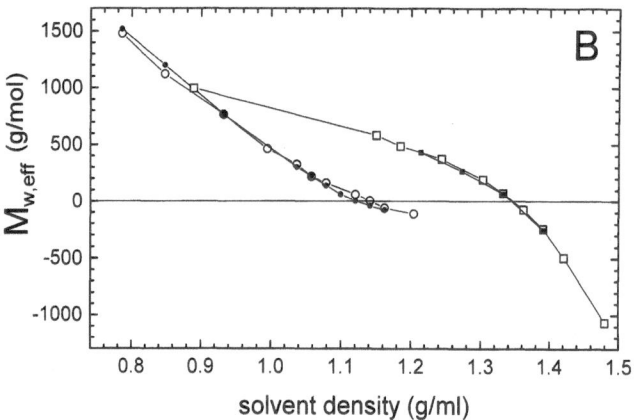

Fig. 2A, B $M_{\text{w,eff}}$ versus solvent density for the PEG derivative of Fig. 1a in different solvents or solvent mixtures. **A** Data collected in salt-free solutions. **B** Comparison of data from **a** with those obtained in the presence of 10 mM tetrabutylammonium hexafluorophosphate (*filled symbols*). Solvent systems: acetonitrile/propylene carbonate (*circles*), tetrahydrofuran/chloroform (*squares*), tetrahydrofuran/propylene carbonate (*crosses*), $H_2O/D_2O/D_2^{18}O$ (*triangles*)

former ones and from each other. The maximum differences in $\bar{v}_{eff}$ are shown by the solvent systems acetonitrile/propylene carbonate (or THF/propylene carbonate), on the one hand, and THF/chloroform, on the other: $\bar{v}_{eff}$ in the former system is 0.870 ml/g if taken directly from the data or 0.900 ml/g if taken from a straight line which fits most of them, whereas the zero crossover of the data for the latter solvent mixture yields $\bar{v}_{eff} = 0.743$ ml/g. The differences between the two $\bar{v}_{eff}$ values thus amount to 15 or 17% of the higher values. Surprisingly, the curve for $H_2O/ D_2O/ D_2^{18}O$ is intermediate between the others, yielding 0.829 ml/g. Figure 2b shows that, in contrast to the findings with a charged supramolecular compound, addition of salt has no or only a small effect on the $\bar{v}_{eff}$ value of the present model compound: in the polar solvent mixture, the difference in $\bar{v}_{eff}$ is only 0.022 ml/g and in the apolar one it is zero. (In the latter case, however, the effective ionic strength is strongly reduced by ion pair formation, as shown by conductivity measurements. Nevertheless, distinct influences of the salt were found with charged compounds [20].)

It should be noted that with the THF/chloroform mixture the plot of $M_{w,eff}$ versus solvent density is strongly nonlinear. As emphasized earlier [20], in such cases it is highly questionable whether the $\bar{v}_{eff}$ value determined from the zero crossover of the plot is applicable to other solvent compositions from the same solvent system, in particular to the corresponding pure solvents. On the other hand, with the other solvent systems the plot is either fully linear or linear over an extended range of solvent densities which includes the pure solvent and needs only slight extrapolation to reach the $M_{eff} = 0$-line. This suggests that the $\bar{v}_{eff}$ values deduced will be applicable to the pure solvent.

Determination of $\bar{v}$ by digital densimetry

To find out whether the $\bar{v}_{eff}$ values determined with solvent mixtures by the buoyant density method agree with those for the constituent pure solvents, additional $\bar{v}$ determinations were performed by applying digital densimetry. The $\bar{v}$ values with acetonitrile, propylene carbonate, THF, chloroform, and H_2O are shown in Table 1, together with the volume determined with the THF/chloroform mixture which corresponds to the zero crossover of the corresponding curve of Fig. 2a. The table also contains the results obtained by the buoyant density method and the densities and dielectric constants of the solvents. In contrast to our experience with charged supramolecular compounds [7], the present $\bar{v}$ measurements did not suffer from poor reproducibility (with the exception of the THF/chloroform mixture) or a strong dependency on solute concentration, which led to relatively small error estimates. A comparison of the digital densimetry data with those obtained by the buoyant density method shows both similarities and discrepancies: whereas the results for samples in acetonitrile or in THF (when propylene carbonate was the second solvent), and in H_2O are virtually identical or at least very close to the corresponding data from equilibrium sedimentation, $\bar{v}$ determined by densimetry in both THF and chloroform strongly differs from that determined by the buoyant density method from THF/chloroform mixtures. Even more strikingly, the $\bar{v}$ value obtained by digital densimetry in the THF/chloroform mixture in which the compound does not sediment differs from the reciprocal density of the solution by approximately 0.14 ml/g or 19%.

We have also tried to calculate the partial specific volume of the PEG derivative in aqueous solutions using established methods [27, 28]. An ab initio calculation led to $\bar{v} = 0.847$ ml/g, which is 1.7–2.2% higher than the experimental results (the same deviation was observed

Table 1 Specific volumes of the poly(ethylene glycol) derivative in different solvents and solvent mixtures, as determined by digital densimetry and the buoyant density method. The density, ρ_0, and the dielectric constant, ε, of the solvents, at 20 °C, are taken from Refs. [29, 30]

Solvent	ρ_0 (g/ml)	ε	$\bar{v}$ (ml/g) (digital densimetry)	$\bar{v}_{eff}$ (ml/g) (buoyant density method)
Acetonitrile	0.78	37.5	0.903 ± 0.002	0.870 (acetonitrile/propylene carbonate)
Propylene carbonate	1.204	65.1[a]	0.862 ± 0.008	0.887 (tetrahydrofuran/propylene carbonate)
THF	0.889	7.6	0.873 ± 0.007	0.743 (tetrahydrofuran/chloroform)
Chloroform	1.483	4.8	0.839 ± 0.005	
H_2O	1.00–1.216	78.5	0.833 ± 0.009	0.829 ($H_2O/D_2O/ D_2^{18}O$
Tetrahydrofuran/ chloroform mixture (23:77 v/v)	1.347		0.886 ± 0.020	0.743

[a] At 25 °C

with unmodified PEG [27]). On the other hand, a "calculus of difference" [7] for the derivative, starting with the experimental value $\bar{v} = 0.838$ ml/g for unmodified PEG [27], yields 0.831 ml/g, which is identical with the value(s) measured by us within the uncertainty of the experimental data.

The state of association of the PEG derivative in different solvents

For future experiments we have planned to use the PEG derivative described here or a related one as a building block for the assembly of supramolecular and macromolecular aggregates; therefore, we investigated its state of association in most of the pure solvents applied in the other studies described here. The analyses were based on the partial specific volume determined in the respective solvents by digital densimetry and shown in Table 1.

In all the solvents used, acetonitrile, THF, chloroform, and H_2O, the predominant form of the PEG derivative was found to be the monomer; however, up to approximately 40% of the material was present in the form of dimers or even larger aggregates (up to approximately octamers). An example is shown in Fig. 3, where the best fit to the data yielded, besides $(60 \pm 10)\%$ monomer, $(40 \pm 10)\%$ dimer (averaged over the sample volume). The exact percentage and nature of the oligomer was barely dependent on the solvent but mainly, for unknown reasons, on the batch of the PEG derivative used. A higher average state of

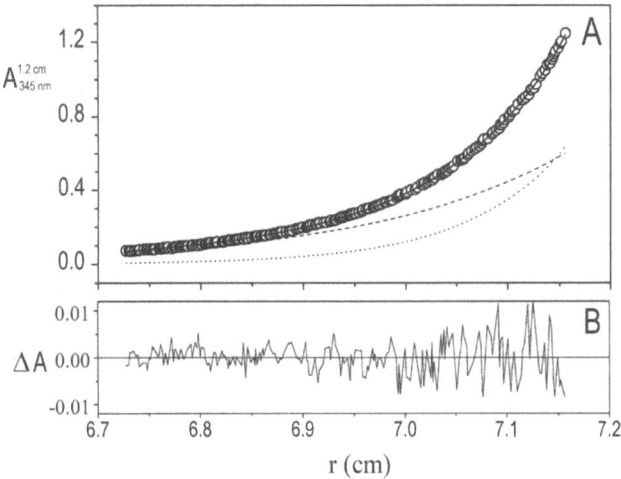

Fig. 3A, B Sedimentation equilibrium analysis of the PEG derivative based on $\bar{v} = 0.903$ ml/g and a monomer molar mass of 3,650 g/mol. **A** Experimental $A(r)$ data collected at 345 nm (*circles*), fitted curve assuming the presence of monomers and dimers of the compound (—), and calculated local contributions of monomers (- - -) and dimers (······). **B** Local differences $\Delta A(r)$ between experimental and fitted data. Solvent: acetonitrile. Initial compound concentration: approximately 0.2 g/l

association of the compound was accompanied by changes in the UV–vis spectra and even in the colour of the solution, indicating that oligomer formation most probably involves stacking of the pyrene constituent of the solute molecules. The oligomer formation was largely irreversible (in agreement with the constant slope over an extended density range of most of the curves of Fig. 2a, which strongly suggests that under the corresponding conditions not only the specific volume but also M_w is virtually independent of solvent composition). The nature of the association suggested was supported by studies on the state of association of the precursor of the PEG derivative used (Fig. 1b): in sedimentation velocity experiments, the experimental data could be perfectly fitted by assuming the presence of a single species which, by determination of the sedimentation coefficient, s, and the diffusion coefficient, D, in H_2O and D_2O [23–25], could be identified as the monomer.

Discussion

In previous reports, our group described the application and tests for the reliability of the buoyant density method for the determination of specific volumes, $\bar{v}_{eff}$, of charged supramolecular compounds [7, 20]. Our efforts stemmed from the observation that in several cases classical digital densimetry seemed to fail to yield correct results, presumably owing to the formation of clusters or microcrystals (at the high solute concentrations required) which had $\bar{v}$ values different from those of monomers or small oligomers of the compounds. The principle of the buoyant density method is to deduce $\bar{v}_{eff}$, for a given compound in a given solvent, by studying mixtures of that solvent and a related solvent of distinctly different density, and by determining the density of that solvent mixture for which the compound does not sediment. A major advantage of the method is that $\bar{v}_{eff}$ is determined at the same low solute concentration as applied during ultracentrifugal analysis of the molar mass or association behaviour of the compound. Another advantage is that much smaller amounts of solute and much lower solute solubilities are required than with digital densimetry. The main disadvantage is the dependency of $\bar{v}_{eff}$ on the solvent, which may lead to significant differences between the specific volume in the solvent of interest and that determined, by the buoyant density method, in a mixture with another solvent [7, 20]. We have suggested a set of criteria which should be fulfilled in order to make a $\bar{v}_{eff}$ value determined by the buoyant density method reliable [20]:

1. A plot of M_{eff} (or $M_{w,eff}$) versus ρ_0 should be linear over an extended range of solvent densities (i.e., M_w and $\bar{v}_{eff}$ should be constant in this range).
2. This density range should include the solvent condition at which the molar mass or the state of

association of the compound will be studied, and it also should include densities where the compound is (at least almost) neutrally buoyant.

3. It should be critically examined whether the linearity of the M_{eff} versus ρ_0 plot is really due to constancy of M_w and $\bar{v}_{eff}$ and not to compensating changes in these parameters.

We have shown previously for a metal coordination array as an electrically charged model compound that suitable solvent systems can be found which fulfil these requirements [20]. With this compound we also observed that $\bar{v}_{eff}$ can vary considerably when the solvent or the solvent system is changed.

In the present work, we have described analogous studies on an uncharged compound of similar mass: a PEG derivative (dye-labelled to allow sedimentation analysis using absorption optics). In contrast to the charged metal coordination arrays studied earlier [7], no indications could be found for a failure of digital densimetry in determining $\bar{v}$ for the present compound; therefore, in the present case digital densimetry appears to be a suitable approach for testing the buoyant density method. In fact, after identifying suitable solvent systems that meet the criteria for the buoyant density method, we obtained specific volume values virtually identical to those from digital densimetry. Interestingly, the distinct solvent dependency of the specific volume was also confirmed for the uncharged PEG derivative. Contrasting the findings with metal coordination arrays [7, 20], we found (with both methods) that the specific volumes in the polar solvents acetonitrile and propylene carbonate and the apolar solvent THF are virtually identical. Such a lack of influence of the solvent polarity on the specific volume may, however, be restricted to uncharged compounds.

The most striking result of the present study concerns a solvent mixture where the application of the buoyant density method for determining $\bar{v}_{eff}$ in the pure constituent solvents apparently failed: the solvent mixture THF/chloroform. In this case, the $\bar{v}_{eff}$ value determined from the compound's buoyant density strongly differed from the values determined by digital densimetry, in both THF and chloroform. In addition, its use in modelling the $A(r)$ data collected in THF did not result in a fit of acceptable quality (whereas the data collected in chloroform could be fitted by applying the specific volumes from both methods). It could be suspected that the divergent $\bar{v}_{eff}$ value is an artefact, caused by demixing of the two solvents of similar polarity but strongly differing density or caused by the formation of a density gradient [31]; however, we think that both effects can be ruled out as an explanation for the discrepancies:

1. We could not find any indication of a macroscopic phase separation in the equilibrium distribution $A(r)$ of the compound.

2. According to calculations by applying the Hermans–Ende equation [32, 33] to a 50:50 v/v THF/chloroform mixture, the density difference between the meniscus and the bottom of the ultracentrifuge cell under the experimental conditions amounts to only approximately 0.009 g/ml; it is distinctly smaller if one of the solvents is in excess. Thus, an apparent shift in the buoyant density of the compound in THF/chloroform mixtures caused by a density gradient will not exceed a few grams per litre and can be neglected. We conclude that the measured $A(r)$ distributions at nearly zero M_{eff} are virtually free of artefacts and that the sedimentation equilibrium of the compound in the corresponding solvent mixtures is in fact described by the $\bar{v}_{eff}$ $\bar{v}$ value obtained by the buoyant density method. This $\bar{v}_{eff}$ value will, however, be applicable only for a limited range of solvent compositions (around the one corresponding to $M_{eff} = 0$). On the other hand, our data show that the $\bar{v}$ value measured in these solvent mixtures by digital densimetry (Table 1) fails to describe the corresponding sedimentation equilibrium data. Probably, this failure is due to neglecting preferential interactions between the compound and chloroform (solvent binding or exclusion from the solute molecules), similar to those described for other systems and studied by other techniques [34].

It should be kept in mind that the extent of the solute/solvent interactions included in measurements of specific volumes is obviously quite different when applying densimetry, on the one hand, and ultracentrifugation utilizing the buoyant density method, on the other. In the present study, digital densimetry was performed under isomolal conditions and yields a true partial specific volume, $\bar{v}$, thereby widely neglecting interactions, whereas the buoyant density method seems to include most of the operative interactions, similar to the isopotential specific volume which may be obtained, by densimetry or conventional sedimentation equilibrium, when working with compounds dialysed to equilibrium prior to the measurements [11, 35]. Preferential binding of one of the solvents to the PEG derivative could also be responsible for the deviation from linearity of the $M_{w,eff}$ versus ρ_0 data observed with the solvent pairs acetonitrile/propylene carbonate and THF/propylene carbonate, at high contents of the second solvent (Fig. 2).

Interestingly, straight lines drawn, in Fig. 2a, through the $M_{w,eff}$ values in acetonitrile and propylene carbonate, THF and propylene carbonate, acetonitrile and chloroform, THF and chloroform, or propylene carbonate and chloroform will be quite close to each other, and their zero crossover will yield $\bar{v}_{eff}$ values relatively close to the $\bar{v}$ values determined, in the pure solvents, by digital densimetry. It remains to be seen whether or to what extent this behaviour is accidental.

It is clear from the foregoing that the buoyant density method for determining $\bar{v}_{\text{eff}}$, though being laborious and requiring considerable ultracentrifuge capacity, should be an important tool for a variety of ultracentrifugal studies in supramolecular and macromolecular chemistry. Its main area of application will be the investigation of compounds with limited availability or with a tendency towards strong self-association and concomitant changes in $\bar{v}$ either of which would prevent the use of classical densimetry. However, additional efforts will have to be made and the existing database will have to be distinctly enlarged before the method can be applied routinely. Above all, suitable solvent pairs will have to be established for a large number of different classes of compounds. These solvent pairs will have to take into account the solubility, polarity, selective solvent binding and, possibly, other molecular properties of the compounds.

Acknowledgements Part of this work was supported by grants from the Deutsche Forschungsgemeinschaft (SFB 266 and SFB 486) and the Dutch Polymer Institute to U.S.S. We are grateful to M.D. Lechner for help concerning the density gradient problems.

References

1. Lehn J-M (1995) Supramolecular chemistry – concepts and perspectives. VCH, Weinheim
2. Lawrence DS, Jiang T, Levett M (1995) Chem Rev 95:2229
3. (a) Semenov A, Spatz JP, Moeller M, Lehn J-M, Sell B, Schubert D, Weidl CH, Schubert US (1999) Angew Chem 111:2701; (b) Semenov A, Spatz JP, Moeller M, Lehn J-M, Sell B, Schubert D, Weidl CH, Schubert US (1999) Angew Chem Int Ed Engl 38:2547
4. Schubert D, van den Broek JA, Sell B, Durchschlag H, Maechtle W, Schubert US, Lehn J-M (1997) Prog Colloid Polym Sci 107:166
5. Schubert D, Tziatzios C, Schuck P, Schubert US (1999) Chem Eur J 5:1377
6. (a) Schütte M, Kurth DG, Linford MR, Cölfen H, Möhwald H (1998) Angew Chem 110:3058; (b) Schütte M, Kurth DG, Linford MR, Cölfen H, Möhwald H (1998) Angew Chem Int Ed Engl 37:2891
7. Tziatzios C, Durchschlag H, Sell B, van den Broek JA, Maechtle W, Haase W, Lehn J-M, Weidl CH, Eschbaumer C, Schubert D, Schubert US (1999) Prog Colloid Polym Sci 113:114
8. (a) Newkome GR, Cho TJ, Moorefield CN, Baker GR, Cush R, Russo PS (1999) Angew Chem 111:3899; (b) Newkome GR, Cho TJ, Moorefield CN, Baker GR, Cush R, Russo PS (1999) Angew Chem Int Ed Engl 38:3717
9. Kurth DG, Lehmann P, Volkmer D, Cölfen H, Koop MJ, Müller A, Du Chesne A (2000) Chem Eur J 6:385
10. Tziatzios C, Durchschlag H, González JJ, Albertini E, Prados P, de Mendoza J, Eschbaumer C, Schubert US, Schuck P, Schubert D (2000) Polym Prepr Am Chem Soc Div Polym Chem 41:934
11. Eisenberg H (1976) Biological macromolecules and polyelectrolytes in solution. Clarendon, Oxford
12. Eisenberg H (1981) Q Rev Biophys 14:141
13. Durchschlag H (1986) In: Hinz H-J (ed) Thermodynamic data for biochemistry and biotechnology. Springer, Berlin Heidelberg New York, pp 45–128
14. Durchschlag H (2001) In: Hinz H-J (ed) Landolt–Börnstein new series VII/2A. Springer, Berlin Heidelberg New York (in press)
15. Durchschlag H (1989) Colloid Polym Sci 267:1139
16. Shima S, Tziatzios C, Schubert D, Fukada H, Takahashi K, Ermler U, Thauer RK (1998) Eur J Biochem 258:85
17. Shima S, Thauer RK, Ermler U, Durchschlag H, Tziatzios C, Schubert D (2000) Eur J Biochem 267:6619
18. Fujita H (1975) Foundations of ultracentrifugal analysis. Wiley, New York
19. Schachman HK (1959) Ultracentrifugation in biochemistry. Academic, New York
20. Tziatzios C, Durchschlag H, Weidl CH, Eschbaumer C, Maechtle W, Schuck P, Schubert US, Schubert D (2001) ACS symposium series. American Chemical Society, Washington, DC (in press)
21. Schuck P (1994) Prog Colloid Polym Sci 94:1
22. Schuck P, Legrum B, Passow H, Schubert D (1995) Eur J Biochem 230:806
23. Schuck P, MacPhee CE, Howlett GJ (1998) Biophys J 74:466
24. Schuck P (1998) Biophys J 75:1503
25. Schuck P, Demeler B (1999) Biophys J 76:2288
26. Kratky O, Leopold H, Stabinger H (1973) Methods Enzymol 27:98
27. Durchschlag H, Zipper P (1994) Prog Colloid Polym Sci 94:20
28. Durchschlag H, Zipper P (1997) J Appl Crystallogr 30:803
29. Merck & Co (1983) The Merck index, 10th edn. Merck & Co, Rayway
30. Weast RC, Selby SM (1967) Handbook of chemistry and physics, 48th edn. CRC, Cleveland
31. Mächtle W (1992) In: Harding SE, Rowe AJ, Horton JC (eds) Analytical ultracentrifugation in biochemistry and polymer science. Royal Society of Chemistry, Cambridge, pp 147–175
32. Hermans JJ, Ende HA (1963) J Polym Sci Part C Polym Symp 1:161
33. Mächtle W, Lechner MD (2001) Prog Colloid Polym Sci
34. Lange H (1964) Kolloid Z Z Polym 199:128
35. Casassa EF, Eisenberg H (1964) Adv Protein Chem 19:287

Progr Colloid Polym Sci (2002) 119:31–36
© Springer-Verlag 2002

POLYMERS, COLLOIDS AND SUPRAMOLECULAR SYSTEMS

Characterization of serum albumin nanoparticles by sedimentation velocity analysis and electron microscopy

Vitali Vogel
Klaus Langer
Sabine Balthasar
Peter Schuck
Walter Mächtle
Winfried Haase
Jacomina A. van den Broek
Christos Tziatzios
Dieter Schubert

V. Vogel · J. A. van den Broek
C. Tziatzios · D. Schubert (✉)
Institut für Biophysik
Johann Wolfgang Goethe-Universität,
60590 Frankfurt am Main, Germany
e-mail: schubert@biophysik.uni-frankfurt.de
Tel.: +49-69-63015835
Fax: +49-69-63015838

K. Langer · S. Balthasar
Institut für Pharmazeutische Technologie,
Johann Wolfgang Goethe-Universität,
60439 Frankfurt am Main, Germany

P. Schuck
Division of Bioengineering and Physical
Science, ORS, National Institutes of
Health, Bethesda, MD 20892-5766, USA

W. Mächtle
Kunststofflaboratorium,
BASF AG, 67056 Ludwigshafen,
Germany

W. Haase
Max-Planck-Institut für Biophysik,
60528 Frankfurt am Main, Germany

Abstract Nanoparticles prepared by desolvation and subsequent cross-linking of human serum albumin (HSA) represent promising carriers for drug delivery. We have studied the particle size distribution and the shape of such HSA nanoparticles in aqueous solutions by sedimentation velocity analysis in the analytical ultracentrifuge and by electron microscopy (EM) of negatively stained samples. The sedimentation of the particles was approximated as ideal and with negligible diffusion, and their particle size distribution was characterized by calculating the sedimentation coefficient distribution, $g^*(s)$, by applying the computer program SEDFIT. Broad distributions were obtained; the s values within any given preparation might vary by as much as a factor of approximately 5. In addition, depending on the preparation the $s_{20,w}$ values of the maxima of the distributions varied over a range of 5,000–20,000 S. Since, according to EM, the particles are spherical, the distributions of the sedimentation coefficients could be transformed into distributions of radii, with the corresponding maxima being located between 85 and 160 nm. The ordinate values in the latter distributions can be corrected for Mie light scattering, which yields true relative concentration versus particle radius plots. The broad distributions described could be fractionated into narrow distributions by preparative sucrose density gradient centrifugation. Such isolated fractions may be useful for studying the relationships between the size of the (loaded) HSA carriers and their biological activity. In addition, the procedure described for the analysis of the size distribution of the HSA nanoparticles may be of significant value for the optimization of the method of preparation.

Key words Serum albumin nanoparticles · Particle size distribution · Sedimentation velocity analysis · Analytical ultracentrifugation · Fractionation of nanoparticles

Introduction

The primary objective of any therapy with pharmacologically active compounds is the controlled delivery of the compound to its site of action. However, frequently only an undesirably small fraction of the drug effectively reaches its target, whereas the remainder may be responsible for causing unwanted or toxic side effects.

A large number of different concepts for enhancing the targeting of the drugs to their specific sites of action have been proposed and developed in attempts to alleviate this situation [1, 2]. One of these is using colloidal carrier systems, such as microemulsions, liposomes, and nanoparticles, as vehicles for drug delivery. The present work is concerned with the last of these systems.

Nanoparticles are solid colloidal particles in a size range between 10 and 1,000 nm. They consist of macromolecular material and can be prepared by polymerization or by dispersion of preformed polymers [3]. In principle, a drug compound can be bound to the surface of nanoparticles after particle preparation or it can be entrapped into the particle matrix during the particle formation. The building blocks of such particles can be of either synthetic or natural origin. Among the latter building blocks, proteins are of special interest. Nanoparticles can be assembled from suitable proteins either by emulsion techniques [4, 5] or by desolvation processes [6, 7].

Besides the surface structure and the functional groups present on the particle surface, the most prominent features of a nanoparticular drug delivery system are its particle size and its particle size distribution [8–10]. These parameters are interesting not only from a physicochemical point of view but have been shown to be of critical influence on both the distribution and the cellular uptake of drug-loaded nanoparticles in biological systems [1, 2]. In pharmaceutics, photon correlation spectroscopy (PCS) is used as the standard method for particle sizing; however, because of the well-known problem of mathematically ill-conditioned analysis of the autocorrelation function [11], the resulting particle size distributions can be strongly biased, being critically dependent on the model used (i.e., unimodal or multi-modal distribution, analysis with or without regularization). We therefore attempted the application of an alternative method: sedimentation velocity analysis in the analytical ultracentrifuge, a standard technique for characterizing particle size distributions in macromolecular chemistry [12–17]. Because the migration of particles in a gravitational field of an ultracentrifuge shows a stronger size dependence than the diffusion coefficients measured by PCS, this method has the potential for significantly higher resolution [17, 18] and, therefore, is much less model-dependent. In the present article, we show that the sedimentation velocity method is ideally suited also for analysing the size distribution of nanoparticles prepared by desolvation and subsequent cross-linking of human serum albumin (HSA) [9]. Since this method, like PCS, only reveals distributions of equivalent Stokes radii, we collected additional shape information on the particles from electron microscopy (EM).

Materials and methods

Materials

HSA (fraction V), glutardialdehyde (8% solution), and fluorescein isothiocyanate (FITC) were obtained from Sigma (Steinheim, Germany). All other reagents were from Merck (Darmstadt, Germany) and were of analytical grade.

Preparation of HSA nanoparticles

HSA nanoparticles were prepared using a previously described desolvation technique [6, 9]. HSA (200 mg) was dissolved in 2.0 ml deionized water followed by the desolvation of the protein solution by the dropwise addition of 6.0 ml ethanol with constant stirring. Directly following the desolvation step, 235 μl 8% aqueous glutardialdehyde solution was added to achieve HSA cross-linking. After stirring for 24 h, the resulting nanoparticles were purified by five cycles of centrifugation (20,000g, 8 min) and redispersion of the pellet in water to the original volume by ultrasonication in a bath-type sonicator. All the preparations were performed at room temperature. The nanoparticle content of the final dispersion after the washing process was determined gravimetrically: A 50.0-μl aliquot of the nanoparticle suspension was transferred to aluminium boats (Lüdi, Flawil, Switzerland) and dried for 4 h at 100 °C; the mass of the particle residue was determined using a Supermicro balance (Sartorius, Göttingen, Germany).

Fractionation of the HSA nanoparticles

Fractionation of the particles was performed in a linear sucrose gradient (20–40% w/w in 20 mM sodium phosphate, pH 7.0, 150 mM NaCl; gradient volume 11 ml). The particles (1.5 mg) in 1.0 ml water were placed on top of the gradient and centrifuged for 90 min at 7,000 rpm (Beckman Spinco L7 ultracentrifuge, rotor SW 41Ti; 20 °C). Fractions (0.5 ml) were harvested using a peristaltic pump and a narrow tube, inserted from above to the bottom of the gradient. The nanoparticles in each fraction were pelleted and resuspended, by sonication, in the buffer containing 23.5% w/v sucrose.

In order to optimize the separation conditions, the initial experiments were performed using HSA particles labelled with FITC [19], which facilitated evaluation of the separation process by visual inspection. In later experiments, the particle content of the fractions was characterized by measuring their turbidity.

Sedimentation velocity analysis

Analytical ultracentrifugation

Prior to study, the HSA nanoparticles were pelleted by preparative centrifugation and resuspended in 20 mM sodium phosphate (pH 7.0), 150 mM NaCl, and 23.5% w/v sucrose with a concentration giving a turbidity between 0.6 and 0.7 at 420 nm in a cuvette with an optical pathlength of 1 cm.

The ultracentrifugal experiments were performed using a Beckman Optima XL-A ultracentrifuge, an An-50Ti rotor, and double-sector charcoal-filled Epon centrepieces of 12-mm optical pathlength. The rotor speed was 3,000 or 5,000 rpm and the rotor temperature was 20 °C. The sample and the reference volumes were 200 and 210 μl, respectively. Apparent absorbance (turbidity) versus radius data were collected at 420 nm, using a radial stepsize of 0.01 mm.

Centrifugal data analysis

For the analysis of the particle size distributions, the sedimentation velocity data were modelled as a distribution of nondiffusing particles based on the observations of very large sedimentation coefficients and of very large particle diameters in electron micrographs. In this limit, the single species solution of the Lamm equation in the sector-shaped ultracentrifugal cell simplifies to the well-known step function:

$$U(s,r,t) = \exp(-2\omega^2 st) \times \begin{cases} 0 & \text{for} \quad r < r_m \exp(\omega^2 st), \\ 1 & \text{else} \end{cases} \quad (1)$$

where s denotes the sedimentation coefficient, r the distance from the center of rotation, r_m the meniscus position, t the time after the start of centrifugation, and ω the rotor angular velocity [20]. Accordingly, the sedimentation of polydisperse mixtures can be described by the differential sedimentation coefficient distribution

$$a(r, t) \cong \int g * (s) U(s, r, t) ds \ , \qquad (2)$$

with $a(r, t)$ denoting the experimentally observed apparent absorbance (turbidity) values at time t and radius r. This integral equation was solved using the method of direct boundary modelling, ls–$g^*(s)$, as described previously [18]. In brief, the sedimentation coefficients were discretized on a grid of s values from 10 to 20,000 S, and the measured turbidity data were fitted by a direct least-squares technique. This approach is well suited to data from absorbance optics, does not generate the artificial "Δt broadening" encountered in the commonly used dc/dt method [21], and provides an undistorted sedimentation coefficient distribution over a broad range of s values [18]. To increase the stability of the inversion of Eq. (2), it was extended by a term for Tikhonov–Phillips regularization, which was implemented to give the most parsimonious distribution on the confidence level of $P = 0.95$ [18, 22]. The area under the differential sedimentation coefficient distribution ls–$g^*(s)$ between values s_1 and s_2 represents the corresponding fraction of the total material loaded, as expressed as the total area under the curve.

In a second step, corrections for solvent density and viscosity contributions were applied to the sedimentation coefficients, and ls–$g^*(s)$ was transformed to standard conditions ls–$g^*(s_{20,w})$ [23]. Since, according to electron microscopic observations, the particles are spherical (see later), we calculated the particle radius, R, corresponding to each sedimentation coefficient, $s_{20,w}$, by using the well-known relationships between $s_{20,w}$, R, and the frictional coefficient, f_0, for spherical particles $[s = M(1 - \bar{v}\rho_0)/f_0, f_0 = 6\pi\eta R, \quad (M/N_A) \times \bar{v} = (4\pi/3)R^3]$ [23]. This leads to a dependence of $s_{20,w}(R) \sim R^2$ and for the differentials $ds \sim 2R dR$, which was used to transform the differential sedimentation coefficient distribution ls–$g^*(s_{20,w})$ to a differential particle size distribution, $a(R)$, preserving the total area under the distribution curve (corresponding to the total loading). As a consequence, the area under this curve gives the fraction of the total material loaded (in units of apparent absorbance) within a given particle radius range.

Because the size of the particles is close to the wavelength of the light used for measuring the turbidity distributions, $a(R)$ will be distorted by the dependency of the scattering intensity on particle size and shape. For spherical particles it can be rescaled with a factor $t(R)$ describing the turbidity signal of a single particle of radius R to form a concentration distribution, $c(R)$. Expressions for $t(R)$ can be derived from the scattering theory of Mie [24] and the correction was implemented by using polynominal interpolation of data tabulated in Ref. [25]. The final differential particle size distribution $c(R)$ in relative concentration units was calculated as $c(R) = a(R)/t(R)$. All numerical procedures were implemented in the software SEDFIT [18].[1]

Auxiliary parameters

The partial specific volume, $\bar{v}$, of the HSA nanoparticles was assumed to be identical to that of the parent HSA, 0.733 ml/g [26]. According to the data in Ref. [26], variations in $\bar{v}$ due to conformational changes (e.g., following desolvation) and addition of sucrose will not exceed ± 0.025 ml/g, which would correspond to a systematic error in M of $\pm 14\%$ and in R of $\pm 4.5\%$.

The density of the buffered sucrose solution used in the sedimentation velocity runs (see earlier) was determined using a

---

[1] Available from P. Schuck on request or from http://www. analyticalultra centrifugation.com

Paar DMA 02 density meter and was found to be 1.0960 g/ml (20 °C). The kinematic solution viscosity was measured using a Ubbelohde viscosimeter, reference no. 50101; the result obtained was 2.13 cP at 20 °C.

Correction of the amplitudes of the size distributions for the dependency of the scattering intensities on particle radius [24, 25] required determination of the values of the refractive indices n_0 and n_1 of the solvent and of the HSA nanoparticles. The former determination used a Bellingham refractometer. For determining n_1, a Shimadzu differential refractometer was used, by applying the relationship $n_1 = n_0 + (dn/dc)\bar{v}$, where dn/dc denotes the specific refractive index increment [27]. Unprocessed HSA (in the sucrose-containing buffer) was used instead of nanoparticles. Its concentration was determined photometrically, based on $A^{1\%}(280 \text{ nm}) = 5.31$ [28]. Measurements of dn/dc were performed at solute concentrations 7, 14, and 21 g/l and at wavelengths 436, 546, and 635 nm and extrapolated to a wavelength of 420 nm. The results obtained were $n_0 = 1.3687$, $dn/dc = 0.164$ ml/g, and $n_1 = 1.592$.

Electron microscopy

The samples, in sucrose-containing buffer (see earlier), were applied to copper grids with a carbon-coated Formvar film. In some experiments, the films were glow-discharged for 30 s before use [29]. The adsorbed sample was washed twice with sucrose-free buffer in order to prevent staining artefacts due to the presence of sucrose. Afterwards, the sample was stained with 2% ammonium molybdate (pH 7.5) filtered though a 0.45-μm Millipore filter prior to use. The morphology of the particles was investigated with a Philips EM 208S electron microscope, at a nominal magnification of 16,000–21,000.

Results

Analysis of the particle size distribution of HSA nanoparticles by sedimentation velocity experiments

In aqueous solutions, the HSA particles rapidly sedimented to the bottom of the ultracentrifuge cell within a few minutes, even at rotor speeds as low as 3,000 rpm. Their sedimentation velocity, therefore, was decreased by adding sucrose, to a final concentration of 23.5% w/v, to the solution in order to increase the time available for data collection. In addition, a single sample only was studied during one run in order to achieve, by the gain in the number of data points which could be collected, high robustness and resolution of the analysis.

As another consequence of the large size of the particles, light scattering strongly dominated the apparent light absorption of the sample. Detection of the boundary movement therefore used sample turbidity at 420 nm (a wavelength of relatively high lamp intensity).

The large size of the HSA nanoparticles also led to virtually complete dominance of sedimentation with negligible diffusion. In combination with the high specific turbidity of the particles, which enabled the use of very low particle concentrations, this allowed treatment of their sedimentation as ideal and nondiffusing. This in turn allowed application of the recently introduced method for calculating apparent sedimentation coefficient distributions, $g^*(s)$, by least-squares boundary modelling, the "ls–$g^*(s)$ method" [18].

Determination of the sedimentation coefficient distributions

Typical sedimentation velocity data on the HSA nanoparticles studied, together with curves fitted to them by application of the ls–$g^*(s)$ method, are shown in Fig. 1. As indicated by the plot of the residuals of the fits, the fits are of reasonable quality, thus confirming the "ideal and nondiffusing" model. The distribution of s values from the analysis is shown in Fig. 2, together with the distribution of another sample. Both distributions are broad and unimodal, the $s_{20,w}$ values centred around 5,000 and 20,000 S, respectively, and varying, within one preparation, by up to a factor of approximately 5.

Of the eight samples studied, the two samples shown in Fig. 2 represented those with the smallest and largest $s_{20,w}$ values, respectively, for the maximum of $g^*(s)$. All of them were prepared under seemingly identical conditions. This indicates that critical parameters in the preparation procedure are unknown and uncontrolled up to now.

Radius distributions and corrections for Mie light scattering

Electron microscopic observations clearly show that the HSA nanoparticles are virtually spherical (see later); thus, the $s_{20,w}$ distributions can be unambiguously converted into radius or mass distributions by applying

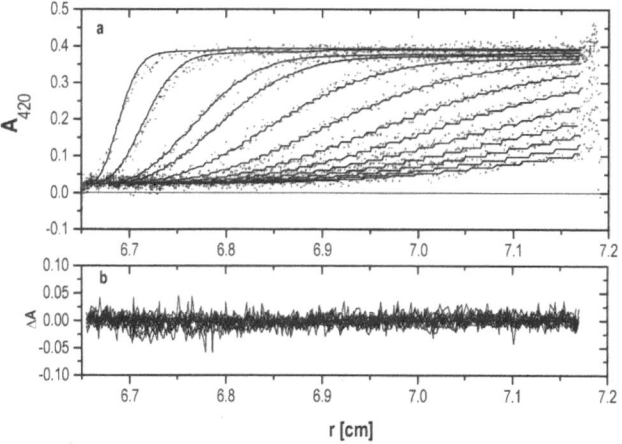

Fig. 1a, b A sedimentation velocity experiment on a sample of human serum albumin (*HSA*) nanoparticles. **a** Experimental turbidity versus radius data (+), and curves fitted to them by direct boundary modelling (–) [18]. **b** Residuals of the fit. The rotor speed was 4,000 rpm and the time interval between subsequent scans was 110–130 s. The discrete steps in the fitted curves of **a** are due to the use of step functions for fitting the particle distributions (Eqs. 1, 2)

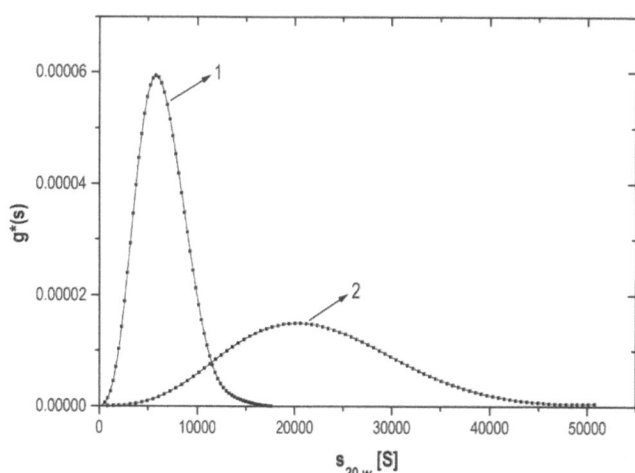

Fig. 2 Sedimentation coefficient distributions: relative contributions to turbidity, $g^*(s)$, versus $s_{20,w}$ for the HSA nanoparticle sample of Fig. 1 (*1*) and for a second sample (*2*)

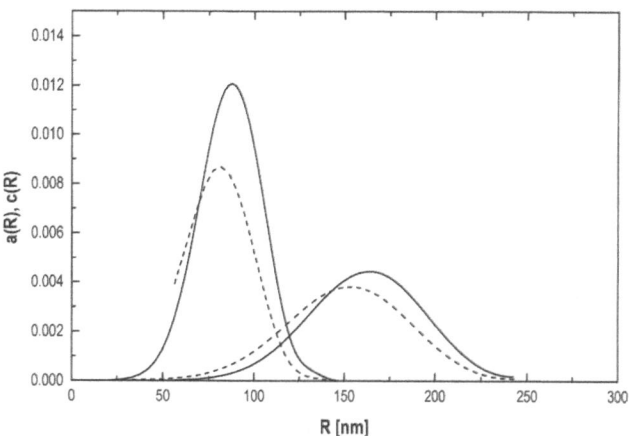

Fig. 3 Transformation of the distributions of Fig. 2 into radius distributions: relative concentration versus particle radius, R, without (–) and with (- - -) corrections for the dependency of the scattering intensity on particle radius

Stokes' law and the assumed partial specific volume of the particles, $\bar{v}$. The distributions of the particle radii corresponding to the curves in Fig. 2 are shown in Fig. 3. Their maxima are located at approximately 85 and 160 nm, respectively. The corresponding particle masses are 2×10^9 and 15×10^9 g/mol, respectively.

Since large particles scatter light more strongly than smaller ones, the turbidity values of Fig. 1 as well as the amplitudes of the distributions of $s_{20,w}$ and R of Figs. 2 and 3 do not represent true relative particle concentrations but are distorted, the mass percentages of the particles at higher s and R values being artificially increased compared to those at lower ones. For spherical particles, this distortion can be corrected by application of Mie's theory, as outlined in detail in Refs. [25, 30]

(Table 2 of Ref. [25] represents a useful collection of data for that purpose). Corrected relative amplitudes for the radius distributions, applying the experimental values for the refractive indices n_0 and n_1 of the solvent and of the dissolved nanoparticles, are included in Fig. 3.

Fractionation of the HSA nanoparticles

Fractionation of mixtures of macromolecules, supramolecular aggregates, organelles, or even cells by sucrose density centrifugation, based either on differences in density or on sedimentation coefficient, is a standard technique in biochemistry. We applied the technique for fractionating the broad distributions shown in Fig. 2 according to their s values. The $s_{20,w}$ distributions of four selected fractions from the same stock solution as the smaller particles of Fig. 2 are shown in Fig. 4. Obviously, the size distribution of most of the particles in a fraction is quite narrow compared to the starting mixture; however, some of the particles both at the leading and at the trailing edge of the peak are much larger or smaller, respectively, than might be expected. We suspect that this effect could be due to interparticle association and dissociation occurring during the preparation steps between fractionation and characterization of the samples, i.e., during pelletting of the nanoparticles which serves for washing and concentrating and during the subsequent ultrasonication of the pellet.

The isolated fractions may be very interesting for studying the relationships between the size of the (loaded) carriers and their biological activity in experiments on drug delivery to cell cultures or isolated cells in solution.

Particle morphology as studied by transmission EM

Both unfractionated samples and selected fractions from the density gradient were studied by transmission EM, by applying negative staining for contrasting the nanoparticles; typical results are shown in Fig. 5. The most striking morphological property of the particles is their nearly perfect spherical shape. The clearest evidence for this is given by Fig. 5c, obtained by glow-discharging the grids before sample application (a procedure which seems to prevent the clustering of the particles on the support film frequently observed with the standard staining procedure; see Fig. 5a, b). The measured particle diameters agree with those determined from the sedimentation analysis.

Discussion

In the foregoing, we described the full characterization of the particle size distribution of spherical nanoparticles

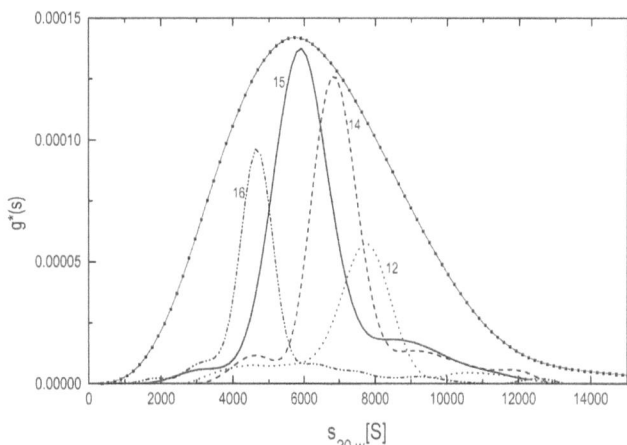

Fig. 4 Fractionation of HSA nanoparticles by sucrose gradient centrifugation: size analysis analogous to Fig. 2, for four selected fractions from a separation of sample 1 of Fig. 2 (the figures correspond to the fraction number). The individual size distributions $g^*(s_{20,w})$ of the fractions are shown together with the original distribution

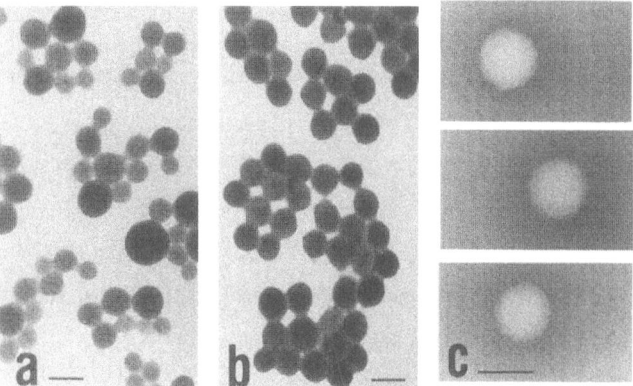

Fig. 5a–c Morphology of the HSA nanoparticles by means of transmission electron microscopy of negatively stained samples. **a** Unfractionated particles, **b** particles in the fraction of highest turbidity from a sucrose gradient, and **c** three arbitrarily selected particles of the sample of **b** on films glow-discharged for 30 s. *Bar length*: 200 nm

prepared from HSA and intended for use as a drug delivery system [6, 9].

Characterizing size distributions of nanoparticles or other polymers by sedimentation velocity analysis in the analytical ultracentrifuge is a standard technique in macromolecular chemistry, in particular in research and development in big chemical companies [12–17]. During the last 2 decades, the methodology applied has become both highly sophisticated and economical [15–17, 30]. The latter aspect was met by the introduction of novel noncommercial apparatus based on commercial preparative ultracentrifuges, and this "hardware" part was supplemented by the development of suitable evaluation programs [14–17, 30]. The present work used a different approach. Following Ref. [18], a commercial analytical ultracentrifuge equipped with absorption optics (as also done in Refs. [13, 31]), combined with a recently introduced linear least-squares method for the calculation of an apparent sedimentation coefficient distribution, was applied. The latter method is based on boundary modelling by a superposition of sedimentation profiles of ideal nondiffusing particles [18]. Our approach certainly lacks the high-throughput capability of the noncommercial systems and lacks the capability of permitting the high-resolution analysis of extremely broad size distributions in a single run, as offered by the "gravitational sweep method" [17, 30]. However, although we have not attempted such optimization so far, our method may be suitable for significantly higher throughput, by use of a longer solution column (thus permitting longer data acquisition times), the direct analysis of the transmitted light data (permitting analysis of different samples in each sector of the ultracentrifuge cell) [32], or the use of the commercial interference optical detection (allowing significantly faster scans and thereby the study of up to eight samples per centrifuge run). Concerning the characterization of very broad distributions, we believe that the combination of the ultracentrifugal size distribution analysis with sucrose density gradient fractionation used by us will lead to results of comparable quality as the "gravitational sweep method", with higher expenditure but with the potentially advantageous side effect of yielding more homogeneous subsamples on a semipreparative scale. We thus consider the approach described as a reasonable alternative to those using special noncommercial apparatus, at least for applications not requiring very high throughput.

With respect to the HSA nanoparticles studied by us, we learned from the analysis described that the size distribution of the unfractionated nanoparticles is much broader and much less reproducible than previously thought. As a consequence, efforts will be made to optimize the preparation procedure, with the aim of assembling HSA nanoparticles with a more predictable and, possibly, adjustable average size and a narrower size distribution. Evaluating and controlling the consequences of modifications in the preparation procedure for the properties of the products should be optimally attained by the methodology described here.

Acknowledgements The authors are grateful to J. Kreuter for helpful discussions, to M. Lewis for helpful comments on the manuscript, and to the German Bundesministerium für Bildung und Forschung for financial support (project 03C0308A).

References

1. Yuan F (1998) Semin Radiat Oncol 8:164
2. Barratt GM (2000) Pharm Sci Technol Today 3:163
3. Kreuter J (1983) Pharm Acta Helv 58:196
4. Scheffel U, Rhodes BA, Natarajan TK, Wagner HN (1972) J Nucl Med 13:498
5. Gallo JM, Hung CT, Perrier DG (1984) Int J Pharm 22:63
6. Marty JJ, Oppenheimer RC, Speiser P (1978) Pharm Acta Helv 53:17
7. Lin W, Coombes AGA, Davies MC, Davis SS, Illum L (1993) J Drug Targeting 1:237
8. MacAdam AB, Shafi ZB, James SL, Marriott C, Martin GP (1997) Int J Pharm 151:47
9. Weber C, Coester C, Kreuter J, Langer K (2000) Int J Pharm 194:91
10. Weber C, Reiss S, Langer K (2000) Int J Pharm 211:67
11. Provencher SW (1979) Makromol Chem 180:201
12. Cantow HJ (1964) Makromol Chem 70:130
13. Aeijelts Averink JW, Reerink H, Boerma J, Jaspers WJM (1966) J Colloid Interface Sci 21:66
14. Scholtan W, Lange H (1972) Kolloid Z Z Polym 250:782
15. Mächtle W (1984) Makromol Chem 195:1025
16. Müller HG (1989) Colloid Polym Sci 267:1113
17. Mächtle W (2000) Biophys J 76:1080
18. Schuck P, Rossmanith P (2000) Biopolymers 54:328
19. Haugland RP (1996) Handbook of fluorescent probes and research chemicals, 6th edn. Molecular Probes, Eugene, Ore
20. Fujita H (1962) Mathematical theory of sedimentation analysis. Academic, New York
21. Stafford WF (1992) Anal Biochem 203:295
22. Schuck P (2000) Biophys J 78:1606
23. Svedberg T, Pedersen KO (1940) Die Ultrazentrifuge. Steinkopff, Dresden
24. Mie G (1908) Ann Phys 25:377
25. Heller W, Pangonis WJ (1957) J Chem Phys 26:498
26. Durchschlag H (1986) In: Hinz H-J (ed) Thermodynamic data for biochemistry and biotechnology. Springer, Berlin Heidelberg NewYork, pp 45–128
27. Mächtle W, Fischer H (1969) Angew Makromol Chem 7:147
28. Means GE, Bender ML (1975) Biochemistry 14:4989
29. Hayat MA (1986) Basic techniques for transmission electron microscopy. Academic, San Diego, Calif
30. Mächtle W (1988) Angew Makromol Chem 162:35
31. Cölfen H, Pauck T (1997) Colloid Polym Sci 275:175
32. Kar SR, Kinsbury JS, Lewis MS, Laue TM, Schuck P (2000) Anal Biochem 285:135

Progr Colloid Polym Sci (2002) 119:37–42
© Springer-Verlag 2002

G. M. Pavlov
N. A. Michailova
E. V. Korneeva
N. P. Yevlampieva
E. I. Rjumtsev
V. S. Khotimsky
E. G. Litvinova
M. B. Chirkova

Properties of a membrane-forming polymer, poly(1-trimethylgermyl-1-propyne), in dilute solutions

G. M. Pavlov (✉) N. A. Michailova
N. P. Yevlampieva · E. I. Rjumtsev
Institute of Physics of St. Petersburg
State University, Ulianovskaya str. 1
198504 St. Petersburg, Russia

E. V. Korneeva
Institute of Macromolecular compounds
of Russian Academy of Sciences
Bolshoi pr. 31
190004 St. Petersburg, Russia

V. S. Khotimsky · E. G. Litvinova
M. B.Chirkova
Topchiev Institute of Petrochemical
Synthesis, Russian Academy of Sciences
Leninskii pr. 29
117912 Moscow, Russia

Abstract Samples and fractions of a membrane-forming polymer, poly (1-trimethylgermyl-1-propyne) (PTMGP), were studied by methods of molecular hydrodynamics in cyclohexane in the molecular weight range $4 < M \times 10^{-4} < 170$. Molecular-weight dependencies of the hydrodynamic characteristics were established and analysed. On the basis of them the equilibrium rigidity of PTMGP chains was evaluated. The influence of volume effects on the size of macromolecules is discussed. A comparison is made with the results published for molecules of poly(1-phenyl-1-propyne) and poly(1-trimethylsilyl-1-propyne).

Key words Polygermylpropyne · Hydrodynamic properties · Macromolecule size

Introduction

Polymers based on acetylene have some interesting and promising physical potential and properties. In particular, they exhibit semiconductor effects and are widely used in membrane technology. Glassy polymers based on silicon-and germanium-containing hydrocarbon monomers exhibit high gas separation levels [1–3]. Among these polymers poly(1-trimethylsilyl-1-propyne) (PTMSP) [1, 4] exhibits extremely high gas permeability, solubility, and diffusion coefficients. Its peculiar feature is the record level of selective gas transport, which exceeds by 1 order of magnitude the values characteristic of highly permeable rubbers [5, 6]. The unique membrane properties of PTMSP are explained by the combination of several features of its molecular structure [7–9]:

1. The presence of double bonds making the chain more rigid.

2. The composition and distribution of $-C=C-$ bonds in cis and trans configurations in the polymer chain.

3. The presence of weakly interacting bulky $Si(CH_3)_3$ side substituents.

All these factors probably ensure a high free volume in the polymer matrix, which is necessary for the unique gas permeability of polymer membranes. Because the use of these polymers in membrane technology is very promising [1, 5, 6] it is urgent to investigate in detail their molecular characteristics, in particular by various methods of molecular physics.

The investigation of various polymer analogues is of considerable interest for understanding the nature of high gas transport in glassy organoelemental polymers. From this viewpoint it is important to study an organogermanium analogue of PTMSP, poly(1-trimethylgermyl-1-propyne) (PTMGP) [3, 10]. In the present work, samples and fractions of this new membrane-forming polymer are studied by methods of molecular hydrodynamics (velocity sedimentation, translational isothermal diffusion, and viscometry).

Methods

A sample of PTMGP was synthesised by using the $TaCl_5$/BuLi catalyst as described in Ref. [10]. According to ^{13}C NMR data, the average ratio of the cis–trans configurations of the $-C=C-$ bonds was 9:91. The repeat unit has the following structural formula:

$$CH_3$$
$$|$$
$$[C=C]_n$$
$$|$$
$$Ge(CH_3)_3$$

Sample fractionation was carried out from solution into CCl_4 by fractional precipitation with ethanol. Seven fractions were obtained (Table 1), fractions 4 and 6 were subsequently refractionated (fractions 8–12), and the 13th one is the unfractionated original sample. The fractionation yield was 97%.

Velocity sedimentation was investigated using an MOM 3180 analytical ultracentrifuge in a single-sector cell at a rotor revolution speed of 40,000 rpm. A Lebedev polarising interferometer was used as the optical recording system [11, 12]. Sedimentation coefficients, s, were calculated from the sedimentation boundary shift, x, with time. The dependencies of $\Delta \ln x$ on Δt are shown in Fig. 1a. Their slope gives the values of $s = \omega^{-2}(\Delta \ln x/\Delta t)$, where $\omega = 2\pi n/60$. The concentration dependence of the sedimentation coefficient which normally satisfies a linear approximation $s^{-1} = s_0^{-1}(1 + k_s c + \cdots)$ was studied. The dependence of s^{-1} on c is shown in Fig. 1b, from which the coefficients s_0 and the Gralen coefficients, k_s, were determined.

Translational diffusion was studied by the classical method of forming a boundary between the solution and the solvent, which was recorded using a polarising interferometer after definite time intervals. The studies were carried out using an in-house-built diffusometer as described by Tsvetkov and coworkers [11, 12]. The tools and calculating algorithm of the diffusion coefficients, D, have also been described elsewhere [11, 12]. The refractive index increments, $\Delta n/\Delta c$, were calculated from the area limited by the diffusion curve.

Intrinsic viscosities, $[\eta]$, were determined from the Hugging plots: $\eta_{sp}/c = [\eta] + k'[\eta]^2 c + \cdots$, where $\eta_{sp} \equiv (\eta-\eta_0)/\eta_0 = (t-t_0)/t_0$, where η and η_0 are the viscosities of the solution and the solvent and t and

t_0 are the times of solution and solvent flow in an Ostwald viscometer.

The values of $[\eta]$, the Huggins parameter, k', and other hydrodynamic characteristics are given in Table 1. The buoyancy factor $(1-v\rho_0)$, or the density increment $\Delta\rho/\Delta c = (1-v\rho_0) = \Delta m/mw$, where $\Delta m = m-m_0$, where m and m_0 are the mass of the same volume of solution and solvent respectively, and w is the weight concentration of the solution, was determined from picnometric measurements of the solution in cyclohexane according to Ref. [13] $(1-v\rho_0) = (0.417 \pm 0.011)$. The corresponding plot is presented in Fig. 2. The data for a sample of PTMSP are shown for comparison in this figure. The cyclohexane viscosity at 25 °C was $\eta_0 = 0.859 \times 10^{-2}$ P.

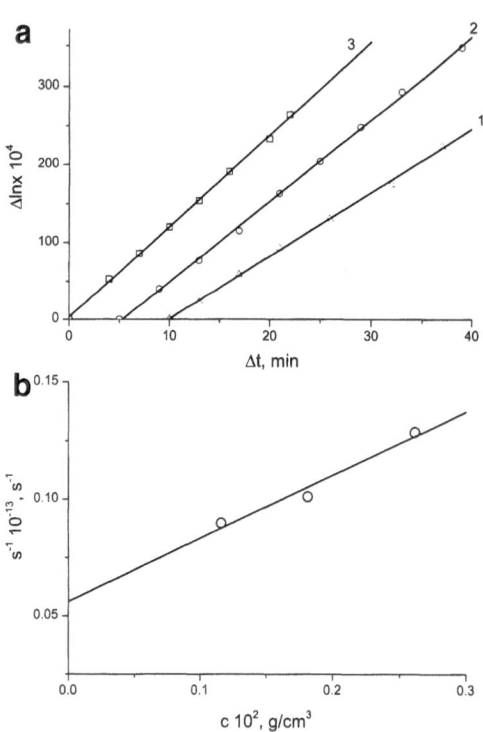

Fig. 1 **a** Dependencies of $\Delta \ln x$ on sedimentation time Δt, where x is the position of the sedimentation peak maximum for solutions of fraction 5 at concentrations $c \times 10^2$ g/cm^3 = 0.261 (1), 0.181 (2), 0.116 (3). **b** The concentration dependence of s corresponding to **a**

Table 1 Hydrodynamic and molecular characteristics of poly(1-trimethylgermyl-1-propyne) molecules in cyclohexane at 25 °C

N	$[\eta]$ (cm³/g)	k'	$D \times 10^7$ (cm²/s)	$\Delta n/\Delta c$ (cm³/g)	$s_0 \times 10^{13}$ (s)	k_s (cm³/g)	$M_{sD} \times 10^{-4}$	$A_0 \times 10^{10}$	$\beta_s \times 10^{-7}$	γ
1	324	0.48	0.77	0.093	15.4	370	119	3.48	1.22	1.37
3	286	0.57	0.72	0.100	15.2	380	126	3.17	1.17	1.33
4	350	0.49	0.81	–	15.0	330	110	3.66	1.20	0.90
5	305	0.54	0.64	0.106	17.8	485	165	3.17	1.24	1.58
6	225	0.53	–	–	10.2	145	40[a]	–	–	0.64
7	49	–	–	–	2.9	–	4.1[b]	–	–	–
8	350	0.39	–	–	16.7	–	150[b]	–	–	–
9	175	–	1.27	0.079	12.7	290	59	3.73	1.48	1.66
10	162	0.40	1.12	0.081	10.0	170	53	3.07	1.05	1.06
11	176	0.52	1.10	0.094	11.4	275	62	3.24	1.26	1.55
12	90	0.76	–	–	11.3	110	42[a]	–	–	1.22
13	375	0.36	0.61	0.097	16.0	540	156	3.17	1.20	1.37

[a] $M_{s\eta}$ and M_{ks}
[b] $M_{s\eta}$

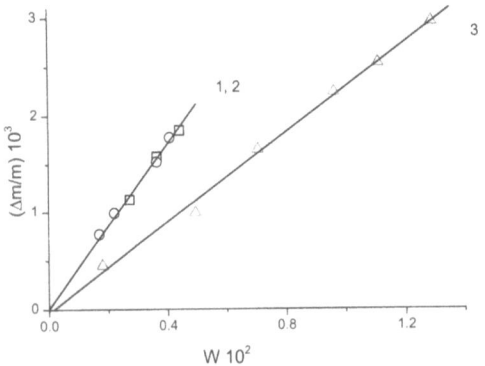

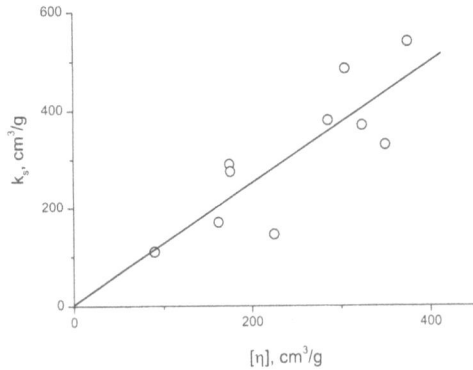

Fig. 2 Dependence of $\Delta m/m$ versus w, where $\Delta m = m - m_0$, m and m_0 are the mass of the same volume of solution and solvent, respectively, w is the weight concentration of the solution, and $\Delta m/mw = (1 - \upsilon\rho_0)$ [13]. *1* and *2* correspond to the different samples of poly(1-trimethylgermyl-1-propyne) (*PTMGP*), $(1 - \upsilon\rho_0) = (0.417 \pm 0.011)$. *3* corresponds to the sample of poly(1-trimethylsilyl-1-propyne) (*PTMSP*), $(1 - \upsilon\rho_0) = (0.232 \pm 0.008)$. All the measurements were made in cyclohexane at 25 °C

Fig. 3 The dependence of k_s versus $[\eta]$ for PTMGP fractions in cyclohexane. The slope is the dimensionless parameter $k_s/[\eta] \equiv \gamma = 1.2 \pm 0.3$, $r = 0.828$

Results and discussion

On the basis of the hydrodynamic characteristics given in Table 1, the following values were calculated: molecular weight (MW) according to Svedberg [13], the hydrodynamic invariant, A_0, [11], and the sedimentation parameter, β_s, [14] according to the equations

$$M_{SD} = R[s]/[D] \ . \tag{1}$$

$$A_0 = \left(R[D]^2 [s][\eta] \right)^{1/3} \ , \tag{2}$$

$$\beta_s = N_A \left(R^{-2}[D]^2[s]k_s \right)^{1/3} \ , \tag{3}$$

where $[s] \equiv s_0\eta_0/(1 - \upsilon\rho)$, $[D] = D_0\eta_0/T$, R is the universal gas constant, N_A is the Avogadro number, and T is the temperature in degrees Kelvin.

The molecular weight and hydrodynamic invariants for PTMGP are listed in Table 1.

The mean value of hydrodynamic invariant is $A_0 = (3.3 \pm 0.3) \times 10^{-10}$ gcm^{2}c^{-2}deg^{-1}mol$^{-1/3}$, and the average value of the sedimentation parameter is $\beta_s = (1.23 \pm 0.08) \times 10^{-7}$ mol$^{-1/3}$. These values are characteristics of linear polymer homologues [12, 15] and were used for calculating MW of samples 6, 7, 8, and 12 (Table 1) from the following equations:

$$M_{s\eta} = (R/A_0)^{3/2} [s]^{3/2} [\eta]^{1/2} \ , \tag{4}$$

$$M_{ks} = (N_A/\beta_s)^{3/2} [s]^{3/2} k_s^{1/2} \ . \tag{5}$$

k_s may be directly compared with the coefficients s_0, and the corresponding scaling relation was obtained: $k_s = 2.56 s_0^{1.85 \pm 0.33}$. The values of k_s may also be correlated with those of $[\eta]$. The corresponding depen-

dence is shown in Fig. 3. The slope of this dependence is the estimation of the average value of a dimensionless parameter $k_s/[\eta] \equiv \gamma$, where $\gamma = (1.2 \pm 0.3)$ for all fractions under investigation.

On the basis of the MW values obtained, the dependencies of the Kuhn–Mark–Houwink–Sakurada (KMHS plot) type were plotted (Fig. 4), and the corresponding scaling coefficients were calculated (Table 2). Within the mean-square errors between the scaling indexes, the following relationships specific for linear polymer homologues are true [11, 17, 18]:

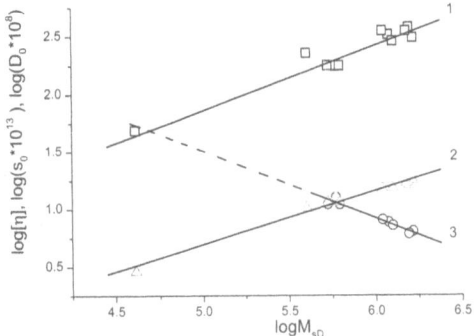

Fig. 4 Kuhn–Mark–Houwink–Sakurada dependencies: *1* $[\eta]$–M; *2* s_0–M; *3* D_0–M

Table 2 Parameters of Kuhn–Mark–Houwink–Sakurada equations for poly(1-trimethylgermyl-1-propyne) molecules in cyclohexane at 25 °C

Relationship	b_i[a]	Δb_i	K_i[a]	R_i[b]	n[c]
$[\eta]$–M	0.56	0.07	0.115	0.9241	12
D_0–M	−0.59	0.06	2.79×10^{-4}	−0.9743	8
s_0–M	0.47	0.03	2.34×10^{-15}	0.9843	12
k_s–s_0	1.85	0.33	2.56	0.9052	9

[a] The coefficients of the type of relationship $D_0 = K_2 M^{b_2}$
[b] The linear correlation coefficients of the relationship $\log k_s = \log K_4 + b_4 \log s_0$
[c] The number of points

$$|b_2| = (1 + b_1)/3 \ , \tag{6}$$

$$|b_2| + b_3 = 1 \ , \tag{7}$$

$$b_4 = (2 - 3b_3)/b_3 \ , \tag{8}$$

where the indices b_1, b_2, b_3, and b_4 are the indices in the KMHS dependencies:

$$[\eta] = K_1 M^{b_1} \ , \tag{9}$$

$$D_0 = K_2 M^{-b_2} \ , \tag{10}$$

$$s_0 = K_3 M^{b_3} \ , \tag{11}$$

$$k_s = K_4 s_0^{b_4} \ . \tag{12}$$

The results given in Table 1 show that sample fractionation takes place not only according to MW. It is probable that uncontrolled variations in the content of cis/trans inclusions of double bonds and different microsequence lengths (cis blocks/trans blocks) in macromolecules at virtually the same composition can also play a certain role.

Bearing in mind the fact that the number of experimental points in the KMHS relationships is different and also taking into account Eqs. (6), (7), and (8), the final KMHS equations were obtained with the correction within the mean-square error:

$$[\eta] = 9.75 \times 10^{-2} M^{0.575} \ , \tag{13}$$

$$D_0 = 1.19 \times 10^{-4} M^{-0.525} \ , \tag{14}$$

$$s_0 = 2.0 \times 10^{-15} M^{0.475} \ . \tag{15}$$

It may be noted that in this case three exponential factors of the KMHS equations are uniquely related to the hydrodynamic invariant A_0:

$$K_\eta^{1/3} K_D = A_0(T/\eta_0) \ , \tag{16}$$

$$K_\eta^{1/3} K_s = A_0[(1 - v\rho_0)/R\eta_0] \ . \tag{17}$$

Using Eqs. (16) and (17) it is easy to estimate the average value of A_0 for PTMGP molecules in cyclohexane. This estimation coincides with the average A_0 value obtained from the parameters given in Table 1.

The preliminary analysis of the PTMGP molecular size can be carried out on the basis of the normalised scaling relationships [19, 20]. This plot is shown in Fig. 5. It is clear that PTMGP molecules are in the intermediate range between rigid-chain and flexible-chain macromolecules. In this case the influence of volume effects on the coil size in solution cannot be completely excluded. A similar situation is common for macromolecules with intermediate rigidity [21, 22]. In particular, it has been observed for the molecules of PTMSP [23].

The value of the volume effects can be characterized by the parameter ε, which is related to $\langle h^2 \rangle \sim M^{1+\varepsilon}$, where $\langle h^2 \rangle$ is the mean-square end-to-end distance of the polymer chain. ε was calculated from the equation $\varepsilon = (2b_\eta - 1)/3 = (2|b_2| - 1)$ [24]. It was estimated that this parameter is in the range $0 \leq \varepsilon \leq 0.08$. It is known [25, 26] that the only theory describing the translational friction of a wormlike necklace with the evaluation of volume effects in analytical form is the Gray–Bloomfield–Hearst theory [27], according to which the following equation is valid for $[s]$:

$$[s]P_0 N_A = [3/(1 - \varepsilon)(3 - \varepsilon)]$$
$$M_L^{(1+\varepsilon)/2} A^{-(1-\varepsilon)/2} M^{(1-\varepsilon)/2} + (M_L P_0/3\pi)$$
$$\times \left[\ln A/d - 1/3(A/d)^{-1} - \varphi(\varepsilon) \right] \ , \tag{18}$$

where d is the hydrodynamic chain diameter and $\varphi(\varepsilon)$ is the function tabulated in Ref. [27]. At $\varepsilon = 0$, Eq. (18) becomes the relationship obtained in the theory [28] for translational friction of a wormlike necklace without volume interactions. In this case we have $\varphi(0) = 1.431$. In the theory of translational friction of a wormlike cylinder the function $\varphi(0)$ takes another value. $\varphi(0) = 1.056$ [29], and this is the reason for greater values of the hydrodynamic chain diameter obtained in the framework of this theory (Table 3). The dependencies corresponding to Eq. (18) for two values of the parameter ε, $\varepsilon = 0$ and $\varepsilon = 0.08$, are shown in Fig. 6a. The estimations following from this plot are listed in Table 3.

Assuming that the size of the macromolecules in the phenomena of translational friction and viscosity are similar and also assuming that A_0 has a constant value in the homologous series, one can write as was done in Refs. [30, 31]:

$$[s]P_0 N_A = \left(M^2 \Phi_0/[\eta] \right)^{1/3} \ . \tag{19}$$

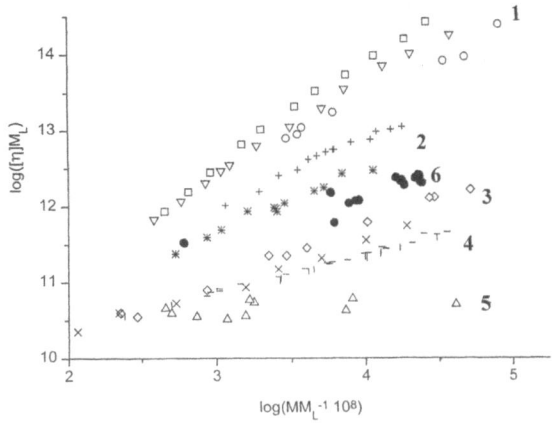

Fig. 5 Normalised double-logarithmic plot of $[\eta]M_L$ against MM_L^{-1} for extra-rigid macromolecules (*1*), rigid (*2*), flexible in thermodynamically good solvents (*3*), flexible in θ-condition (*4*), globular (*5*) [19, 20], and data from Table 1 (*6*)

Table 3 Values of equilibrium rigidity, A, and hydrodynamic diameter, d, of poly(1-trimethylgermyl-1-propyne) molecules in cyclohexane at 25 °C

ε/method	Slope	$A \pm \Delta A \times 10^8$ (cm)	Intercept	$d \pm \Delta d \times 10^8$ (cm) [26, 27]	$d \pm \Delta d \times 10^8$ (cm) [28]	r
$0/s$	$(12.3 \pm 0.9) \times 10^{-16}$	112 ± 16	$(1.85 \pm 0.8) \times 10^{-13}$	2 ± 1	3 ± 2	0.9752
$0/[\eta]$	1.46 ± 0.09	73 ± 6	100 ± 90	5 ± 5	8 ± 7	0.9802
$0.08/s$	$(22.7 \pm 1.5) \times 10^{-16}$	81 ± 11	$(1.1 \pm 0.8) \times 10^{-13}$	4 ± 3	$-$	0.9780
$0.08/[\eta]$	2.7 ± 0.17	51 ± 5	12 ± 90	15 ± 14	$-$	0.9811

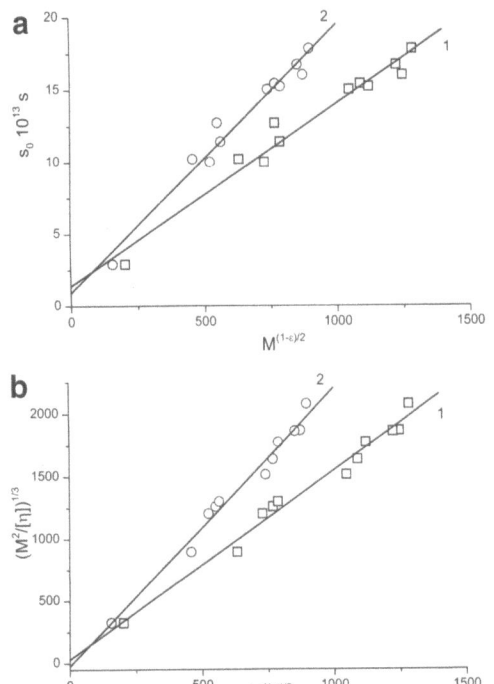

Fig. 6 Dependencies of **a** s_0 and **b** $(M^2/[\eta])^{1/3}$ on $M^{(1-\varepsilon)/2}$ at $\varepsilon = 0$ (*1*) and $\varepsilon = 0.08$ (*2*)

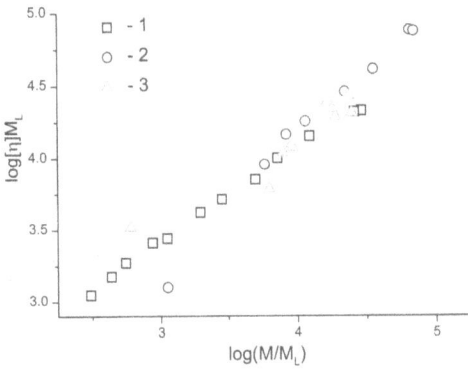

Fig. 7 Dependence of $\log[\eta]M_L$ on $\log(M/M_L)$ for PTMGP and PTMSP: data from Ref. [32] (*1*); data from Refs. [7, 23] (*2*); data from Tables 1 and 3 (*3*)

Consequently, viscometric data can be considered in a system of coordinates, $(M^2/[\eta])^{1/3} = f(M^{(1-\varepsilon)/2})$, where the function type is given by the right-hand side of Eq. (18). The authors of Ref. [26] were the first to use the dependence $(M^2/[\eta])^{1/3} = f(M^{(1-\varepsilon)/2})$; at $\varepsilon = 0$ this dependence is transformed into the well-known Bushin–Tsvetkov plot [12] for a wormlike cylinder in the absence of volume interactions in the polymer chain. The plots corresponding to Eqs. (18) and (19) are shown in Fig. 6b. The results obtained from these plots are given in Table 3.

Under the objective conditions the investigated range of MW for **PTMGP** is not very wide, and the main data are situated in the range of high MW ($M > 400 \times 10^3$); therefore, the determination reliability of the hydrodynamic diameter of macromolecules is not high, although the linear correlation coefficients, r, of the corresponding plots have high values ($r > 0.97$). However, the evaluations of the Kuhn segment length, A, are sufficiently reliable (Table 3). According to the data of translational

diffusion, the range of A values is $80 < A_f \times 10^8$ cm < 110. According to viscometric data this range is $50 < A_\eta \times 10^8$ cm < 70. These quantitative evaluations are in good agreement with the qualitative consideration based on normalised scaling relationships (Fig. 5). The differences in the evaluations of A_f and A_η are usually observed for linear polymers and are explained by the disagreement between the theories of translational and rotational friction of polymer chains [12].

In conclusion, these results are compared with those obtained earlier for PTMSP [7, 32] in which the Si atoms are presented and the ratio of cis–trans inclusions of $-$C$=$C$-$ bonds is different. The comparison is also made with poly(1-phenyl-1-propyne) studied in cyclohexane [33]. This comparison was also carried out using normalised scaling relationships. The results are shown in Fig. 7. It follows from this figure that the sizes of the polymers compared are similar. It may be the consequence of the competing effect of the side-group structure; the composition and/or microsequence length difference of the polymers being compared as well as due to the difference in the hydrodynamic quality of solvents in the polymer–solvent systems being compared. The separation of these effects is not a trivial experimental problem and requires additional investigations over a wide range including relatively low MW.

Acknowledgement The authors are grateful for financial support from INTAS (grant N 2000-00230).

References

1. Masuda H, Higashimura T (1987) Adv Polym Sci 81:121
2. Plate N, Antipov E, Tepliakov V, Khotimskii V, Yampolskii Y (1990) Vysokomol Soedin 32:1123
3. Litvinova E, Khotimsky V (1994) Abstract book of the 2nd international symposium "Progress in membrane science and technology". Enschede, The Netherlands, p 57
4. Masuda H, Isobe E, Higashimura T, Takada K (1983) J Am Chem Soc 105:7473
5. Takada K, Matsuya H, Masuda H, Higashimura T (1985) J Appl Polym Sci 30:1605
6. Plate N, Bokarev A, Kaliuzhnyi N, Litvinova E, Khotimskii V, Volkov V, Yampolskii Y (1991) J Membr Sci 60:13
7. Savoca A, Surnamer A, Tieu C (1993) Macromolecules 26:6211
8. Shtennikova I, Kolbina G, Korneeva E, Khotimsky V, Litvinova E. (1998) J Appl Chem 71:1350
9. Rjumtsev E, Yevlampieva N, Shtennikova I, Khotimsky V, Litvinova E, Plate N (1999) Vysokomol Soedin 41:1169
10. Khotimsky V, Chirkova M, Litvinova E, Rebrov A, Antipov E (2000) Vysokomol Soedin 42:1169
11. Tsvetkov V, Frenkel S, Eskin V (1970) Structure of macromolecules in solution. Butterworths, London
12. Tsvetkov V (1989) Rigid-chain polymers. Consultants Bureau, New York
13. Svedberg T, Pedersen KO (1940) The ultracentifuge. Oxford University Press, Oxford
14. Pavlov G, Frenkel S (1986) Vysokomol Soedin 28:353
15. Pavlov G, Frenkel S (1995) Prog Colloid Polym Sci 99:101
16. Wales M, van Holde K (1954) J Polym Sci 14:81
17. Budtov V (1992) Physical chemistry of polymer solutions. Chemistry, St Petersburg
18. Pavlov G, Frenkel S (1982) Vysokomol Soedin 24:178
19. Pavlov G, Rowe A, Harding S (1997) Trends Anal Chem 16:401
20. Pavlov G, Harding S, Rowe A (1999) Prog Colloid Polym Sci 113:76
21. Pavlov G, Korneeva E, Michailova N, Ananieva E (1992) Carbohydr Polym 19:243
22. Pavlov G, Ivanova N, Korneeva E, Michailova N, Panarin E (1996) J Carbohydr Chem 15:417
23. Shtennikova I, Kolbina G, Yakimansky A, Plate N, Khotimsky V, Litvinova E (1999) Eur Polym J 35:2073
24. Ptizin O, Eizner Yu (1959) Zh Tekh Fiz 29:1105
25. Bushin S, Astapenko E (1986) Vysokomol Soedin 28:1499
26. Pavlov G, Panarin E, Korneeva E, Kurochkin E, Baikov V, Uschakova V (1990) Makromol Chem 191:2889
27. Gray H, Bloomfield V, Hearst J (1967) J Chem Phys 46:1493
28. Hearst J, Stockmayer W (1962) J Chem Phys 37:1425
29. Yamakawa H, Fujii M (1974) Macromolecules 7:128
30. Tsvetkov V, Lavrenko P, Pavlov G, Bushin S, Astapenko E, Boikov A, Shildiaeva N, Didenko S, Malishenko B (1982) Vysokomol Soedin 24:2343
31. Pavlov G, Selunin S, Shildiaeva N, Yakopson S, Efros L (1985) Vysokomol Soedin 27:1627
32. Masuda H, Isobe E, Higashimura T (1985) Macromolecules 18:841
33. Hirao T, Teramoto A, Sato T, Norisuye T, Masuda H, Isobe E, Higashimura T (1991) Polym J 23:925

Progr Colloid Polym Sci (2002) 119:43–49
© Springer-Verlag 2002

Zoltán Bozóky
Lajos Fülöp
Gyõzõ Jánoki

Combination of preparative and analytical ultracentrifugation methods for the investigation of low-density lipoprotein aliquots to prepare radioactive-labelled lipoproteins

Z. Bozóky (✉) · G. Jánoki
National Research Institute for
Radiobiology and Radiohygiene
P.O.B. 101, 1775 Budapest, Hungary
e-mail: bozoky@hp.osski.hu

L. Fülöp
Regional Blood Centre, Gödöllõ
Hungary, Hungary

Abstract Low-density lipoprotein (LDL) can be radiolabelled with different techniques and various tracers (99mTc, 123I, 125I, 131I, 111In, 67 Ga) and can be used as a tool for the noninvasive exploration of a variety of disorders of lipoprotein metabolism. Density-gradient centrifugation allows the simultaneous isolation of the major lipoprotein density classes, i.e. very low density lipoproteins, LDLs, high-density lipoproteins, forming discrete bands of lipoproteins in preparative tubes. The analysis of the principal lipid constituents was done in isolated fractions. The cholesterol and triglyceride values in the fractions were determined by enzymatic methods. The fraction limits of the bands were defined by means of prestaining the serum lipoproteins. Aliquots of the fractions were taken for Schlieren analysis after adjusting the density and underlayering with a salt solution, in a spinning ultracentrifugation capillary band-forming cell. We obtained quantitative results by measuring the Schlieren areas between the sample curves and the reference baseline curve by computerised numerical and graphical techniques. The decomposition of the integrated curve was carried out using a nonlinear regression program followed by deconvolution algorithm analysis in order to determine the parameters of the constituent Gaussian subclasses. With the LDL particle concentration technique, we measured the concentration of LDL and analysed the polydispersity of the aliquots for the preparing of the radioactive-labelled lipoproteins. LDL is the major transport protein for endogenous cholesterol in human plasma. Radioiodination of LDL was performed using the iodogen method and LDL was radiolabelled with 99mTc using sodium dithionite as a reducing agent. The development of intracellularly trapped ligands permitted the sites of in vivo degradation of LDL to be identified. The methods described proved to be useful for a clear and immediate visual presentation of the concentration values of lipoproteins in the whole spectra and for the identification of the heterogeneity of lipoproteins subfractions. Combination of preparative and analytical ultracentrifugation methods allow lipoprotein aliquots to be investigated before and after radioactive labelling with isotopes to identify labelled materials.

Key words Single-spin density - gradient ultracentrifugation · Preparative isolation of lipoproteins · Investigation of lipoproteins by Schlieren curves · Radioactive-labelled lipoproteins

Introduction

Epidemiological studies indicate that development of coronary heart disease is due to increased plasma concentration of low-density lipoproteins (LDLs) and decreased plasma concentration of high-density lipoproteins (HDLs) [1, 2].

Furthermore, characteristic patterns of the different lipoprotein density classes or the appearance of additional lipoprotein fractions in the serum may lead in certain metabolic situations to the development of atherosclerosis [3, 4]. A characteristic feature of atherosclerotic lesions is the accumulation of LDLs which undergo oxidative modification [5, 6]. Uptake of oxidized LDL via scavenger receptor(s) on macrophages can lead to the intracellular accumulation of lipids. Radiolabelled LDL may be used as a tracer to study in vivo the distribution and metabolism of LDL as well as to allow the detection of atherosclerotic lesions in animal models and in human studies. Preparative ultracentrifugation analyses which are based on the density of lipoproteins permits the investigation of lipoproteins in their natural state. A number of preparative procedures are based on the principle of selective lipoprotein flotation at different salt densities and the lipoprotein classes are sequentially isolated by increasing the serum density in steps by the addition of salts such as NaCl, KBr, and NaBr or a combination of these [7]. Among the preparative procedures density-gradient ultracentrifugation employing single-spin techniques is widely used. A rapid method has been applied for the complete separation of very low density lipoproteins (VLDLs), LDLs, HDLs, and free protein by a single discontinuous-gradient ultracentrifugation in a radial rotor [8]. In the present study we employed a combination of simple discontinuous density-gradient method and analytical ultracentrifugation to determin quantitatively the lipoprotein aliquots before radioactive labelling.

Materials and methods

Serum samples

The blood samples were taken from the veins of normal individuals after 1 night of fasting. The blood was allowed to clot for 40 min at room temperature and the serum was then separated by centrifugation at 1,200g and 20 °C for 30 min. The serum was stored at 4 °C in the presence of 0.01 g/l ethylenediaminetetraacetate (EDTA) salt. The classification of the problems of metabolism according to World Heath Organisation categorisation was performed by the determination of the serum total lipids, total cholesterol, triglyceride, and agarose gel electrophoresis [9].

Preparative density-gradient ultracentrifugation

The preparative isolation of serum lipoproteins was carried out in a fixed angle rotor of 20° on a type of 3180 ultracentrifuge made by the Hungarian Optical Works (Budapest).The conditions of the gradient formation and lipoprotein separation were as follows. Serum (3 ml) was adjusted to a density of 1.3 g/ml with solid KBr (0.49 g/ml) and (in some cases 60 μl 1 g/l solution Sudan Black in ethylene glycol was added to aid the visualisation of the lipoprotein bands) placed in the bottom of a cellulose acetate centrifuge tube and overlayered with 8.5 ml 1.00 6 g/ml NaCl solution containing 0.01 g/l EDTA salt made to a volume of 11.5 ml.

Tubes were loaded into a P50 angular rotor of 20° for 130 min at 50,000 rpm. In the rotor the g force at the top of the tube was 95,030 and at the bottom of the tube it was 187,265. At the end of the run the tubes were removed from the rotor and 23 fractions were collected from each tube by puncturing the bottom using a fraction collector. Fractions of approximately 0.5 ml were taken from the tubes by drop flow. By means of prestaining the serum lipoproteins the fraction limits of the bands were defined. The Sudan Black solution was prepared using a method described by Terpstra et al. [10]. Analysis of the principal lipid constituents was done on the whole serum and the isolated lipoprotein fractions. The cholesterol and trigyiceride values in the fractions were determined by enzymatic methods [11, 12]. In the case of the control density gradient the solutions were removed in the same manner. At the end of the run the tubes were removed according to the localisation of the lipoprotein bands 1.5, 4, 2.5, 1.5, 2 ml, corresponding to free serum protein, HDL, LDL, intermediate-density lipoprotein (IDL), and VLDL, respectively, and analytical ultracentrifugation runs and radioactive labelling were performed using these aliquots.

Analytical density-gradient ultracentrifugation

The runs were carried out using an A 65-2 analytical rotor with a type 3180 ultracentrifuge made by the Hungarian Optical Works (Budapest) operating in refractometric Schlieren mode at 546 nm with a high-pressure mercury vapour light source lamp (HBO2000). We applied band-forming, capillary underlaying type single-sector centrepieces furnished with special holes. The 4° band-forming and 12-mm optical path length centrepieces were assembled with negative-angle wedge windows (−1°40′) on the bottom of the cell. The setting of the optical system was selected at a constant Philpot angle of 20°. Solid KBr was used to adjust the density of the sera or other preisolated lipoprotein fractions. The coordinates of the Schlieren distribution curves of the samples were obtained from the photograph, input into the computer, interpolated using piecewise cubic polynomials, and displayed. The interpolation of the Schlieren distribution curves of gradient baseline was found in the same way, the curves were subtracted, and the result was integrated exactly for the ordinates of all pixels. Then, this integral curve was displayed on the screen and one Gaussian curve was adapted to its main peak, guessing and inputting the three parameters of the Gaussian (mean value, standard deviation, height of the maximum) [13, 14].

The optimisation procedure is specialised to minimise the sums of squares, i.e. in our case:

$$F(\alpha_1,\beta_1,h_1,\ldots,\alpha_N,\beta_N,h_N)=\sum_x\left\{y(x)-\sum_{i=1}^N h_i\exp[-(x-\alpha_i)2/4\beta_i]\right\}^2,$$

(1)

where α is the mean value, β is the standard deviation, h is the height of the maximum, h_i is the maximum value of the function, α_i is the place of the maximum value, and β_i is the deviation.

Here the summation is taken over all x values (ordinates of pixels on the screen) and $y = y(x)$ is the integral curve. The optimisation procedure employs gradient vectors (obtained by a finite-difference formula) and Gaussian–Newtonian vectors, which are used to define a plane on which the functional F is minimised locally. The minimum value of F is called the fitting error. The lipoprotein particle concentrations were calculated from the area under the integral of the Gaussian curve using a calibration data constant.

LDL investigation method

LDL aliquots (2×50 μl) were injected into the holes of the centrepiece from the upper window with a microsyringe. The cell was assembled and 0.5 ml NaCl ($\rho = 1.042$ g/ml) solution containing 0.1 g/L EDTA was put into the sector of the centrepiece. Runs were performed at 50,000 RPM and 20 °C. When the rotor was accelerated to 2000–4000 RPM; the increasing hydrostatic pressure forced the adjusted-density sample from the holes of the centrepiece through capillaries under the salt solution in the sector. The maximum speed was obtained after 5 min. Photographs of the Schlieren pattern were taken at 20 min after reaching full speed.

HDL investigation method

Sera or isolated HDL aliquots fractions (2×50 μl) were adjusted to a density of 1.3 g/ml with solid KBr and were injected into the holes of the centrepiece from the upper window with a microsyringe. The cell was assembled and 0.5 ml NaCl ($\rho = 1.042$ g/ml) solution containing 0.1 g/l EDTA was put into the sector of the centrepiece. Runs were performed at 50,000 rpm and 20 °C. Photographs were taken after reaching full speed at 80 min.

VLDL investigation method

Isolated VLDL aliquots (2×50 μl) were injected into the holes of the centrepiece from the upper window with a microsyringe. The cell was assembled and 0.5 ml NaCl ($\rho = 1.006$ g/ml) solution containing 0.1 g/l EDTA was put into the sector of the centrepiece. Runs were performed at 50,000 rpm and 20 °C. Photographs were taken after reaching full speed at 20 min.

IDL investigation method

Isolated IDL aliquot fractions (2×50 μl) were adjusted to a density of 1.063 g/ml with solid KBr and were injected into the holes of the centrepiece from the upper window with a microsyringe. The cell was assembled and 0.5 ml NaCl ($\rho = 1.006$ g/ml) solution containing 0.1 g/l EDTA was put into the sector of the centrepiece. Runs were performed at 50,000 rpm and 20 °C. Photographs were taken after reaching full speed at 20 min.

Lipoprotein labelling with 99mTc

Native and copper-oxidized LDL were labelled with 99mTc according to the method described by Lees et al. [15]. Briefly, LDL (4–8 mg protein) was mixed with 40–60 mCi 99mTc-pertechnetate (TcO^{4-}) and 10 mg sodium dithionite, which was dissolved just before use in 0.5 M glycine buffer, pH 9.8, and the mixture was incubated for 30 min. 99mTc-LDL was separated from free 99mTc by Sephadex G25 chromatography. The purified 99mTc-LDL was dialysed against 150 mM NaCl, 1 mM EDTA, and 3 mM.

NaN$_3$ in tris(hydroxymethyl)aminomethane (Tris) buffer, pH 7.4, at 4 °C for 2 h, sterilised through a 0.22-μm filter.

Lipoprotein labelling with 125-I or 131-I

Aliquots of these lipoproteins were labelled with 125-I or 131-I: 2–4 g/l apo-B-containing lipoproteins were incubated with iodogen and 100 μCi 125-I at 9.5 pH. The radioiodine and lipoprotein were mixed together in the tube for 10–15 min, and the mixture was removed by decantation. The unreacted iodine was separated by column chromatography using Sephadex gel. The purified 125-I or 131-I was dialysed against 150 mM NaCl, 1 mM EDTA, and 3 mM NaN$_3$ in Tris buffer, pH 7.4, at 4 °C for 2 h, sterilized through a 0.22-μm filter [16].

LDL oxidation

Oxidised LDL was obtained by copper oxidation: LDL in 0.15 M NaCl was oxidized by incubation for 30 min at 37 °C with copper sulphate (2 μM). Oxidation was stopped by adding EDTA (2 μM) to the solution.

Results and discussion

In order to verify the lipoprotein assignment we examined the lipoprotein samples both obtained from normal subjects and from patients with different types of HLP. Analysis of the principal lipid constituents was done on the whole serum and the isolated lipoprotein fractions. The bands of the migrated lipoprotein were constant or only very small differences were noted. The localisation of the lipoprotein bands in the tubes indicated that three fractions at the bottom correspond to free serum protein. Lipoprotein bands with properties of HDL were present in the gradient fractions from 4 to 10. Bands with the properties of LDL were present in fractions 12–17. In fractions 18 and 19 IDL was present in type III HLP samples. The band at the top of the tube isolated in fractions 20–23 corresponded to VLDL.

Measurement of cholesterol

We determined VLDL, LDL, and HDL cholesterol in normal sera and in sera of patients of type IIB, type III, type IV, and type V HLP. The distributions of cholesterol were taken from the tubes in the 23 fractions after single-spin density centrifugation. Fractions were collected on the basis of the appearance of the bands.

In the normolipidemic sample 210–220 mg/dl (5.42–5.68 mmol/l) was the upper limit for serum cholesterol: VLDL cholesterol contained less than 40 mg/dl (1.03 mmol/l), LDL cholesterol was up to 160 mg/dl (4.13 mmol/l), and HDL cholesterol was in the range up to 60 g/l (1.55 mmol/l) (Fig. 1). In the type II serum LDL cholesterol was greater than 200 mg/dl (5.16 mmol/l) ((Fig. 2). In the type III, type IV, and type V VLDL cholesterol was isolated greater than 90 g/l.

Figures 1–3 show the values of cholesterol measured by the enzymatic method: the fractions (0.5 ml) were taken from preparative tubes by drop flow.

The most common enzymatic method employs the Trinder reaction, which includes the breakdown of cholesterol esters to free cholesterol by cholesterol esterase. Cholesterol oxidase, in the presence of oxygen, oxidizes free cholesterol to form cholest-4-ene-3-one and hydrogen peroxide. The hydrogen peroxide reacts with phenol and 4-aminoantipyrine in the presence of peroxidase to form a quinoneimine dye. The resulting colour is measured by photometry.

The amount of dye formed, determined by its absorption at 500 nm, is directly proportional to the total cholesterol concentration.

Measurement of triglyceride

The values of triglycerides measured by the enzymatic method are shown in Figs. 4–6: the fractions (0.5 ml) were taken from preparative tubes by drop flow.

The present procedure involves hydrolysis of triglycerides by lipase. The glycerol concentration was determined in a coupled assay that terminates in the formation of a quinoneimine dye. The amount of dye formed,

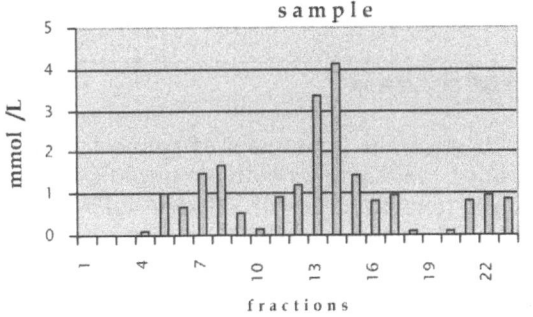

Fig. 1 Cholesterol distribution in a normal sample

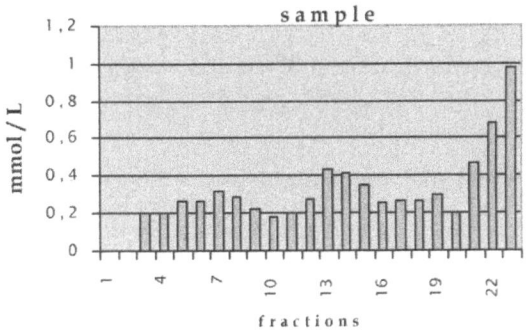

Fig. 4 Triglyceride distribution in a normal sample

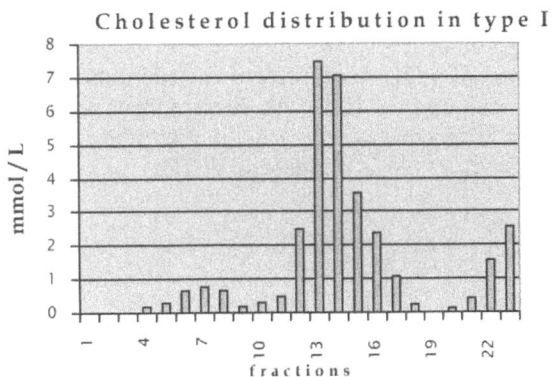

Fig. 2 Cholesterol distribution in a type II high-density lipoprotein (*HLP*) sample

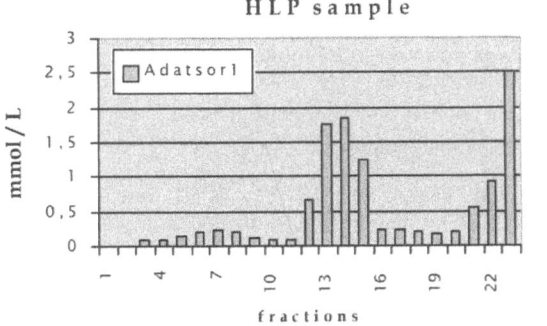

Fig. 5 Triglyceride distribution in a type II HLP sample

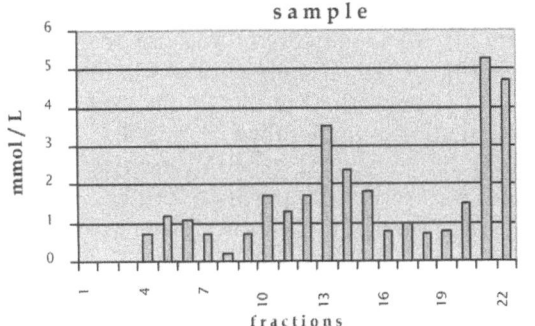

Fig. 3 Cholesterol distribution a type IV HLP sample

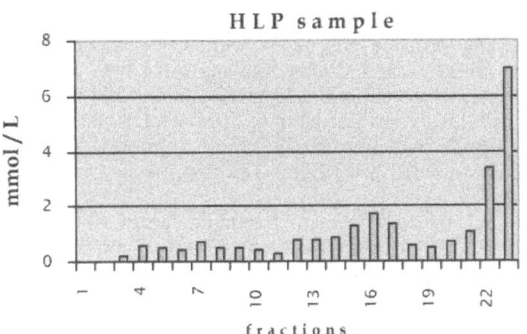

Fig. 6 Triglyceride distribution in a type IV HLP sample

determined by its absorption at 500 nm, is directly proportional to the triglyceride concentration.

The distributions of the triglycerides were taken also from the tubes in the 23 fractions after single-spin density centrifugation. Triglyceride isolated from normolipidemic sera containing less than 100 g/l (1.14 mmol/l) VLDL triglyceride (Fig. 4). Triglyceride was measured in a subject with mildly elevated VLDL triglyceride of 120 g/l (1.37 mmol/l) (Fig. 5) and in a subject with severe hypertriglyceridemia greater than 250 g/l (2.85 mmol/l) VLDL triglyceride (Fig. 6).

Prestaining sera

The prestained normal serum contained a trace of VLDL at the top of the tube, an intensely stained band containing LDL at the middle of the tube, and a faintly stained band of HDL at the lower region of the tube. Patients with type II HLP had an intensely stained board band (LDL) at the middle of the tubes. Prestained samples from patients of type III HLP revealed a diffusely stained band between VLDL and LDL due to the presence of floating IDL. In serum from patients with type IV HLP and type V HLP there was an intensely stained band at the top of the tube (the prestained samples are not shown).

LDL Schlieren curves

Deconvolution analysis of these integrated density curves revealed the presence of three fractions in the normolipidemic and type II HLP samples and of four fractions in the type IV HLP sample.

HDL Schlieren curves

Deconvolution analysis of these integrated density curves revealed the presence of two and three fractions in the samples of healthy subjects and patients with HLP.

VLDL Schlieren curves

Deconvolution analysis of these integrated density curves revealed the presence of three fractions in the normolipidemic and four fractions in the type II HLP sample and in the type IV HLP sample.

IDL Schlieren curves

Deconvolution analysis of these integrated density curves revealed the presence of three fractions in the sample.

Radiolabelling lipoproteins

The efficiency of the radiolabeling of LDL with technetium was more than 98%.

Rabbits were positioned under the collimator of a γ-ray camera after 10 mCi 99 mTc-LDL had been injected in vivo in a bolus. At 1 and 6 h postinjection the liver uptake was detected. The time of maximum liver uptake was at 1 h postinjection (figures not shown). 99mTc can be linked directly in the presence of reducing agent, and the Tc-LDL bonds are strong enough for receptor uptake. Our results show that 99 mTc-LDL can be used to assess the organ distribution pattern of LDL in rabbits by external imaging. The present results are an

Fig. 7 Low-density lipoprotein (*LDL*) Schlieren curves: **a** normal serum with 130 mg/dl LDP cholesterol; **b** type II HLP with 192 mg/dl LDL cholesterol; **c** type IV HLP serum with 88 mg/dl LDL cholesterol

 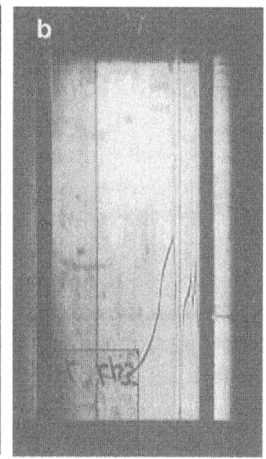 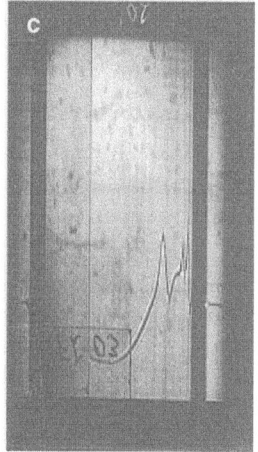

Normal serum with 130 mg / dl LDL chol. Type II HLP with 192 mg/ dl LDL chol Type IVHLP serum with 88 mg / dl LDL chol

Fig. 8 HDL Schlieren curves:
a normal serum with 60 mg/dl
HDL cholesterol from aliquots;
b type II HLP with 40mg/dl
HDL cholesterol from aliquots;
c type IV HLP serum with
45mg/dl HDL cholesterol from
aliquots; **d** normal serum with
65 mg/dl HDL cholesterol from
whole sera together with the
LDL peak

HDL Schlieren curves

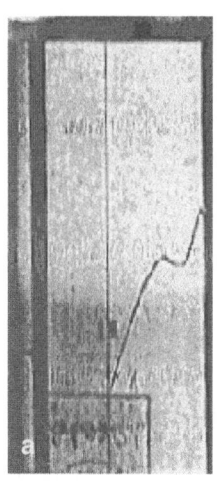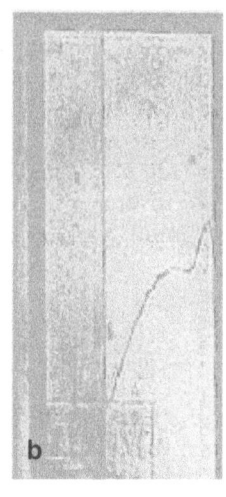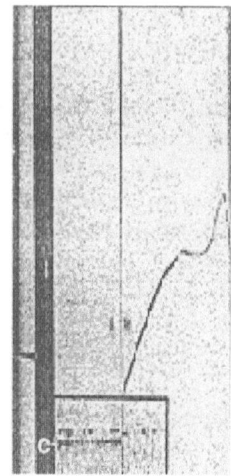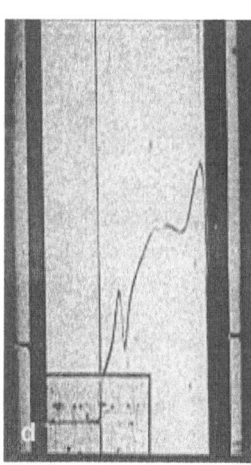

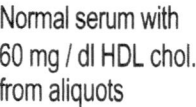

Normal serum with
60 mg / dl HDL chol.
from aliquots

Type II HLP with
40 mg/ dl HDL chol
from aliquots

Type IV HLP serum
with 45mg / dl HDL
chol from aliqouts

Normal serum with 65 mg/dl
HDL chol from whole sera
together with LDL peak.

VLDL Schlieren curves

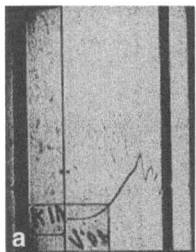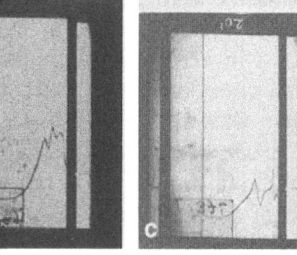

Type III HLP with   Type II HLP with
120 mg / dl VLDL chol. 105 mg/ dl VLDL chol

Type IVHLP serum with
135 mg / dl V LDL chol

Fig. 9 Very low density lipoprotein (*VLDL*) Schlieren curves: **a** type
III HLP with 120 mg/dl VLDL cholesterol; **b** type II HLP with
105 mg/dl VLDL cholesterol; **c** type IV HLP serum with 135 mg/dl
VLDL cholesterol

IDL Schlieren curves

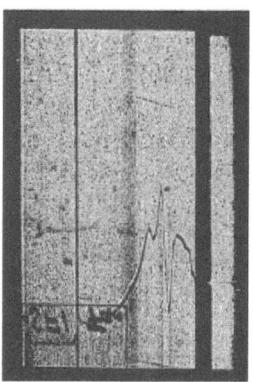

Type III HLP with
140 mg/dl IDL chol

Fig. 10 Intermediate-density lipoprotein (*IDL*) Schlieren curve for
type III HLP with 140 mg/dl IDL cholesterol

important advance towards the goal of external imaging
of atherosclerosic lesions in cholesterol diet rabbits.

The present study describes the construction of a
discontinuous density-gradient ultracentrifugation in a
fixed-angle rotor which permits the investigation of the
major lipoprotein classes (VLDL, IDL, LDL, HDL) of a
whole serum sample. We developed new analytical
ultracentrifuge micromethods for the investigation of
serum lipoproteins from ultracentrifuge Schlieren scans
after preparative isolation. We obtained quantitative
results by measuring the Schlieren areas between the
sample curves and the reference baseline curve by using
computerised numerical and graphical techniques.
Combination of preparative and analytical ultracentrif-

ugation methods can be used for the investigation of
lipoprotein (VLDL, IDL, LDL, HDL) aliquots to
determine concentration values and the presentation of
lipoprotein Schlieren data to gain an insight into the
profile, allowing comparison of the serum lipoprotein
concentration. We determined the qualitative and quantitative characteristics of lipoprotein particles in normal
subjects and in patients defined as presenting type IV,
type III, and type II hyperlipoproteinemia. The aim of
this work was to develop a rapid and reproducible
ultracentrifugation method to obtain lipoprotein aliquots for radiolabelling and to evaluate lipoprotein label-

ling techniques which are suitable for the scintigraphic delineation of experimental atherosclerotic lesions.

Conclusions

The present study describes the construction of discontinuous preparative and analytical density-gradient ultracentrifugation methods which permit the investigation of major lipoprotein classes (VLDL, LDL, HDL) and separated lipoprotein fractions. We designed and applied band-forming, capillary-underlayering-type single-sector centrepieces furnished with special holes. We used special software for the analysis of this type of single-spin density-gradient ultracentrifugation. Besides Schlieren optics, UV absorption optics can be used at 254 nm to record the flotation of the lipoproteins in the cell at the same time.

Because of its rapidity and convenience, lipoproteins probably remain fully native. Lipoprotein profiles of serum from hyperlipoproteinemia phenotypes examined by our methods were qualitatively different from each other and from profiles of normolipidemic subjects. There is a relationship between the enzymatic cholesterol quantity of the separated fractions in different HLP types and the estimation of lipoprotein Schlieren curves on the basis of concentration values. The lipoprotein Schlieren curves were more often characterised by a preponderance of less dense and denser subfractions in samples of patients with hyperlipoprotenemic disease than in controls.

This method may be used for monitoring the separated subfraction of major lipoprotein classes before and after radioactive labelling and oxidative modification of lipoprotein subfractions and can detect the heterogeneity of lipoproteins.

References

1. Fischer WR (1983) Metabolism 32: 283–91
2. Crouse JR, Parks JS, Schey HM (1985) J Lipid Res 26:566–74
3. Swinkels DW, Hak-Lemmers HLM, Demacker PNM (1987) J Lipid Res 28:1233–1239
4. Austin MA, Breslow JL, Hennekens CH, Buring ME, Willet WC, Krauss RM (1988) J Am Med Assoc 260:1917–21
5. Witztum JL (1994) Lancet 344:793–795
6. Morel DW, DiCorleto PE, Chilson GM (1984) Arteriosclerosis 4:357–364
7. Havel RJ, Eder HA, Bradgon HJ (1955) J Clin Invest 34:1345–1353
8. Chung BH, WilkinsonT, Greer JC, Segrest JP (1980) J Lipid Res 21:284–291
9. (a) Fredrickson DS, Levy RI, Lees RS (1967) N Engl J Med 276:34–44; (b) Fredrickson DS, Levy RI, Lees RS (1967) N Engl J Med 276:94–103; (c) Fredrickson DS, Levy RI, Lees RS (1967) N Engl J Med 276:48–150; (d) Fredrickson DS, Levy RI, Lees RS (1967) N Engl J Med 276:218–225; (e) Fredrickson DS, Levy RI, Lees RS (1967) N Engl J Med 276:273–281
10. Terpstra AHM, Sanchez-Muniz FJ, West CE, Woodward CJH (1982) Comp Biochem Physiol B 71:669–763
11. Burstein MS, Savaille S (1960) Clin Chim Acta 5:609–615
12. Bucolo G, David H (1973) Clin Chem 22:98–101
13. Dixin WJ (1981) BMDP Statistical Software. University of California Press, Berkeley
14. Medgyessy P (1977) Decomposition of superpositions of density functions and discrete distributions. Publishing House of the Hungarian Academy of Science, Budapest, p 90
15. Lees RS, Garabedian HD, Lees AM (1985) J Nucl Med 26:1056–1062
16. Matthews CME (1957) Phys Med Biol 2:36–56

Progr Colloid Polym Sci (2002) 119:50–57
© Springer-Verlag 2002

Gisela Berth
Helmut Cölfen
Herbert Dautzenberg

Physicochemical and chemical characterisation of chitosan in dilute aqueous solution

G. Berth (✉) · H. Cölfen · H. Dautzenberg
Max Planck Institute for Colloids and
Interface Research, 14424 Potsdam
Germany
e-mail: berth@mpikg-golm.mpg.de

Abstract Static and dynamic light scattering as well as analytical ultracentrifugation and viscosity measurements were used to investigate the chain conformation of chitosans in salt-containing solutions (pH 4.5; ionic strength about 0.12 M). The samples of various degrees of acetylation were chemically homogeneous. The molecular-weight dependence of the radius of gyration has given clear evidence for a relatively flexible wormlike chain with a persistence length of about 6 nm irrespective of the degree of acetylation, where excluded-volume effects and the polydispersity of the samples were taken into account. In contrast, the interpretation of the hydrodynamic data via a "whole-body approach" according to the Wales–van Holde ratio suggested a strongly elongated chain conformation. The failure of the latter to properly reflect the chain conformation was ascribed to the high extent of draining. A nearly free-draining case can also account for the high scaling exponent of the relationship between intrinsic viscosity and molecular mass.

Key words Chitosan · Wormlike chain · Draining effects · Light scattering · Analytical ultracentifugation

Introduction

In previous studies we gained experience in the field of macromolecular and chemical characterisation of several (ionic) polysaccharides such as pectins, xanthans, xylinans and chitosans (reviewed in Ref. [1]). The application of static light scattering (SLS) has been at the centre of our attention and the studies were occasionally complemented by ultracentrifugation analyses. Regarding the average molecular weights, we could often achieve respectable agreement but some disagreement remained with respect to the chain conformation. In this contribution we take up this topic using chitosans as a model substance to discuss potentials and limitations of different approaches for an estimate of the chain conformation of a polymer in solution.

Chitosans are produced on a large scale from chitin. The chitin normally comes from crustaceans. The main source is the shells of shrimp, lobster and crabs. Major fields of application are in the cosmetic and pharmaceutical industry. In the USA chitosan is also used in slimming cures. Our interest in chitosan was stimulated by possible applications in the field of self-organising or nanosized materials. We wanted to characterise chitosans in terms of their chemical and macromolecular parameters, where the interest was focussed, in particular, on the chain conformation in solution. This seemed to be a challenging task. The literature that already existed when we started was full of contradictions. Proposals ranged from Gaussian coils to almost rodlike structures and there was no conclusive explanation for the diversity of findings. Our own results have already been published [2–4]. In the following, some selected results are introduced but major attention is focussed on methodical problems which we were faced with when we tried to make sense out of the data.

Chemically chitosans are a family of linear copolymeric polysaccharides which are built up of glucosamine and N-acetyl glucosamine. The monomeric units are linked via β-1.4-glycosidic bonds. There is a close structural relationship to cellulose.

An important chemical parameter is the degree of acetylation (DA). The DA relates the number of acetylated units to the total number of monomer units and is given in percent. The majority of commercial chitosans has average DAs between 20 and 30%. For special applications DAs up to about 60% are of interest. The bulky acetyl groups have occasionally been claimed to affect the chain conformation for steric reasons but in our judgement there is no convincing evidence for that in the literature. So we had to include this aspect in our own investigations [4].

Chitosans are soluble in aqueous acidic solution. Then the free amino groups bind protons and the macromolecules become polycations whose charge density depends on the chemical composition. Like other polysaccharides, chitosans have broad molecular-weight distributions.

From the viewpoint of a polymer scientist, the most characteristic structural features can be resumed in three items:

– Linear copolymers of various chemical composition.
– (Cationic) polyelectrolytes (in acidic aqueous solution).
– Polydisperse materials (at least) with respect to the molecular mass.

Each of these features was of importance for the strategy of our studies.

Experimental

We used three commercial samples with DAs between 20 and 30% and five laboratory-made chitosans with DAs between 5 and 55%. The DAs were determined by colloid titration [4]. Acetate buffer (0.02 M) of pH 4.5 was used as the solvent. The ionic strength was chosen to be about 0.12 mol l^{-1} by adding sodium chloride.

The methods used are as follows:

– SLS and dynamic light scattering.
– Analytical ultracentrifugation (sedimentation equilibrium and sedimentation velocity).
– Membrane osmometry.
– Capillary viscometry.
– Colloid titration with toluidine blue as indicator.
– Semipreparative gel permeation chromatography on Sepharose and Sephadex gels.
– High-pressure size-exclusion chromatography on TSK columns plus multiangle laser light scattering detection.

All experimental details have been reported in previous publications [2–4].

Results and discussion

Checking the chemical homogeneity of the samples

For polydisperse copolymers, the chemical composition can vary within the molecular-weight distribution, so it seemed useful to check the chemical homogeneity of the samples used in these studies. This could be done by analysing the DA distribution across the molecular-weight distribution. Therefore, we fractionated the samples by gel permeation chromatography (GPC) and determined both the DA (by means of colloid titration) and macromolecular parameters such as molecular weight and reduced viscosity in the fractions. A typical set of results is shown in Fig. 1.

One can see the polymer elution covers a broad elution volume range. The molecular weights in the fractions decrease continuously over a broad range with increasing elution volume and so do the reduced viscosities (which are nearly identical with the intrinsic viscosity because of the low polymer concentration in the eluant). Both quantities combined lead to the straight universal calibration line on the bottom which is indicative of a successful separation. The respective DA values are evenly distributed over more than 90% of the population, apart from the low percentage of low-molecular-weight but highly acetylated material in the dark-shadowed area. Such small amounts of low-molecular-weight "impurities" were not supposed to be relevant for the experiments we are discussing here and therefore it seemed to be reasonable to proceed with the unfractionated parent samples.

Static light scattering

Data evaluation was done in the framework of the Rayleigh–Debye theory. For polymers of the chitosan type, a Zimm plot is an appropriate data presentation. A typical Zimm plot is shown in Fig. 2. The angular dependence of the scattered light at zero concentration provides $R_{G,z}$ from the z-average of the square of the radius of gyration. The concentration dependence at zero angle provides the second virial coefficient, B, and the joint point of interception of the two straight lines with the ordinate gives the weight-average molecular weight, M_w.

Having done such measurements for all the samples in this study, it is useful to consider the molecular-weight dependence of the radius of gyration. From the theory of macromolecules in solution it follows that the scaling exponent in a double-logarithmic plot is 0.5 for Gaussian coils in the unperturbed state and 1.0 for rigid rods. All values in-between are possible.

Our own data along with data from the literature [5–8] are plotted in Fig. 3. Although data for rather different chitosans in somewhat different solvents have been mixed up, we find our own values in good agreement with the others. The majority of points straddle a straight line, the scaling exponent of which is 0.55. This is indicative of a rather flexible wormlike

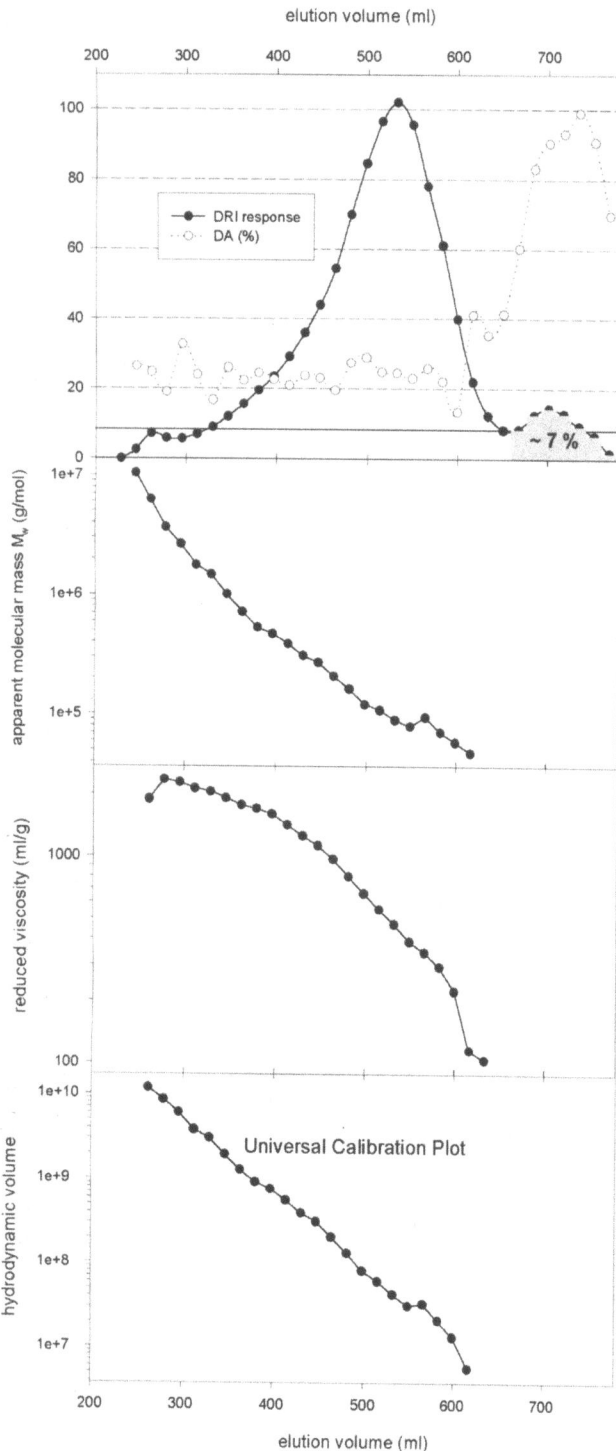

Fig. 1 Check of the chemical homogeneity after gel permeation chromatography on Sepharose CL-2B/ Sephadex G-75 by analysing the degree of acetylation (*DA*) distribution across the molecular-weight distribution. Sample description: DA 25–30%; $M_w = 140,000 \, \mathrm{gmol}^{-1}$; $[\eta] = 650 \, \mathrm{mlg}^{-1}$

chain and we now turn to the details of the interpretation.

The characteristic stiffness parameter of the worm-like-chain model is the persistence length, L_P. For a monodisperse system of unperturbed wormlike chains, L_P is related to the radius of gyration and the contour length, L, according to Eq. (1):

$$R_{\mathrm{G},\theta}^2 = \frac{L L_\mathrm{P}}{3} - L_\mathrm{P}^2 + \frac{2 L_\mathrm{P}^3}{L} - \frac{2 L_\mathrm{P}^4}{L^2} \left(1 - e^{-L/L_\mathrm{P}} \right) \ . \tag{1}$$

L is obtained from the experimentally measured molecular weight divided by the mass per unit length, M_L. In our case, M_L varies with the DA and is between 335 and 370 g(molnm)$^{-1}$. M is in fact M_w and consequently L means precisely L_w – the weight-average contour length.

For polyelectrolytes, L_P results from an electrostatic contribution, $L_{\mathrm{P,e}}$, and the intrinsic persistence length, $L_{\mathrm{P,i}}$, which means the stiffness of the same chain in the absence of any charges:

$$L_\mathrm{P} = L_{\mathrm{P,i}} + L_{\mathrm{P,e}} \ . \tag{2}$$

$L_{\mathrm{P,e}}$ depends on the ionic strength of the solution via the Debye–Hückel screening length and can be calculated according to the Odijk approach by means of the following equation:

$$L_{\mathrm{P,e}} = \frac{\lambda_\mathrm{B} \lambda_\mathrm{D}^2}{4 b^2 \xi_\mathrm{M}^2} \ , \tag{3}$$

where

$$\lambda_\mathrm{D} = \left(\frac{1000}{8 \pi \lambda_\mathrm{B} N_\mathrm{A} I} \right)^{1/2} \ ,$$

where λ_D is the Debye–Hückel screening length in centimetres, N_A is Avogardro's constant, I is the ionic strength of the solvent (moles per litre), λ_B is the Bjerrum length in centimetres (about 7 Å in water at room temperature), b is the charge distance along the chain and ξ_M is the effective charge according to the Manning theory. At a glance one can see that the higher, the smaller λ_D and hence $L_{\mathrm{P,e}}$.

Under our conditions, the electrostatic contribution was calculated to be on the angstrom scale. This is within the experimental error, so the electrostatic term of the persistence length becomes negligible. What was measured by experiment is $L_{\mathrm{P,i}}$ and this is what we are normally interested in when, for instance, we want to judge possible steric effects of the DA on the chain conformation [2, 4].

Before L_P can be calculated from the experimentally measured radii of gyration and L_w, two corrections have to be made.

Firstly, the experimentally obtained radius of gyration comes from a *z*-average value and cannot be related to the weight-average contour length

Fig. 2 Zimm plot of chitosan in acetate buffer of pH 4.5 and an ionic strength of about 0.12 moll^{-1}. $M_w =$ 139,000 gmol^{-1}; $R_{G,z} = 58$ nm; $B = 4.78 \times 10^{-3}$ mlmolg^{-2}

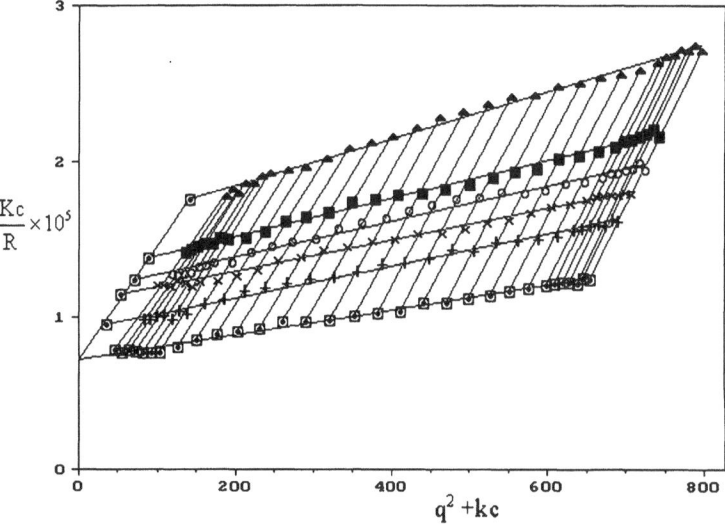

Fig. 3 Molecular-weight dependence of the radius of gyration

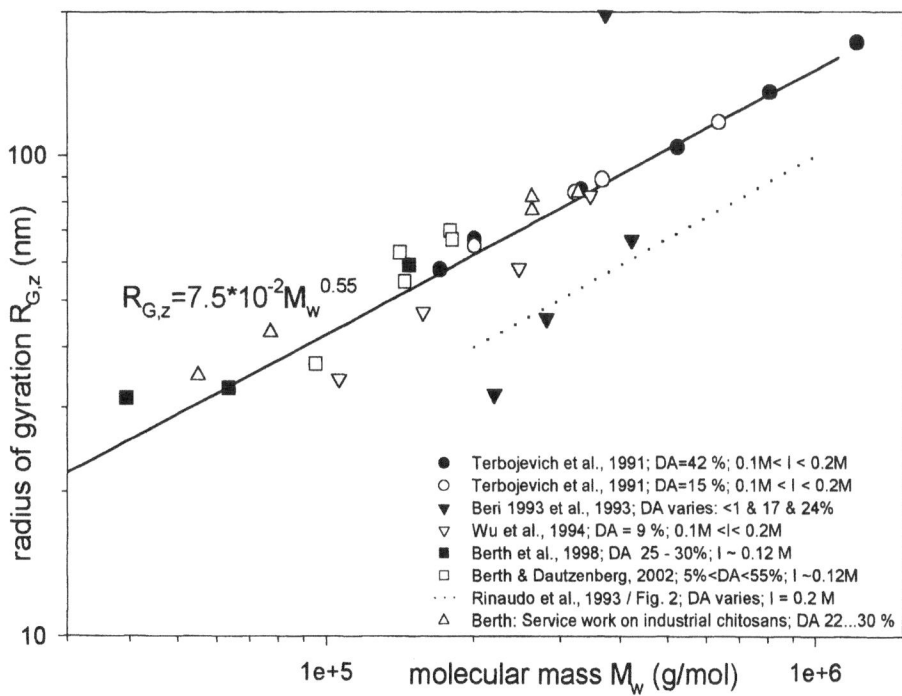

without any correction for polydispersity. Since the number-average molecular weights of our samples were known by membrane osmometry measurements [2], the polydispersity, σ_M, could be estimated. Moreover, a special logarithmic distribution function approximated the molecular-weight distribution obtained by GPC sufficiently well [1]. So we applied Eqs. (4) and (5)

$$\langle R_G^2 \rangle_z = \frac{1}{M_w} \int_0^\infty M R_G^2 p_w(M) \mathrm{d}M , \qquad (4)$$

$$p_w(M) \frac{= M^{-3/2} \exp\left[-(\ln M - \ln M_w)/2\sigma_M^2\right]}{(2\pi)^{1/2} \sigma_M M_w^{-1/2} \exp\left[\sigma_M^2/8\right]}$$

$$\text{and} \quad \frac{M_w}{M_n} = e^{\sigma_M^2} \qquad (5)$$

to relate the z-average of R_G^2 to the weight average of the molecular mass.

Secondly, the Zimm plot in Fig. 2 reveals a highly positive B value. This holds for all the samples in this study and means the measurements were done in a thermodynamically good solvent, so excluded-volume effects had to be considered. The excluded-volume concept takes into account the fact that in a thermodynamically good solvent segment–segment interactions lead to an expansion of the chain dimensions. This is described by the famous formula introduced by Flory

$$R_G^2 = \alpha^2 R_{G,0}^2 \ , \tag{6}$$

with α being the expansion factor, where

$$\alpha^5 - \alpha^3 = (134/105)z \quad \text{and} \quad z = \left(\frac{3}{2\pi l_K^2}\right)^{3/2} \beta N_K^{1/2} \ , \tag{7}$$

with β being the excluded volume and N_K and l_K being the number and length of the Kuhn segments.

To estimate α, we assumed the electrostatic term of the excluded volume to be dominant,

$$\beta = \beta_{el} = 8\pi\lambda_D L_P^2 \ . \tag{8}$$

With $N_K = \frac{L}{l_K} = \frac{L}{2L_P}$, one obtains

$$z_{el} = \frac{3^{3/2}}{4\pi^{1/2}}L^{1/2}\lambda_D L_P^{-3/2} \ . \tag{9}$$

Starting with the persistence length that was obtained assuming an unperturbed state, one can calculate z and α. An iterative procedure leads to the final value of $L_P \cong 6$ nm. The value would have been significantly higher without the corrections for polydispersity and excluded-volume effects (Table 1).

To resume the result, the SLS results lead to a single-stranded wormlike chain with a persistence length of about 6 nm in the observed range of ionic strengths between 0.1 and 0.3 mol l^{-1} (when the data from the literature in Fig. 3 are involved). This applies irrespective of the DA between 5 and 55%. According to this, a chain with a molecular weight of 140,000 gmol^{-1} corresponding to a contour length of about 400 nm consists of about 30 Kuhn segments. This means the chains are relatively flexible and resemble Gaussian coils rather than rigid rods.

Hydrodynamics

Viscosity

In the framework of the Kirkwood–Riseman theory [9], the intrinsic viscosity, $[\eta]$, can be calculated according to

$$[\eta] = 6^{3/2}\Phi_0 \frac{\langle R_G^2 \rangle^{3/2}}{M} \tag{10}$$

for the nondraining case with $\Phi_0 = 2.87 \times 10^{23}$ mol^{-1} for the Flory constant.

Using our data, the calculated values in Table 2, column 4 are twice or three times as high as the measured ones in column 3. These findings made us think about draining effects. Compared with the non-free-draining case, partial draining is known to reduce the intrinsic viscosity but to increase the diffusion coefficient as well as the sedimentation coefficient. Partial draining would explain the failure of approaches that have been developed for the non-free-draining case only.

The extent of draining can be estimated by comparing the actually measured intrinsic viscosity with the theoretical value for the non-free-draining case. Φ_0 varies with the draining parameter, X, according to the function

$$\Phi_0(X) = \left(\frac{\pi}{6}\right)^{3/2} N_A [XF(X)] \ , \tag{11}$$

where $\Phi_0 = 2.87 \times 10^{23}$ mol^{-1} and $X = \infty$ for the non-free-draining case and $X = 0$ for the free-draining case. $XF(X)$ is a decreasing function of X (see the tabulated values in Ref. [9], p. 269). The low X values obtained for our samples (Table 2, column 6) suggest the case of a nearly free draining wormlike chain.

The next item to be considered is the $[\eta]$–M_w relationship in Fig. 4. Again we have mixed our own data with those from other laboratories [8, 10, 11] on a variety of samples under various conditions. The points straddle a joint regression line, the scaling exponent of which is found to be 0.92. Such high values are

Table 1 Selected results of static light scattering (*SLS*) (M_n by membrane osmometry)

Samples	M_w (gmol^{-1})	$B \cdot 10^3$ (mlmolg^{-2})	$R_{G,z}$ (nm)	M_w/M_n	L_w (nm)	α	$\langle R_{G,0}\rangle_z$ (nm)	$\langle R_G\rangle_w$ (nm)	$\langle R_{G,\theta}\rangle_w$ (nm)	L_P (nm)
Chitosan A	63,830	3.55	33.0	2.05	182	1.18	27.9	23	19.5	6.5
Chitosan C	147,000	6.76	59.3	2.56	439	1.26	47.1	37	29.4	6.0

Table 2 Estimate of the extent of draining based on viscosity and SLS measurements

Sample	M_w (gmol^{-1})	$R_{G,w}$ (nm)	$[\eta]_{exp}$ (mlg^{-1})	$[\eta]_{NDC}$ (mlg^{-1})	$\frac{[\eta]_{exp}}{[\eta]_{NDC}} = \frac{\Phi_0(X)}{\Phi_0(\infty)}$	X
Chitosan A	64,000	23	333	834	~0.4	~1
Chitosan B	39,300	22	150	1,483	~0.1	~0.2
Chitosan C	147,000	37	650	1,453	~0.4	~1

commonly ascribed to a high chain stiffness. Gaussian coils under theta conditions would give an exponent of 0.5 according to Eq. (10) and, dependent on increasing chain stiffness and excluded-volume effects, the exponent would increase to values close to unity. The maximum value is reached at 2.0 for rigid rods. This is one way of interpreting the scaling exponent but the suggested high chain stiffness would be in conflict with the model based upon the light scattering data. The draining effect can resolve the apparent contradiction again. Following Kirkwood and Riseman [9], one obtains the equation

$$[\eta] = \frac{N_A n \varsigma}{6 \eta_0 M} \langle S^2 \rangle_0 \qquad (12)$$

for the free-draining case ($X = 0$), with S being the end-to-end distance, n the number of segments per chain, ς the translational friction coefficient of the segment, η_0 the solvent viscosity and N_A Avogadro's number. Since S^2 is proportional to M and n is proportional to M, it follows that $[\eta] \sim M^{1.0}$ (scaling exponent 1.0). So, in our example, the high experimental scaling exponent in the $[\eta]$–M relationship reflects the nearly free draining situation rather than a high chain stiffness.

Analytical ultracentrifugation

Sedimentation equilibrium experiments were done on only one sample (Table 3, chitosan A). The weight-average molecular weight obtained is in reasonably good agreement with the light scattering results. Sedimentation velocity measurements were performed on several samples and we used the Svedberg equation as well as an approach proposed by Scheraga and Mandelkern for an estimate of the average molecular masses. (Both the sedimentation coefficients and the diffusion coefficients were related to water and 20 °C in the common way, index 20,w.) The results are compiled in Table 3.

At a glance one notices that almost all of the hydrodynamically estimated molecular masses are significantly lower than the light scattering and sedimentation equilibrium data. The lowest ones result from the Scheraga–Mandelkern approach based upon intrinsic viscosities and sedimentation coefficients for the samples specified. Furthermore, when we used the diffusion coefficient $D^0_{20,w}$ (derived from dynamic Zimm plots) to calculate the ratio of the radius of gyration to the hydrodynamic radius (via Stokes' law), we obtained somewhat higher values than the anticipated value of about 2.0 which is predicted for polydisperse Gaussian coils in a thermodynamically good solvent. The failure of the Scheraga–Mandelkern approach becomes plausible in the light of the draining effects discussed.

So there was interest in what would result from a so-called "whole-body approach" if we interpreted our hydrodynamic data in terms of conformations, i.e., in terms of the axial ratio of ellipsoids of revolution.

A plot of the reciprocal sedimentation coefficients ($1/s_{20,w}$) versus concentration is shown in Fig. 5. There is a strong scattering of points owing to the low magnitude of s but, after a critical evaluation of all details, we

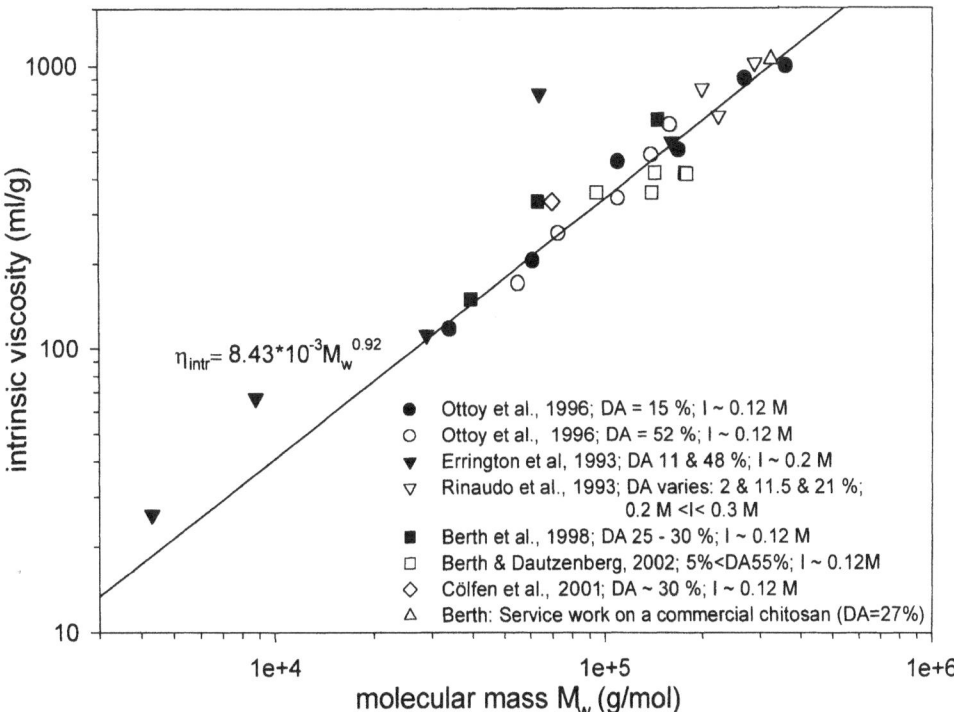

Fig. 4 $[\eta]$–M_w relationship for chitosans of various DA

Table 3 Average molecular masses obtained from different approaches and ratios of the radius of gyration to the hydrodynamic radius

Sample	M_w by SLS (gmol^{-1})	$[\eta]$ (mlg^{-1})	M_w by sedimemtation equilibrium (gmol^{-1})	M_w (gmol^{-1}) via Svedberg[a]	M_w (gmol^{-1}) via Scheraga–Mandelkern[b]	R_G/R_h[c]
Chitosan A	63,800					
	72,000	333	70,000	52,300	33,500	2.3
	64,900					
Chitosan B	39,300	150	n.d.	44,200	21,300	
	44,100					2.4
Chitosan C	147,000					
	134,000	650	n.d.	110,000	76,300	2.5
	180,000					

[a] $M_w = \left(s^0_{20,w}RT\right)/\left[D^0_{20,w}(1-\bar{v}\rho_0)\right]$

[b] $\beta = \frac{N_A s^0_{20,w}[\eta]^{1/3}\eta_0}{M^{2/3}(1-\bar{v}\rho_0)100^{1/3}}$

[c] R_h from $D^0_{20,w}$ (from dynamic Zimm plots) via Stokes' law

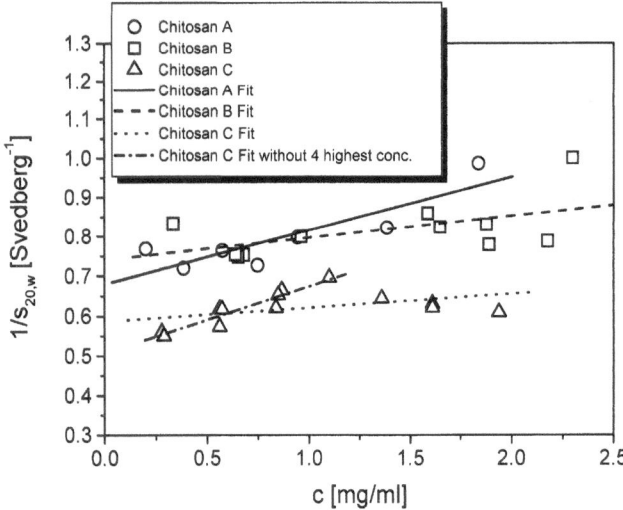

Fig. 5 Reciprocal sedimentation coefficients versus concentration for three industrial samples of various average molecular weight

which allows an estimate of the gross conformation of a polymer in solution. $\bar{v}$ and v_s are the partial specific volume and the total specific volume of the solute, respectively. v is the Simha factor. f/f_0 describes the deviation of the frictional coefficient from that of a sphere with a partial specific density $\bar{\rho} = \frac{1}{\bar{v}}$. There is a simple relation between R, the Perrin factor, P, and v:

$$R = \frac{2}{v}\left[1+P^3\right] . \qquad (14)$$

R is 1.6 for compact spheres whereas a more asymmetric or a stiffer conformation lowers the value. R can be converted into an axial ratio (a/b) of oblate or prolate ellipsoids of revolution by means of a special inversion algorithm. The ratio a/b then describes the shape of that hypothetical ellipsoid of revolution which models the observed hydrodynamic behaviour best. For several reasons discussed in Ref. [3] we took an average value of $R\sim0.4$ as most reasonable for all our samples, irrespective of their molecular masses, and this led to an axial ratio as high as 25.

If taken as a reflection of the real shape, such a strongly elongated shape would be in severe contradiction to the observed molecular-weight dependence of R_G and $[\eta]$. If we stopped at this point we would have to conclude on extremely controversial results from light scattering and ultracentrifugation regarding the chain conformation. In any case, the light scattering results

proceeded with the underlined values in Table 4, which reflect the general facts well [3]. The slope of the straight lines gives k_s, the concentration dependent regression coefficient. k_s divided by $[\eta]$ gives the Wales–van Holde ratio, R,

$$R = \frac{k_s}{[\eta]} = \frac{2\bar{v}}{v v_s}\left[\frac{v_s}{\bar{v}} + \left(\frac{f}{f_0}\right)^3\right] , \qquad (13)$$

Table 4 Results of sedimentation velocity experiments and Wales–van Holde ratio, R

Sample	$S^0_{20,w}$ (S)	k_s (mlg^{-1})	k_s (gl^{-1})[b]	$R = k_s/[\eta]$
Chitosan A	1.43 ± 0.05	140 ± 30	127	0.42/0.38[b]
Chitosan B	1.35 ± 0.082	56 ± 29	110	0.37/0.73[b]
Chitosan C	(1.72 ± 0.06)	(30 ± 20)	(199)	(0.05)
	1.97 ± 0.06[a]	166 ± 23[a]	283	0.26/0.44[b]

[a] Neglecting the data points at higher concentrations

[b] Calculated according to Lavrenko et al. [12]

Table 5 Comparison of size parameters derived from SLS and hydrodynamic data (for details see the text)

	Chitosan A	Chitosan B	Chitosan C
M_w (gmol^{-1})	64,000	39,300	147,000
$R_{G,w}$ (nm)	23.0	22.7[a]	37.1
b (nm)	1.83	1.19	3.00
R_G (nm)	20.5	13.3	33.6

[a] Questionable because of uncertainties especially in the SLS measurement

would have to be given preference since SLS is the only method that provides information on the particle mass and size without any assumptions. In contrast, the hydrodynamic behaviour of a particle is known to depend on three factors that cannot easily be separated from each other. These are the particle shape, solvation and draining. A comment made by Wales and van Holde [13] is supportive of this judgement. These authors explicitly confined their concept to the non-free-draining case and mentioned that cellulose derivatives did not match the pattern.

As an attempt to make sense of the hydrodynamic data we introduced [3] the following proposal. The aim was to check whether size information from hydrodynamic data would be consistent with R_G from SLS.

From the axial ratio (a/b) of an ellipsoid of revolution one obtains the respective Simha factor, v, from tables. v together with [η] allows one to calculate the total specific volume of the solute, v_s, where solvation is included according to the equation

$$[\eta] = v v_s , \tag{15}$$

where v_s is given by

$$v_s = \frac{V_E N_A}{M} , \tag{16}$$

with V_E being the volume of a single ellipsoid of revolution and M the molecular mass.

The volume of a single prolate ellipsoid of revolution (V_E) is defined by

$$V_E = \frac{4\pi}{3}\tau b^3 , \tag{17}$$

where $\tau = a/b$ for the axial ratio and b is the short axis of the ellipsoid of revolution. Equation (17) combined with Eq. (16) and rearranged gives

$$b = \left(\frac{3 v_s M}{4\pi N_A \tau}\right)^{1/3} , \tag{18}$$

so the length of axis b can be calculated from M, [η] and k_s, all of them accessible by experiment.

The corresponding radius of gyration of such an ellipsoid of revolution is obtained from

$$R_G^2 = \frac{1}{5}b^2(2 + \tau^2) \tag{19}$$

and, to our surprise, the values obtained are not far from the weight-average R_G derived from SLS (Table 5).

In other words, these data interpreted in terms of a wormlike chain according to Eq. (1) would lead to (almost) the same stiffness parameter that was derived from SLS and there was no longer a basic contradiction regarding the chain conformation.

Conclusions

According to light scattering results, many polysaccharides such as chitosans, cellulose derivatives, alginates, hyaluronic acids, heparin, pectins, etc. behave as relatively flexible chains in salt-containing aqueous solutions. Nevertheless, their hydrodynamic behaviour analysed by means of the "whole-bodyapproach" is modelled best by a highly asymmetric shape. This, however, does not necessarily mean a high chain stiffness or rodlike shape in reality.

Acknowledgements G.B. thanks the Deutsche Forschungsgemeinschaft and Wella AG, Germany, for their interest and financial support. H.C. thanks the Herrmann Schnell Foundation for financial support.

References

1. Berth G, Dautzenberg H (1998) Recent Res Dev Macromol Res 3:225–248
2. Berth G, Dautzenberg H, Peter MG (1998) Carbohydr Polym 36:205–216
3. Cölfen H, Berth G, Dautzenberg H (2001) Carbohydr Polym 45:373–383
4. Berth G, Dautzenberg H (2002) Carbohydr Polym 47:39–51
5. Terbojevich M, Cosani A, Conio G, Marsano E, Bianchi E (1991) Carbohydr Res 209:251–260
6. Beri RG, Walker J, Reese ET, Rollings JE (1993) Carbohydr Res 238:11–26
7. Wu C, Zhou SQ, Wang W (1995) Biopolymers 35:385–392
8. Rinaudo M, Milas M, Dung PL (1993) Int J Biol Macromol 15:281–285
9. Yamakawa H (1971) Modern theory of polymer solutions. Harper and Row, New York
10. Ottøy MH, Vårum KM, Christensen BE, Anthonsen MW, Smidsrød O (1996) Carbohydr Polym 31:253–261
11. Errington N, Harding SE, Vårum KM, Illum L (1993) Int J Biol Macromol 15:113–117
12. Lavrenko PN, Linow KJ, Görnitz E (1992) In: Harding SE, Rowe AJ, Horton JC (eds) Analytical ultracentrifugation in biochemistry and polymer science. The Royal Society of Chemistry, Cambridge, pp 517–531
13. Wales M, van Holde KE (1954) J Polym Sci 14:81–86

Progr Colloid Polym Sci (2002) 119:58–63
© Springer-Verlag 2002

N. Errington
P. Mistry
A. J. Rowe

Protein hydration varies with protein crowding and with applied pressure: a sedimentation velocity study

N. Errington · A. J. Rowe (✉)
NCMH Business Centre
School of Biosciences
University of Nottingham
Sutton Bonington
Leicestershire LE12 5RD, UK
e-mail: arthur.rowe@nottingham.ac.uk
Tel: +44-115-9516156.
Fax: +44-115-9516157

P. Mistry
Department of Biochemistry
University of Leicester
Leicester LE1 7RH, UK

Abstract We have recently shown, using Derjaguin–Landau–Verwey–Overbeek theory, that the stability of protein solutions can be accounted for primarily in terms of the energy barrier presented by the bound water surrounding the protein, rather than by net repulsive forces. In further work we demonstrated experimentally, using precision densimetry and dynamic light scattering, that the amount of this water bound to proteins varies with temperature. We have now used the analytical ultracentrifuge (AUC) to study possible effects of crowding and applied hydrostatic pressure on water binding and hence on the sedimentation velocity of proteins. We show that whilst self-cancelling of effects minimises changes in the s values, there are predictions, which we are able to confirm experimentally, that both high protein concentration and elevated hydrostatic pressure at levels found in the AUC will lead to effects attributable to additional hydration. It is concluded that protein hydration is, over a range of conditions, a variable rather than a constant quantity. This finding is significant in relation to the stability and formulation of protein solutions.

Key words Hydration · Sedimentation velocity · Pressure · Crowding · Stability of protein solutions

Introduction

The solubility of proteins in aqueous salt solutions and the stability with respect to time of such solutions is a topic of long-standing interest and importance. A fundamental approach to cell biology requires an appreciation of protein solubility properties in a functional context: it can be as important for a specific protein (e.g. a cytoskeletal or motility protein) to be insoluble under defined conditions as for it to be soluble. Equally, formulation of biological products for pharmaceutical or food use can only be achieved successfully if we are aware of the factors which control stability of the product and, hence, influence effective shelf life.

We have embarked upon an attempt to revisit these issues. Our starting point has been to view the solubility of proteins in aqueous solution in terms of Derjaguin–Landau–Verwey–Overbeek (DLVO) theory. This theory has over the past few decades found very wide application to classical colloidal systems, but almost no application to proteins. We have shown [1] that application of DLVO theory using a modern value for the effective Lifshitz–Hamaker constant leads to a very simple finding for average globular proteins of typical surface potential: the net repulsive potential between such molecules is negligible, being of the order of kT only. From this it follows that the presence of a hydration layer is essential for both the solubility of these proteins and for the stability of their solutions. This concept agrees well [1] with many studies over the years which have shown that under physiological levels of pH and salt concentration soluble proteins behave as "hard spheres".

A new formalism has also been developed [1] which shows that experimentally determined dwell times for water molecules vicinal to protein surfaces are long enough to account, in principle, for protein monomers being sterically unable to approach to smaller separation distances at which net attractive forces are dominant. In a further study we explored in some detail the consequences of the proposed functionality of this hydration layer so far as temperature variation of protein solutions is concerned. A new high-precision densimetric technique has been defined (volume–temperature dispersion, VTD) which enables the apparent thermal expansion coefficient of proteins to be monitored under a range of conditions, particularly as a function of temperature. For many of the proteins studied, this coefficient was found to fall in value as the temperature was raised. This effect can be both qualitatively and quantitatively explained in terms of the progressive loss of water, bound (as has long been appreciated) with an average density higher than that of bulk solvent water. In the case of one protein (ovalbumin) studied in some detail, this loss was verified by a careful study of the temperature dependence of the Stokes radius, measured by dynamic light scattering.

This result is of some importance. If a function of the "bound" water surrounding a protein in solution is to maintain a minimal separation between protein monomers, it follows that conditions which favour the presence of bound water will favour protein solution stability, and vice versa. Storage of protein solutions at subambient temperature will favour stability, if the amount of bound water is thereby increased. Our VTD results imply that this is so for most of the proteins which we have studied: in accord with known facts and long established laboratory practice.

Other conditions may also, however, change the level of hydration of proteins in solution and thus affect stability. Recalling that water is, on average, bound with a significant density excess compared to bulk solvent (see previously), and hence with volume contraction of the system, then by Le Chatelier's principle we can predict that any conditions which impose a volume constraint on a protein solution will inevitably favour additional binding of water. Two such conditions are crowding of the solution, i.e. by progressive addition of solute to high concentrations, and application of moderate external pressure. Both conditions are known from experience to favour protein solution stability. In the present study we investigate, using the analytical ultracentrifuge (AUC), the extent to which these conditions are associated with additional hydration levels. To achieve this, we monitor the change in effective swollen specific volume (V_s) over a wide range of concentration, and we investigate, using the highest precision attainable, the possible presence of a (small) predicted regression of the s value of a sedimenting solvent–solution interface upon applied hydrostatic pressure. In this latter context it was necessary to perform prior computation to demonstrate the small, but by no means negligible, effects arising from compressibility of solute and solvent components and from the effects of pressure on the viscosity of water.

Materials, methods and theory

Materials

The proteins and chemicals were purchased from Sigma.

Methods

Sedimentation velocity analysis was carried out using (for the study of pressure effects) a Beckman XL-I (interference optics) or a Beckman XL-A AUC (absorption optics), or for studies of packing volume an MSE Mk II AUC, using schlieren optics with an analogue multiplexer, all at 20 °C and at appropriate speeds. Care was taken to ensure that the rotor was at the selected temperature prior to commencement of the run.

For the purpose of the study of pressure effects, data was captured every 1 min (interference optics) or every 3.5 min (absorption optics). Sedimentation coefficients were evaluated over various time intervals, using the program SEDFIT [2]. For studying packing volumes, data was logged photographically and prints evaluated using a digitising tablet with local software to capture and process data. A single value for the sedimentation coefficient was returned together with a value for the radial dilution averaged over the time period studied. To ensure reproducibility between runs, one of the (lower) concentrations used was designated as a calibration standard and included in one channel in the rotor in each run.

Theory

Studying effects of crowding at high solute concentration

The specific volume occupancy associated with the presence of given mass of solute species in solution is given by V_s (millilitres per gram). It is argued in the Introduction that the crowding of the macromolecular species at high solute concentration should (by Le Chatelier's principle) result in additional water becoming bound. From this it follows that as the protein concentration, c (grams per millilitre) increases, so V_s will also increase. If not, then a single value for V_s will suffice over the entire concentration range. These alternatives can be tested by the use of sedimentation–concentration dependence theory [3, 4], as V_s is an integral term in this treatment.

The general form of the equation for sedimentation–concentration dependence is given by

$$s^c = s^0 \left(1 - \frac{2c(V_s + vF^3) - (cV_s)^2(2\phi_p - 1)/\phi_p^2}{2cvF^3 + 1} \right), \qquad (1)$$

in which s^c and s^0 are the sedimentation coefficients at solute concentration c and infinite dilution, respectively, F is the conventionally defined frictional ratio (more usually written as f/f_0), v is the partial specific volume (millilitres per gram) and ϕ_p is the dimensionless maximum packing fraction of the solute. For perfect spheres the latter has the value 0.64: for real, approximately spherical particles it is more usually taken as 0.60.

Equation (1) has been rearranged from the form given in Ref. [4] to eliminate the limiting concentration-dependence parameter k_s [3]. It was been fitted to experimental data using nonlinear least-squares analysis via the program ProFit (Quantum Soft, Zurich, Switzerland). All parameters are fixed, other than for V_s, being precalculated after the extrapolation of the (reciprocal) s values from lower concentration data to yield s^0. The value of V_s obtained is designated $(V_s)_{av}$ and is the value which best fits the whole range of the data.

The quantity $(V_s)_0$, operative at infinite dilution, is found by application of the basic theory of the concentration dependence of s at high dilution [3] using in-house software BIOMOLS, with the constraint that the correct (known) M value must be echoed. A comparison of $(V_s)_0$ with $(V_s)_{av}$ indicates the presence or otherwise of a variation in V_s as indicated by the range of concentration studied.

A limitation in this treatment is that a constant value for the partial specific volume, v, is assumed over the entire concentration range. From our arguments (Introduction and Ref. [1]) it is clear that a slight but significant decrease in v is expected under conditions where the hydration increases; however, as discussed further later, this reduction in v acts to increase s values, in contrast to the effect produced by a larger V_s value. Hence, our approach, in which we compare $(V_s)_{av}$ with $(V_s)_0$ is qualitatively sound, but will tend to underreport any effects seen.

Studying effects of pressure on sedimentation rates

It might at first sight be expected that the effect of additional water of hydration would be to increase the Stokes radius (r_H) and hence because of elevated friction to reduce s values; however, since the additionally bound water is deemed to have a density higher than that of bulk solvent water, it follows that the partial specific volume of the protein will decrease, tending to cause an increase in sedimentation velocity.

We computed what the net effect arising from these two opposing effects might be. Our treatment has of necessity to be simple and approximate. We assumed that water already bound (at 1 bar) retains the same density under pressures applied (to above 100 bar), and that the additionally bound water shares the same density value. Neither assumption is likely to be exact.

Let us take a protein of partial specific volume 0.739 ml/g (the exact value is of minimal importance) of which it is assumed that -0.02 ml/g is attributable to "striction", i.e. excess density of bound water. A V_s/v ratio of 1.4 (typical for many proteins) is assumed. Calculation then shows that the transition from the anhydrous state to the hydrated state is predicted to be accompanied by an increase in r_H, leading to a reduction in s of 10.6%, whereas the extra density of the bound water relative to bulk solvent should cause an increase in s of 8%. It follows that for a change in hydration defined by the entire water bound to the protein, a fall in s is predicted, but only of the order of 3.4%. Thus, even a quite substantial change in the amount of bound water under applied hydrostatic pressure might only result in a decrease in s of less than 1%. Nonetheless, with careful work changes of less than 1% ought to be detectable.

It is, however, important to be certain as to what the behaviour would be of a reference system, in which no changes in hydration with pressure occurred. Small changes in s will result from the change in density and viscosity of dilute aqueous salt solutions (the solvent) under applied pressure, from the effect of pressure on the hydrodynamic particle (compressibility of protein and of bound water) and from consequent changes in friction arising from a smaller Stokes radius of the particle under pressure. The last issue appears not to have been addressed previously, so far as we can ascertain. It is very small in magnitude (less than 0.1% for a pressure change of 100 bar), but would be more significant in the case of a temperature change.

We computed the expected change in s at 20 °C over the range 0–100 bar excess pressure for a hypothetical globular protein, of partial specific volume $v = 0.73$ ml/g, using current estimates for the compressibility of water and protein, and for the effect of pressure on water viscosity. The last named effect is very temperature sensitive, the specific reduction in viscosity decreasing by a factor of 50 over the range 0–30 °C.

Using a value of 0.453 g/Pa for the compressibility of water [5], 0.05 g/Pa for the compressibility of protein [6] and -0.4 ps for the

derivative of the viscosity of water with respect to pressure [7], we find that over the given pressure range the sedimentation coefficient will change in value by around -0.4%, on the assumption that the degree of hydration stays unchanged. Thus, a change in s differing significantly (from -0.4%) is required to confirm a change in hydration level.

Results

Effects of crowding

A plot of s versus c for apoferritin and ovalbumin is shown in Fig. 1. Lines of best fit using Eq. (1) are shown. These describe the data well over the range studied. A fixed value for the maximum packing fraction, $\phi_p = 0.60$, was employed. The values for $(V_s)_{av}$ estimated are shown in Table 1 and are contrasted with $(V_s)_0$ values, operative at infinite dilution. For both proteins, but especially for ovalbumin, there is a very considerable increase in the former value, in comparison to the latter one. Ovalbumin, a glycoprotein, might be more liable to these effects than a simple protein, but even for apoferritin the suggestion from these figures (Table 1) is that around 40% more water may become bound under crowded conditions. For apoferritin an attempt was made to model the gain in hydration by making the parameter V_s into a simple linear function of solute concentration. The resulting fit (Fig. 1) was good, possibly better than the simple fit using an average value of V_s, but there could be a wide choice of type of function incorporated in this

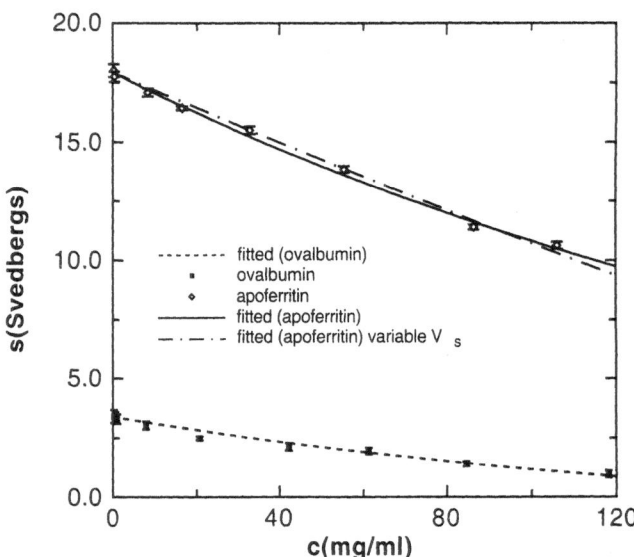

Fig. 1 Plot of sedimentation coefficient, $s_{20,w}$ against average solute concentration for apoferritin and ovalbumin. The fitted lines are obtained by nonlinear least-squares fitting of the data values to Eq. (1), fixing all values other than for the swollen specific volume, V_s. For the apoferritin data an alternative fit is shown, in which V_s is treated as a simple linear function of concentration (*dotted line*)

Table 1 Average values for V_s returned by curve-fitting experimental data, compared to the values computed at infinite solute dilution

Protein	$V_{s,0}$	$V_{s,av}$
Apoferritin	1.00	1.39
Ovalbumin	1.25	3.10

way, and it is not felt that the precision of the data justified this approach being pursued further.

An attempt was made in fitting the data to float both V_s and ϕ_p, following previous practice [4]. This yielded aphysical results, with negative values for ϕ_p.

A limitation of the data is that pressure effects, as detailed later, were not specifically considered; however, all the s values were computed over a similar (albeit not identical) region of the cell radius, so this matter should contribute to "noise" only, rather than introducing any systematic error. An apparent difference between the present and earlier results obtained using other proteins is discussed in the Discussion.

Effects of excess hydrostatic pressure

The analysis of the data from the six proteins studied using SEDFIT is shown in Fig. 2, in which the s values are plotted against the excess pressure at the mid-point of the sedimenting boundary. It is important that the position of the meniscus is determined with precision. Almost invariably the early points lie away from the regression line defined by the majority of the points, probably owing to small uncertainties in the position of the meniscus. An extreme case is shown for a sample of ovalbumin studied by absorption optics (Fig. 2), where it

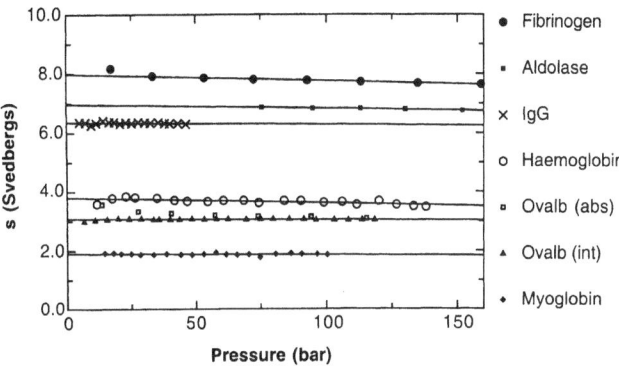

Fig. 2 Plots of estimated sedimentation coefficient against excess hydrostatic pressure at the mid-point of the boundary for the proteins fibrinogen, aldolase, IgG, haemoglobin, ovalbumin, and myoglobin dissolved in 100 mM NaCl phosphate (50 mM) buffered to pH 7. Absorption optics were used in all cases except for ovalbumin, where data from both absorption optics and interference optics are plotted. The regression lines shown are linear least-squares fits to the data points

is noted that the plot approaches asymptotically the slope of the line obtained using interference optics, in a different run. Overall, we find a small but highly significant regression of s with pressure of around -2% (Table 2), which is greater than the baseline prediction of -0.4% for no change in hydration (see earlier).

Discussion

Our aim in this study was to explore the extent to which protein hydration, which we have recently shown to vary with temperature, may also vary in a predictable fashion with molecular crowding and with applied hydrostatic pressure. We used analytical velocity sedimentation as a prime tool for this purpose, as both protein concentration and the level of applied hydrostatic pressure are readily usable as variable parameters. This is not the case with alternative methods. In particular, dynamic light scattering poses problems arising from both multiple scattering effects, from limited theory of concentration dependence at high c values, and from the lack of high-pressure apparatus. It is important to assess, however, the extent to which a variation in the level of bound water could actually affect the sedimentation coefficient.

The predicted range of variation of s on the basis of variable hydration

Owing to partial cancellation of effects, the percentage variation of s for a given change in hydration will be around 3 times smaller than the variation in r_H. Dynamic light scattering will thus be a much more sensitive probe for variable hydration, but the precision of modern analytical ultracentrifugation technology, with massive data capture and advanced analytical software, is such that any changes ought to be detectable with some certainty.

Table 2 The experimentally observed pressure dependence of s for six proteins (for conditions see legend to Fig. 2). The standard error of the regression indicates that for all but one of the proteins studied (IgG, data logged only over a small range of pressure) the regression differs significantly or highly significantly from zero. For all cases other than ovalbumin the regression is also significantly greater than the regression (-0.4%) predicted on the basis of solvent and solute compressibility and pressure effects on solvent viscosity

Protein	% change/100 bar	$\pm$
Fibrinogen	-2.8	0.12
Aldolase	-2.0	0.35
IgG	-1.1	0.97
Haemoglobin	-4.8	1.01
Ovalbumin	-1.1	0.24
Myoglobin	-2.4	1.49

The variation of s with high solute concentration as a probe for a variation of V_s

The approach which we adopted in our attempt to estimate V_s values under realistic concentration conditions involved the use of an extended s–c dependence theory [3, 4]. The theoretical equation involved is here rearranged in a manner which makes it evident that other than for the frictional ratio, which is determinable empirically, and the maximum packing fraction, which for approximately spherical particles can be assumed, it is V_s which determines the concentration dependence of sedimentation over the full range of c. Since it is known that the extended equation gives an accurate fit to numerical data both for ideal spheres and for real particles [4], one can with some confidence use this equation to estimate V_s for particles which fall within this definition. Indeed, even for a particle (fibrinogen) which plainly cannot be defined as close to spherical, it has been found that a good fit to empirical data is obtained, with the not unreasonable finding that ϕ_p (floated in the fit) has a lower value (0.40) than for spheres [4].

Results from our earlier published work [4] on bovine serum albumin (BSA) and fibrinogen showed that a single estimate for V_s described the data fairly well over the full range of c, albeit with a slightly lower value for ϕ_p (0.49) than expected for BSA (the lower value for fibrinogen accords with expectation for an symmetric particle). As has been remarked, this agrees well with a prediction made for these particular proteins from VTD. These latter studies showed little change of hydration as the temperature was varied, and it may be that this stability of the hydration level is also found under other conditions (crowding).

The variation of s with hydrostatic pressure

Our results demonstrate experimentally – we believe for the first time for globular proteins – that there is a small but significant variation of s during the course of a run, which clearly exceeds in magnitude the small (−0.4%) effect predicted on the basis of direct pressure effects on solvent and solute (see earlier). This decrease cannot be attributed to radial dilution. Concentrations of protein in the region of 1 mg/cm^3 were employed. For a typical globular protein, concentration-dependence coefficient, k_s, of around 4 cm^3/g [3], a positive change of the order of 0.05% only is predicted. Shear effects (which again would give a positive change) need not be considered for globular proteins. The presence of a neutral salt should avoid charge effects. So far as the evaluation of the data is concerned, we found the SEDFIT routine to give results which are both stable and reliable. For control purposes we evaluated a run using SVEDBERG software [8], which like SEDFIT fits to the Lamm equation, and

DC/DT+ [9], which computes the $g(s^*)$ profile via the time derivative. Similar results were obtained, but with somewhat less stability and, in the case of DC/DT+, where critical decisions concerning groups of data to be selected must be taken, less objectivity. The results from different runs using different instruments using different optical systems are in good agreement (Fig. 1).

The only factor which clearly varies significantly – and indeed over a wide range – during the course of a run is the hydrostatic pressure operative across the sedimenting boundary. At 50,000 rpm, this pressure increases by around 20 bar/mm from the meniscus, so attaining a value in excess of 100 bar before the boundary becomes too near to the cell base for the data collected to be valid. The size of the decrease in s observed with respect to the s value extrapolated to the meniscus, and after subtraction of the expected "baseline" effect, is of the order of −1.7%, which probably implies that at least 10–20% of additional bound water is acquired as the pressure is raised.

An interesting consequence of this finding is that ideally all the measured s values should be back-extrapolated to zero run time (effectively, to the boundary at the meniscus position). This approach does in any case have the advantage that one knows the precise solute concentration to which the measured s value refers. Procedures which involve extrapolation to infinite time, such as the van Holde–Weischet analysis [10], may in some cases introduce significant error as a result of the (admittedly small) pressure dependence seen in the present study.

Can variable hydration explain variable stability of proteins?

The hydration level of a protein can vary when the chemical nature of the solvent is altered. This has recently been described by Timasheff [11] for lysozyme in the presence of a range of cosolvents. We sought to investigate whether, under fixed conditions of the chemical environment, the hydration level can equally vary, but in response to physical changes alone.

In this context, it has long been appreciated by formulation chemists that protein solutions are stabilised against aggregation by low temperature, high solute concentration, and modestly elevated (10^2–10^3 bar) pressure. Our results provide a unified approach to the underlying reasons for these observations, in terms of hydration being a variable quantity, even under defined, fixed chemical conditions. We have considered the effect of lowered temperature in increasing hydration elsewhere. We now show that under both high solute and moderately elevated pressure conditions, the level of hydration is also elevated, to an extent which is clearly significant. This accords fully with the basic hypothesis which we have advanced [1] regarding the stability of protein solutions: the layer of bound water is the most important factor in determining stability and the level of spatial extension of that bound water is not – as has been widely assumed – a constant, but is a variable quantity.

Acknowledgement We are grateful to BBSRC/EPSRC for grant support of this work.

References

1. Rowe AJ (2001) Biophys Chem (in press)
2. Schuck P (1998) Biophys J 75:1503–1512
3. Rowe AJ (1977) Biopolymers 16:2595–2611
4. Rowe AJ (1992) In: Harding SE, Rowe AJ, Horton JC (eds) Analytical ultracentrifugation in biochemistry, polymer science. Royal Society of Chemistry, Cambridge, pp 394–406
5. Chaplin M (2001) http://www.sbu.ac.uk/water/data.html
6. Gross M, Jaenicke R (1994) Eur J Biochem 221:617–630
7. Cho CH, Urquidi J, Robinson GW (1999) J Chem Phys 111:10171–10174
8. Philo JS (1977) Biophys J 72:435–444
9. Philo JS (2000) Anal Biochem 279:151–163
10. van Holde KE, Weischet WO (1978) Biopolymers 17:1387–1403
11. Timasheff SN (1999) Adv Protein Chem 51:355–432

Progr Colloid Polym Sci (2002) 119:64–69
© Springer-Verlag 2002

Determination of thermodynamic properties of sodium alginate from bacteria and seaweeds in aqueous solutions

A. Straatmann
W. Borchard

A. Straatmann · W. Borchard (✉)
Institut für Physikalische und
Theoretische Chemie der
Gerhard-Mercator-Universität Duisburg
47048 Duisburg, Germany

Abstract Alginates are biopolymers produced by seaweeds as well as by some bacteria like *Pseudomonas aeruginosa* or *Azotobacter vinelandii*. They consist of copolymers containing 1,4-β-D-mannuronate and its 5-epimer α-L-guluronate. Their composition and physical properties depend on the source of the alginate. Different alginates from seaweeds and bacteria have been investigated using analytical ultracentrifugation. The samples were prepared in sodium chloride solution to suppress charge effects of the polymers. From the concentration gradient in equilibrium centrifugation the molar masses and the second virial coefficient are derived. The results obtained by this method are in good agreement with the results of light scattering measurements. The properties of different alginates are compared and discussed.

Key words Bacterial alginate ·
Sedimentation-diffusion
equilibrium · Aggregation

Introduction

Alginates are used in a wide range of industrial applications, for example, as stabilising or gelling agents in the food industry. For these purposes they are usually extracted from brown seaweeds, for example, from *Laminaria hyperborea* or *Ascophyllum nodosum* [1, 2]. Alginates are also produced as extracellular polymer substances (EPS) by some bacteria like *Pseudomonas aeruginosa* or *Azotobacter vinelandii*. These bacteria live in microbial aggregates, the so-called biofilms, which grow in every aqueous environment. They can cause great damage on filters, pipelines, ship hulls and other technical systems. The EPS determine the stability, the structure and the functionality of these microbial aggregates [3]. Recently a lot of research has been done to investigate the physical properties of EPS compounds. Since the EPS is a rather complicated gel-like system of polysaccharides, proteins, nucleic acids (phospho)lipids and other organic matter there is need for comparable model systems [4]. *P. aeruginosa* is one strain of bacteria where alginate is the main EPS product. The physical properties of seaweed alginate in comparison to bacterial alginate were investigated. In this contribution it is examined whether commercially available alginate can be used in model systems for the EPS matrix.

Alginates consist of a long unbranched chain of 1,4-β-D-mannuronate (M) and its 5-epimer α-L-guluronate (G), see Fig. 1.

The equatorially linked mannuronate residues are rather flexible, while the diaxially linked guluronate residues are fairly rigid. Both monomers can be found as blocks as well as in alternating sequences. Depending on the source the M/G ratio varies and, therefore, the physical properties and the gelling behavior of the alginates, which are in turn determined by the composition and the sequence of the monomers, differ [2, 5]. For example, the guluronate blocks form gels in the presence of bivalent cations especially Ca^{2+} ions. The Ca^{2+} ions can build stable complexes with the carboxylic groups of the guluronate, whereas the carboxylic groups of the mannuronate are sterically hindered from forming these complexes [1].

Fig. 1 **a** 1,4-β-D-Mannuronic acid, **b** α-L-guluronate

In contrast to seaweed alginates, some mannuronate residues of alginates from bacteria are acetylated in the O-2 or in the O-3 position [1, 6]. The acetylation reduces the net negative charge of the polyanions and decreases the capacity and selectivity of cation interaction [1, 3]. The viscosity of alginate solutions from acetylated alginate and the water-retention capacity of the alginate is increased [7].

A peculiarity of alginate from *P. aeruginosa* is the lack of blocks of guluronic acid; therefore, no gelation via complexation with Ca^{2+} ions is possible. For macromolecules exceeding a molar mass of 150,000 g/mol entanglements are expected and a nonpermanent network can be formed [8]. The polymer can then undergo a phase transition from the solution to a gel. The alginate gels were found to be permanent on a short time scale, but under stress the cross-linking junctions were loosened on a long time scale [9].

Several alginates from seaweeds have been studied recently [10–13]. In our work two industrial alginates and one bacterial alginate are investigated and the results are compared. The measurements were performed in sodium chloride solutions to suppress charge effects of the polyelectrolyte. Harding et al. [14] recommended to use buffers with an ionic strength of at least 0.1. Horton et al. [11] showed that above an ionic strength of 0.3 no further depression of the Donnan effect could be detected. In the present experiments sodium chloride solutions with a concentration of 0.14 mol/l were used.

Theory

Equilibrium measurements

Solutions in ultracentrifugal fields have to be described as continuous systems. They are determined by the gradient of the potential of the external field, which causes a

gradient of the chemical specific potential $\tilde{\mu}_i$. When the equilibrium between diffusion and sedimentation is reached the following condition applies for each solute i, with $i = 2, 3, \ldots, N$, at each radial distance from the axis of rotation, r:

$$\left(1 - \tilde{V}_i^0 \rho\right)\omega^2 r \mathrm{d}r = \left(\frac{\partial \tilde{\mu}_i}{\partial \rho_i}\right)\mathrm{d}\rho_i \ , \tag{1}$$

where $1 - \tilde{V}_i^0 \rho$ is the buoyancy term, ρ the density of the solution, ω the angular velocity and ρ_i the partial density or mass concentration of the solute.

The specific chemical potential depends on the partial density in the following way [15]:

$$\tilde{\mu}_i = \tilde{\mu}_i^0 + \frac{RT}{M_i}\ln \rho_i \gamma_i \ , \tag{2}$$

where R is the gas constant, T the thermodynamic temperature in Kelvin and γ_i the activity coefficient of compound i on the partial density scale.

The logarithm of the activity coefficient can be expanded into a series using the virial coefficients of the osmotic pressure [16]:

$$\ln \gamma_i = \frac{2M_i}{RT}A_2\rho_i + \frac{3M_i}{2RT}A_3\rho_i^2 + \cdots \ , \tag{3}$$

with M_i being the molar mass of compound i and A_2 and A_3 being the second and third virial coefficients.

Using Eqs. (1), (2), and (3) the following relation is obtained:

$$\left(1 - \tilde{V}_i^0 \rho\right)\omega^2 r \mathrm{d}r$$
$$= \frac{RT}{M_i}\left(\frac{1}{\rho_i} + \frac{2M_i}{RT}A_2 + \frac{3M_i}{RT}A_3\rho_i + \cdots\right)\mathrm{d}\rho_i \ . \tag{4}$$

By multiplying Eq. (4) by $M_i\rho_i$, Eq. (5) is derived:

$$M_i\rho_i\left(1 - \tilde{V}_i^0 \rho\right)\omega^2 r \mathrm{d}r$$
$$= RT\left(1 + \frac{2}{RT}A_2M_i\rho_i + \frac{3}{RT}A_3M_i\rho_i^2 + \cdots\right)\mathrm{d}\rho_i \ . \tag{5}$$

For dilute solutions where the partial density of the solute, ρ_2, is smaller then the partial density of the solvent, ρ_1, the density of the solution, ρ, can be substituted by the density of the pure solvent, ρ_{01}. Assuming that $\tilde{V}_i^0 = \tilde{V}_2^0$ and remembering the definition of the weight-average molar mass, M_w,

$$M_w = \frac{\sum_{i=2}^{N} m_i M_i}{\sum_{i=2}^{N} m_i} = \frac{\sum_{i=2}^{N} \rho_i V M_i}{\sum_{i=2}^{N} \rho_i V} = \frac{\sum_{i=2}^{N} \rho_i M_i}{\sum_{i=2}^{N} \rho_i} \ , \tag{6}$$

subsequent summation over all species i and integration leads after some rearrangements to

$$\omega^2 \int_{r_{\text{ref}}}^{r} \rho_2 \left(1 - \tilde{V}_2^0 \rho_{01}\right) r \, dr$$

$$= RT \left(\frac{1}{M_w} \rho_2 + A_2^* \rho_2^2 + 3A_3^* \rho_2^3 + \cdots\right) . \quad (7)$$

The buoyancy term can be obtained by measuring the density of the solvent and the polymer solutions with respect to the weight fraction of the polymer, w_2. From the schlieren patterns of the solution the concentration profile in the ultracentrifugal cell can be determined and the integral on the left-hand side of Eq. (7) can be calculated numerically.

The apparent molar mass, $M_{w,\text{app}}$, can be obtained from a plot of the integral versus the partial density. This apparent value depends on the concentration of the solution and has to be extrapolated to the initial partial density zero ($\rho_{2,0} \rightarrow 0$).

$$\frac{1}{M_{w,\text{app}}} = \frac{1}{M_w} + 2A_2 \rho_{2,0} + 3A_3^2 \rho_{2,0} + \cdots \quad (8)$$

From Eq. (8) the weight average of the molar mass of the polymer and the osmotic virial coefficients can be obtained.

Materials and methods

Sodium alginate Protanal LF 20/60 supplied by Pronova with an M/G ratio of 0.4, Manucol DM supplied by Monsanto, M/G 2.1, and bacterial alginate SG81 from *P. aeruginosa*, M/G 2.5, were used. The alginate was dialyzed against water for 24 h and then freeze-dried. The purified bacterial alginate contained 1% of proteins, which could not be extracted. The sodium alginate was dissolved in 0.14 mol/l sodium chloride solution and stirred overnight to prepare a stock solution. This was diluted to produce solutions with concentrations between 0.05 and 0.5 wt% for the alginate from Pronova and between 0.05 and 0.25 wt% for the other alginates. The measurements were done with an analytical ultracentrifuge (model E from Beckman) with a modified schlieren optics [17]. A six-hole An-G titanium rotor with sector-shaped cells was used. The pictures were taken with a charge-coupled-device camera, transmitted to a "framegrabber" card and digitized with the program Autoscope (Digithurst 1991). The schlieren gradients were redrawn with a graphic tablet from Wacom. For the equilibrium runs the integral in Eq. (7) was calculated with a visual basic program developed in our research group and the data were saved as ASCII files. In case of the velocity runs the position of the hypothetical sharp boundary according to Goldberg [18] was calculated using a similar visual basic program.

After the equilibrium runs had commenced pictures were taken twice a day until the schlieren pattern remained constant for 2 days. This was done for each alginate for three different speeds between 6,000 and 18,000 rpm. Equilibrium was reached after 24–36 h. The velocity runs were performed at 30,000 rpm at 20 °C and pictures were taken every 30 min.

The densities of the alginate solutions were measured with a DMA 60 density meter from Paar (Graz, Austria) with two DMA 602 comparative measuring cells. The partial specific volumes were 0.48 ml/g for the seaweed alginates and 0.76 ml/g for the bacterial alginate.

Scanref monocolor from NFT was used to measure the refractive index increment. A refractive index increment of 0.26 ml/g was obtained for the seaweed alginates and 0.14 ml/g for the bacterial alginate at 633 nm.

Results and discussion

During the equilibrium runs an unusual course of the schlieren pattern was obtained (Fig. 2).

From the meniscus onwards the first part of the equilibrium pattern of the refractive index gradient is as expected from solutions [19]. At a certain distance from the axis of rotation, there is, however, no further increase in the refractive index gradient. The specific pattern could be seen for all speeds between 9,000 and 18,000 rpm. To prove that a real equilibrium was established, the distinct velocities were employed by starting from a lower speed one time and slowing down from a higher speed another time. By overlaying the schlieren pictures obtained both times, it could be shown that there was no difference in the concentration profiles and that they were independent of the way. The higher the centrifugal acceleration the more pronounced the step in the schlieren pattern and the further it was moved to the bottom of the cell. It was not possible to find a low enough speed such that this effect did not occur. At 6,000 rpm no gradient of the seaweed alginate could be resolved. In the case of the bacterial alginate a small step was obtained at this velocity, but the resolution was too small for further analysis.

A linear dependence of the refractive index gradient versus radial distance has been found for κ-carrageenan

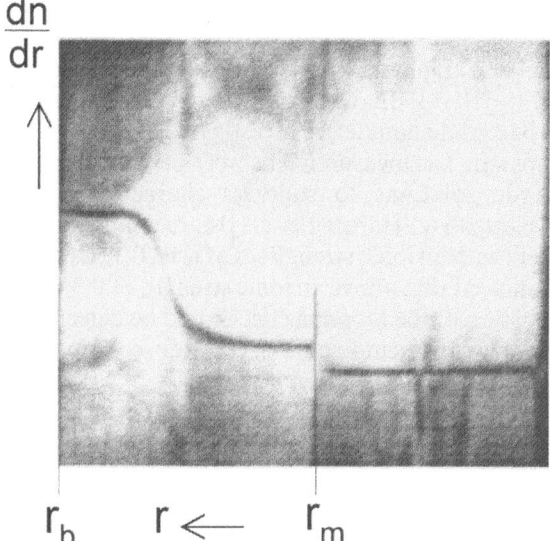

Fig. 2 Typical picture of the schlieren pattern (Pronova LF 20/60) obtained in sodium chloride solution ($c = 0.15$ mol/l) with an initial weight fraction of the polymer of $w_{2,0} = 0.4\%$ and a rotational speed of $v = 18,000$ rpm, 20 °C)

gels [20]. In the case of a continuous gel there must be a phase boundary, which can be seen as a meniscus between these phases. In the case of a solution no constant refractive index gradient would be expected; however, experimentally it is found that neither a new phase boundary nor a continuous refractive index gradient occurs. Therefore, we presume that at a distinct concentration microgels are reversibly formed in the centrifugal field. The determination of the concentration at which the schlieren line runs into the plateau zone showed that it was independent of the starting concentration.

For the seaweed alginates the concentration range at which a microgel is probably formed was 2.3 ± 0.3 g/l (Table 1). For the bacterial alginate a lower critical concentration of 1.2 ± 0.2 g/l was found.

When the alginate solutions were not exposed to the ultracentrifugal field anymore, the microgels were dissolved evenly again.

It is remarkable that the partial specific volumes of the diverse alginates are so different. The literature value for the specific volume of seaweed alginate is 0.44 ml/g [11], which is comparable to the value of 0.48 ml/g which was found for the seaweed alginate in this contribution. For the bacterial alginate a much higher value of 0.76 mg/ml was determined. The partial specific volume of proteins is about 0.7 l/g [21], but the small amount of about 1% protein in the bacterial alginate is not enough to raise the specific volume of the alginate considerably. The reason for the different partial specific volumes might be that the bacterial alginate contains a considerable number of acetyl groups. The second reason to mention is that the partial specific volumes are extrapolated values of density measurements performed in an extremely dilute region, so that the uncertainty in the data is ±0.06 ml/g.

For the seaweed alginate the measurements at 18,000 rpm were used for further evaluations. At this velocity a good resolution of the schlieren pattern was obtained. In the case of the bacterial alginate the

measurements at 9,000 rpm were used owing to extensive meniscus depletion at higher velocities.

An example for the concentration profile of the integral from Eq. (7) is shown in Fig. 3 for an alginate solution from Pronova. The plot could be approximated using two virial coefficients to represent nonideality. Using further virial coefficients did not improve the results.

The profile of the reciprocal apparent weight-average molar mass, $1/M_w$, as a function of the concentration of Pronova LF 20/60 shows the expected linear increase below 2.7 g/l and a decrease above this concentration (Fig 4). As described earlier, this is approximately the critical concentration at which a microgel is built. The decrease is typical of self-aggregating systems [22].

Though the curve could be described by the use of three virial coefficients in the whole concentration range, an extrapolation is only possible with a linear function. Since the values of $1/M_w$ can be approximated linearly in the dilute region, only concentrations below 2.7 g/l were considered in the present analysis.

The plots for the three different alginates are shown in Fig. 5.

For both seaweed alginates $1/M_w$ versus the partial density ρ_2 could be described as a linear function with a positive slope. In the case of the bacterial alginate a descending slope was found for the whole concentration range. No significant change in the slope was obtained at a partial density of $\rho_2 = 1.2$ g/l, corresponding to the beginning of the plateau zone of the schlieren pattern. To determine a change in the slope at this concentration probably more measurements at lower concentrations are necessary, which is not possible with the equipment used.

Table 1 Plateau concentration of sodium alginate solutions from Pronova

ρ_2^0 (g/l)	$\rho_{2,p}$ (g/l)
0.51	2.57
1.06	1.85
1.48	2.13
2.03	2.83
2.53	2.66
3.09	2.34
3.50	2.23
4.01	2.23
5.03	2.07
6.02	2.24
Mean	2.3

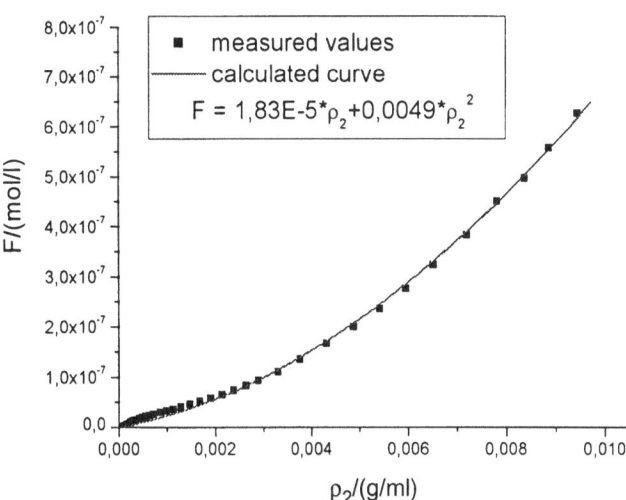

Fig. 3 $F = \omega^2 \int_{r_{ref}}^{r} \rho_2 (1 - \tilde{V}_2^0 \rho_{01}) r dr$ versus ρ_2 for a solution of Pronova LF 20/60 in sodium chloride solution ($c = 0.14$ mol/l) with $\rho_{2,0} = 2$ g/l, $v = 18,000$ rpm, 20 °C

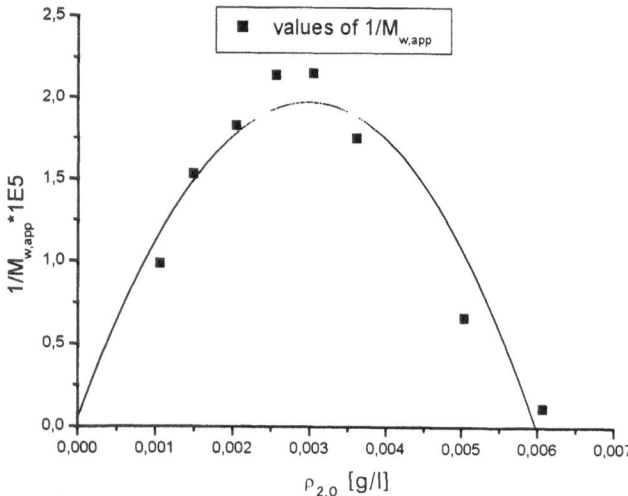

Fig. 4 $1/M_{w,app}$ versus the initial partial density of the polymer. $\rho_{2,0}$. for Pronova LF 20/60

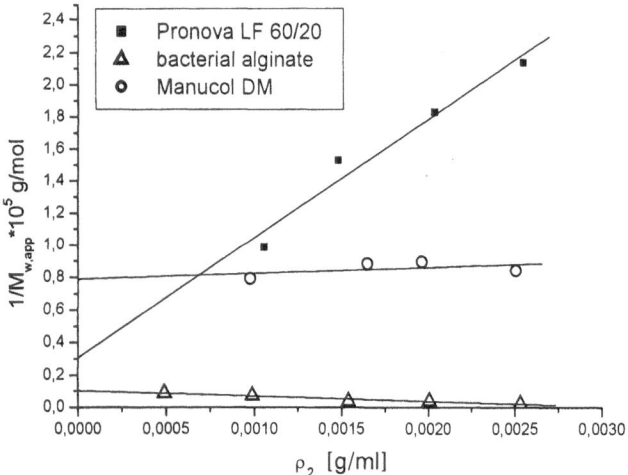

Fig. 5 Comparison of $1/M_w$ versus $\rho_{2,0}$ for different alginates

The results obtained from ultracentrifugal equilibrium are summarized in Table 2. Though the error range of the molar masses is rather high, the results obtained for the seaweed alginates are in good agreement with the results from other authors. A molar mass of 240,000 g/mol was determined for the Pronova sample by

Horton et al. [11] using a low-speed sedimentation–diffusion equilibrium technique with interference optics. For Manucol DM a molar mass of 130,000 g/mol was obtained, which is the same value as that reported earlier by Ball et al. [12]. With light scattering techniques the same value was determined by Windhues and Borchard.

The molar mass of the bacterial alginate was 5 times higher than that of the seaweeds. This is in agreement with the high viscosity of solutions from bacterial alginate and the lower critical concentration for the formation of microgels. These microgels can, on the one hand, be built by electrostatic interactions caused by the charged groups. On the other hand, more entanglements are expected at higher molar masses than in the case of short chains. These entanglements may act as junction points for network formation.

A molar mass of 1,100,000 g/mol for the bacterial alginate was determined using light scattering techniques, which is in the same range as the molar mass detected with the sedimentation–diffusion equilibrium.

For both seaweed alginates small but positive virial coefficients of the order of magnitude of 1×10^{-3} mlmol/g^2 were found. This is in good agreement with the results of Horton et al. [11], who found a second virial coefficient of 4×10^{-3} mlmol/g^2 for sodium alginate from Pronova.

The second virial coefficient of the bacterial alginate was found to be negative. This might be due to the higher tendency of aggregation at higher polymer concentration. Another cause might be the hydrophobic acetyl groups, which cause a lower solubility of the alginate in water.

Windhüs and Borchard found negative second virial coefficients for all alginates using light scattering. The reason may be that their measurements were performed with solutions in the concentration range from 1 to 5 g/l, which are higher than those used in this work. Above a concentration of 2.7 g/l, higher virial coefficients are necessary to describe the system. Measurements at concentrations above the critical concentration where microgels are built yield negative virial coefficients.

It could be shown that all alginates may form gels above a critical concentration. The complexation with bivalent cations is not necessary for this gelling.

Table 2 Results obtained from equilibrium measurements

	1,4-β-D-Mannuronate /5-epimer α-L-guluronate	M_w (g/mol)	A_2 (mlmol/g^2)
Pronova LF 60/20	0.4	180,000 ± 80,000	$2.9 \times 10^{-3} \pm 1 \times 10^{-4}$
Manucol DM	2.1	130,000 ± 40,000	$1.8 \times 10^{-3} \pm 2 \times 10^{-4}$
Bacterial alginate	2.2	980,000 ± 80,000	$-1.6 \times 10^{-4} \pm 3 \times 10^{-5}$

The sedimentation coefficients for infinite dilution at 20 °C were not corrected for viscosities because the viscosities of the salt dilution and water are nearly the same. The values of s_2 were 2.3 S for Pronova LF 20/60 and 3.9 S for Manucol DM and the bacterial alginate. For Manucol DM the calculated sedimentation coefficient is higher then the values found by Harding et al. (2.2 S) [13] and by Wedlock et al. 2.5 S [10]. This might be due to the different values of ionic strength. Harding et al. used a buffer with a ionic strength of 0.3 and Wedlock et al. used one with an ionic strength of 0.2. If there are not enough low-molecular-weight ions to screen the charges of the polyelectrolyte, aggregates are formed and the sedimentation coefficients of these partially aggregated alginate macromolecules are measured, which are expected to be larger. The shape of the molecule therefore depends on the ionic strength as well and the frictional coefficient changes. This might lead to a higher sedimentation coefficient. The high value of s_2 for the bacterial alginate is in agreement with the higher molar mass compared to the seaweed alginate. As the negative second virial coefficient indicates, aggregation cannot be excluded in this case as well and further investigations at higher salt concentrations or lower alginate concentrations might be necessary to confirm this value.

Another reason for the discrepancy of the sedimentation coefficients might be the low velocity at which the measurements were done. In comparison with the higher velocities usually used, the variance in the sedimentation coefficients is higher.

Conclusion

The physical properties of the seaweed alginate and the bacterial alginate are similar though water or aqueous salt solutions are poor solvents for the bacterial alginate. Algal alginate may be used in model systems for the EPS matrices of biofilms, if bacterial alginate is not available. For more similar solution properties the commercial alginate should probably be acetylated.

Aqueous solutions of seaweed alginates and a bacterial alginate can be described by the molar mass and the second virial coefficient in the dilute regions below a critical concentration of 2.7 g/l. This concentration was found to be approximately the concentration at which microgels are built in the case of seaweed alginates. For the bacterial alginate the critical microgel building concentration is 1.2 g/l; therefore a negative virial coefficient is obtained even at concentrations below 2.7 g/l. Further investigations have to be made at lower polymer concentrations to find out whether positive virial coefficients can be obtained. The sedimentation coefficients of the three alginates were determined in sodium chloride solution with an ionic strength of 0.14 mol/l. The aggregation tendency of the Manucol DM is too high at this ionic strength to determine the sedimentation of the nonaggregated polymer in solution.

Acknowledgement The authors thank Pronova Biopolymer AG for providing material of high quality and J. Wingender for providing the bacterial alginate extracted from pseudomonas Adruyinosa.

References

1. Rehm BHA, Valla S (1997) Appl Microbiol Biotechnol 48:281
2. Fasihuddin BA, Wedlock DJ, Omar S, Phillips GO (1988) In: Philips GO, Wedlock DJ, Williams PA (eds) Gums and stabilisers for the food industry. IRL, Oxford pp 89–96
3. Flemming HC, Wingender J, Mayer C, Körstgens V, Borchard W (2000) In: Lappin-Scott H, Gilbert P, Wilson M, Allison D (eds) Community structure and co-operation in biofilms. SGM symposium, vol 59. Cambridge University Press, Cambridge, p 87
4. Wingender J, Neu TR, Flemming H-C (1999) Microbial extracellular polymeric substances, characterization, structure and function. Springer, Berlin Heidelberg New York
5. Rees DA, Welsh EJ (1977) Angew Chem 89:228
6. Lee JW, Ashby RD, Day DF (1996) Carbohydr Polymers 29:337
7. Skjåk-Bræk G, Zanetti P, Paoletti S (1989) Carbohydrates 185:131
8. Graessly WW (1975) In: Advances in polymer science 16. Springer, Berlin Heidelberg New York, p 3
9. Körstgens V, Borchard W (2001) J Microbiol Methods 46:9
10. Wedlock DJ, Fasihuddin BA, Phillips GO (1986) Int J Biol Macromol 8:57
11. Horton JC, Harding SE, Mitchell JR, Morton-Holmes DF (1991) Food Hydrocolloids 5:125
12. Ball A, Harding SE, Mitchell JR (1988) Int J Biol Macromol 10:259
13. Harding SE, Vårum KM, Stokke BT, Smidsrød O (1991) Adv Carbohydr Anal 1:63
14. Harding SE (1995) Carbohydr Polym 28:227
15. Fujita H (1962) Mathematical theory of sedimentation analysis. Academic, New York
16. Meyerhoff G (1960) Angew Chem 72:699
17. Coelfen H (1994) Dissertation. Duisburg
18. Goldberg RJ (1952) J Phys Chem 57:194
19. Schachmann HK (1959) Ultracentrifugation in biochemistry. Academic, New York
20. Hinsken H, Borchard W (1995) Colloid Polym Sci 273:913
21. Durchschlag H, Christl P, Jaenicke R (1991) Colloid Polym Sci 86:41
22. Cölfen H, Harding SE (1997) Eur Biophys J 25:333

Progr Colloid Polym Sci (2002) 119:70–76
© Springer-Verlag 2002

INTERACTING SYSTEMS

Natalia A. Chebotareva
Iraida E. Andreeva
Valentina F. Makeeva
Boris I. Kurganov
Nataliya B. Livanova
Stephen E. Harding

Self-association of phosphorylase kinase from rabbit skeletal muscle in the presence of natural osmolyte, trimethylamine *N*-oxide

N. A. Chebotareva (✉) · I. E. Andreeva
V. F. Makeeva · B. I. Kurganov
N. B. Livanova
A.N. Bach Institute of Biochemistry
Russian Academy of Sciences
Leninskii pr.33 117071 Moscow, Russia
e-mail: natalia@cheb.phys.msu.su
Fax: +7-95-9542732

S. E. Harding
University of Nottingham
NCMH Unit, School of Biosciences
Sutton Bonington, LE12 5RD, UK

Abstract The effect of trimethyl-amine *N*-oxide (TMAO) on self-association of phosphorylase kinase (PhK) has been studied by analytical ultracentrifugation and turbidimetry in 40 mM *N*-(2-hydroxyethyl)piper-azine-*N*′-ethanesulfonic acid buffer, pH 6.8 and 8.2. PhK is a hexade-camer $(\alpha\beta\gamma\delta)_4$ with a molecular mass of 1,300 kDa. The oligomeric state of the native enzyme is dependent on the protein concentration and the concentrations of Ca^{2+} and Mg^{2+}, which are essential for the enzymatic activity. In the absence of Ca^{2+} and Mg^{2+} the enzyme exists in the monomeric and dimeric forms (with $s_{20,w} = 23$ and 36.5 S); however, the addition of 0.1 mM Ca^{2+} and 10 mM Mg^{2+} results in the appear-ance of associates of higher order. TMAO (0.6–1.0 M) was found to favor greatly self-association of PhK. In the presence of TMAO, apart from the association products, consisting of a rather low number (n) of PhK molecules ($n = 2$, 3, 4,...), two distinct rapidly moving bound-aries with $s_{20,w} = 189$ and 385 S are registered on the sedimentation profiles. These boundaries corre-spond to 24-mers and 70-mers (the molecular masses of associates were estimated using the Mark–Houwink–Kuhn–Sakurada equa-tion, assuming a spherical form. The kinetics of TMAO-induced association of PhK was monitored by following the increase in the turbidity of the enzyme solution at varying concentrations of the pro-tein and TMAO. The initial rate of the association is proportional to the enzyme concentration squared, suggesting that the initial step of PhK association is the stage of dimerization.

Key words Analytical ultracentrifu-garion · Self-association · Phosphor-ylase kinase · Trimethylamine *N*-oxide · Crowding

Introduction

Phosphorylase kinase (PhK; EC 2.7.1.38) catalyzing phosphorylation and activation of glycogen phosphory-lase *b* is one of the largest and most structurally complex protein kinases known [1, 2]. The enzyme molecule is a hexadecamer with the subunit composition $(\alpha\beta\gamma\delta)_4$ and molecular mass 1,320 kDa [3]. Electron microscopy studies of PhK showed that the tight complex of the catalytic γ subunits (44.7 kDa) with regulatory δ subunits (16.7 kDa) identical to the Ca^{2+}-binding protein cal-modulin forms a compact nucleus of the structure, whereas the regulatory α (138.4 kDa) and β (125.2 kDa) subunits form ridges on the surface of the molecule, giving it the shape of a "butterfly" or "chalice" [4–6].

Ca^{2+} and Mg^{2+} ions stimulate the PhK activity by inducing tertiary and subsequent quaternary structural changes in the enzyme molecule [7–9].

Using the method of sedimentation in an analytical ultracentrifuge Cohen [3] showed that at rather high concentration of protein (3 g/l), besides "monomeric" enzyme sedimenting at 23 S (corresponding to a molecular mass of 1,300 kD), dimeric ($s_{20,w} = 37$ S) and trimeric ($s_{20,w} = 48$ S) forms were observed. Data on Sepharose 4B column chromatography of PhK also suggest the existence of aggregated forms with low catalytic activity compared with the activity of the main peak [3, 10]. The solutions in those studies did not contain added Ca^{2+} and Mg^{2+}.

Carlson and King [11] showed that turbidimetry is a convenient method for studying self-association of PhK. Ca^{2+} and Mg^{2+} act totally synergistically in promoting the self-association and chelation of the metals, with excess ethylenediaminetetraacetate reversing the process.

The rates, equilibria, and mechanisms of biochemical reactions in dilute solutions differ significantly from those under crowding conditions within the cell [12]. Crowding theory [13, 14] predicts that the protein assembly should be favored by environmental effects arising from substantial volume occupancy.

It is known that trimethylamine N-oxide (TMAO), a naturally occurring osmolyte, favors the formation of supramolecular structures [15, 16]. The space-filling effect of a concentrated solution of TMAO should displace the isomerization equilibrium in favor of the more compact protein conformation [17–24] and a self–association equilibrium in favor of the polymeric state [25–28].

In this connection, of special interest is a more detailed study of the effect of concentrated solutions of TMAO on self-association of various proteins and such a complex enzyme as PhK to better appreciate the mechanisms of protein oligomerization and the behavior of enzymes under conditions which mimic those in the cell.

Materials and methods

Sedimentation velocity experiments were carried out in Optima XL-I and model E analytical ultracentrifuges (Beckman Instruments, Palo Alto, USA), the latter instrument being equipped with a photoelectric scanner and a monochromator, and using 12-mm double-sector cells. Sedimentation profiles were recorded by measuring the absorbance of the enzyme at 280 nm. Sedimentation coefficients were calculated according to the formula

$$s = \ln(r/r_m)w^2 t, \tag{1}$$

where r and r_m are radial positions of the boundary and meniscus, respectively, w is the angular velocity, and t is time of sedimentation. Sedimentation coefficients were also estimated from an apparent sedimentation coefficient distribution, [$g(s^*)$ versus s^*], which was determined using the so-called time-derivative program (Beckman). Sedimentation coefficients were corrected to solvent density and viscosity (20 °C, water) in the standard way.

Assuming a globular form for the PhK association products the following relation between the sedimentation coefficients of monomer (s_1) and n-mer (s_n; $n = 2, 3, 4, ...$) was used:

$$s_n = s_1 n^{2/3}, \tag{2}$$

where 2/3 is the Mark–Houwink–Kuhn–Sakurada coefficient for a sphere.

The dimerization constant for the monomer–dimer equilibrium of PhK was calculated according to the equation: $K_{ass} = (1-f)/(2f^2 c)$, where f is a monomer fraction, which was estimated from the sedimentation profiles and corrected for radial dilution, and c is the molar enzyme concentration calculated on the monomer.

PhK was isolated from rabbit skeletal muscle according to Cohen [3] using ion-exchange chromatography on (diethylamino)ethyl-Toyopearl at the final step of purification [29]. The concentration of PhK was determined spectrophotometrically at 280 nm using an extinction coefficient of $\varepsilon^{1\%}_{1cm} = 12.4$. TMAO and N-(2-hydroxyethyl) piperazine-N'-ethanesulfonic acid (Hepes) were purchased from Sigma Chemical Co. (USA). The other reagents (of high purity) were from Russian suppliers.

The kinetics of self-association of PhK was followed by an increase in the optical absorbance at 360 nm using a Hitachi 557 spectrophotometer (Japan) equipped with a thermostated cell holder. The kinetics experiments were performed at 20 °C in 40 mM Hepes buffer, pH 6.8 and 8.2, containing 1 mM β-mercaptoethanol. Prior to the sedimentation and kinetics experiments PhK was dialyzed against 40 mM Hepes buffer, pH 6.8 or 8.2, at 4 °C for 3 hours.

The initial rates of self-association of PhK were calculated by setting the kinetics curves using a polynomial of the second degree with the Microcal Origin 5.0 program.

Results

Analytical ultracentrifugation

The sedimentation patterns of PhK in 40 mM Hepes buffer, pH 6.8, and the $g(s^*)$ distribution functions without added Ca^{2+} and Mg^{2+} and in the presence of 0.1 mM Ca^{2+} and 10 mM Mg^{2+} are shown in Fig. 1. In the absence of ions the enzyme exists mainly in the monomeric and dimeric forms (with $s_{20,w} = 23.7$ and 34.8 S, respectively) (Fig. 1a, b). The addition of 0.1 mM Ca^{2+} and 10 mM Mg^{2+} results in the appearance of associates of higher order (Fig. 1c, d). A comparison of the absorbance values corresponding to the plateaus in Fig. 1 a and c indicates that in the presence of 0.1 mM Ca^{2+} and 10 mM Mg^{2+} a part of the enzyme (about 40%) precipitates during rotor acceleration. Analysis of the sedimentation behavior of PhK by plotting the $g(s^*)$ distribution and fitting the $g(s^*)$ profile with Gaussian functions (Fig. 1d) gives three species with $s^* = 23.2$, 33.8, and 47.7 S. Assuming a globular character for the PhK associates, we can conclude that the sedimentation coefficients correspond to monomer, dimer, and trimer.

Corresponding analysis of sedimentation of PhK at pH 8.2 in the presence of 0.1 mM Ca^{2+} and 10 mM Mg^{2+} yields a $g(s^*)$ distribution for PhK of the form shown in Fig. 2. The $g(s^*)$ distribution may be also resolved into species with sedimentation coefficients corresponding to monomer, dimer, and trimer. Under the conditions used the enzyme solutions contain large associates which precipitate over several minutes after the commencement of the centrifugation run. A decrease in the Mg^{2+} concentration results in the disappearance

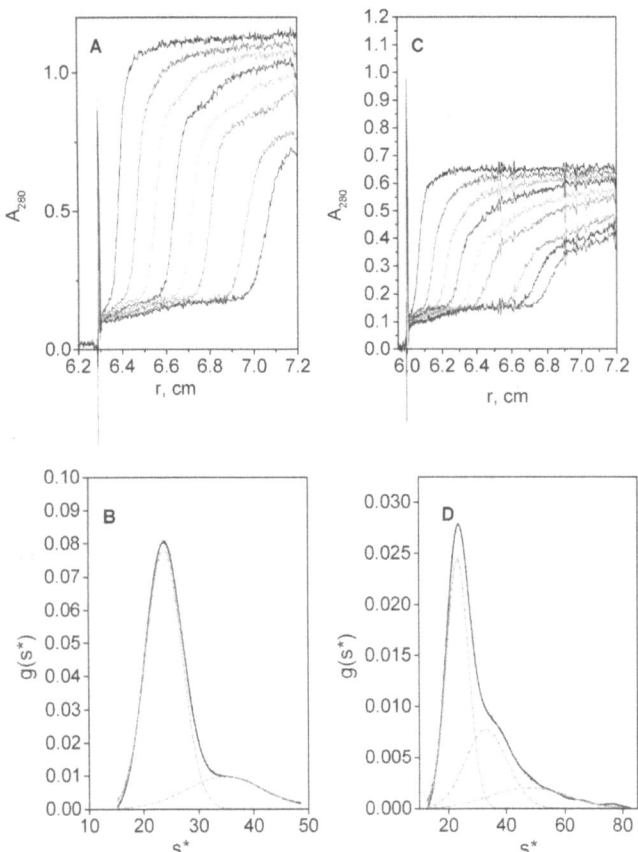

Fig. 1a–d Sedimentation velocity analysis of self-association of phosphorylase kinase (*PhK*) (0.86 g/l) in 40 mM *N*-(2-hydroxyethyl) piperazine-*N'*-ethanesulfonic acid (Hepes) buffer, pH 6.8, at 20 °C. Sedimentation was carried out at 30 000 rpm. The time interval between scans was 3 min. **a** The sedimentation patterns of PhK in the absence of Ca^{2+} and Mg^{2+} ions. **b** The $g(s^*)$ versus s^* plots obtained from **a**. **c** The sedimentation patterns of PhK in the presence of 0.1 mM Ca^{2+} and 10 mM Mg^{2+}. **d** The $g(s^*)$ versus s^* plots obtained from **c**

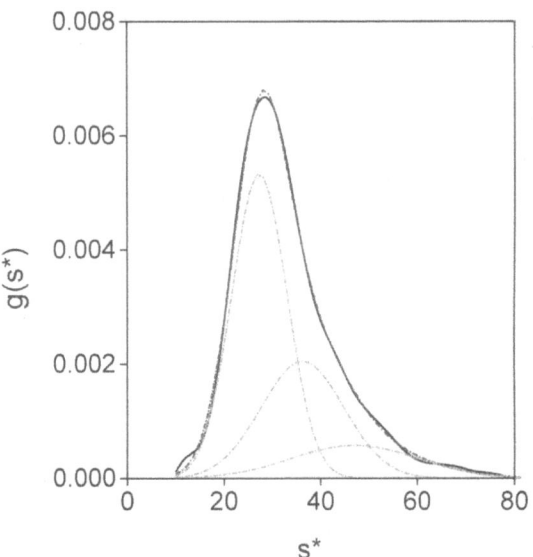

Fig. 2 Sedimentation of PhK (0.2 g/l) in 40 mM Hepes, pH 8.2, supplemented with 10 mM Mg^{2+} and 0.1 mM Ca^{2+}. Plots of $g(s^*)$, versus s^*

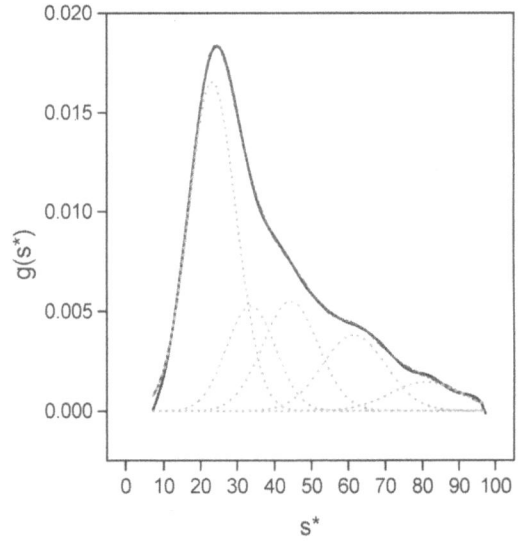

Fig. 3 Sedimentation of PhK (0.41 g/l) in 40 mM Hepes, pH 6.8, supplemented with 2 mM $MgCl_2$ and 0.1 mM $CaCl_2$. Plots of $g(s^*)$ versus s^*

of the large associates able to precipitate during rotor acceleration (Fig. 3, pH 6.8). Analysis of the $g(s^*)$ distribution indicates that the monomer–dimer–trimer mixture ($s_{20,w} = 23.2$, 33.7, and 46.5 S) is supplemented by a tetramer ($s_{20,w} = 61.2$ S) and a further species with $s_{20,w} = 80$ S (possibly a hexamer).

The effect of TMAO on the association of PhK was studied in the presence of 0.1 mM Ca^{2+} and 2 mM Mg^{2+}. The first ten sedimentation profiles in the presence of 0.6 M TMAO and in the absence of this osmolyte are shown in Fig. 4. TMAO was found to favor greatly self-association of PhK. Two distinct rapidly moving boundaries with $s_{20,w} = 385$ and 189 S are detected in the sedimentation patterns. These boundaries correspond to 24-mers and 70-mers (the molecular masses of associates were estimated using the Mark–Houwink–Kuhn–Sakurada relation: $s_{20,w} \sim M^b$ with $b = 0.667$, assuming a globular conformation). Such

large associates are absent from the sedimentation patterns obtained in the absence of TMAO (Fig. 4b).

It was of special interest to study the effect of TMAO on the self-association of Phk without added Ca^{2+} and Mg^{2+}. In this case only boundaries corresponding to the monomeric and dimeric forms were detected both in the presence and in the absence of TMAO (Fig. 5a, b, respectively). It is worth stressing that large associates are lacking. The boundaries corresponding to the individual oligomeric enzyme forms in Fig. 5a are well resolved,

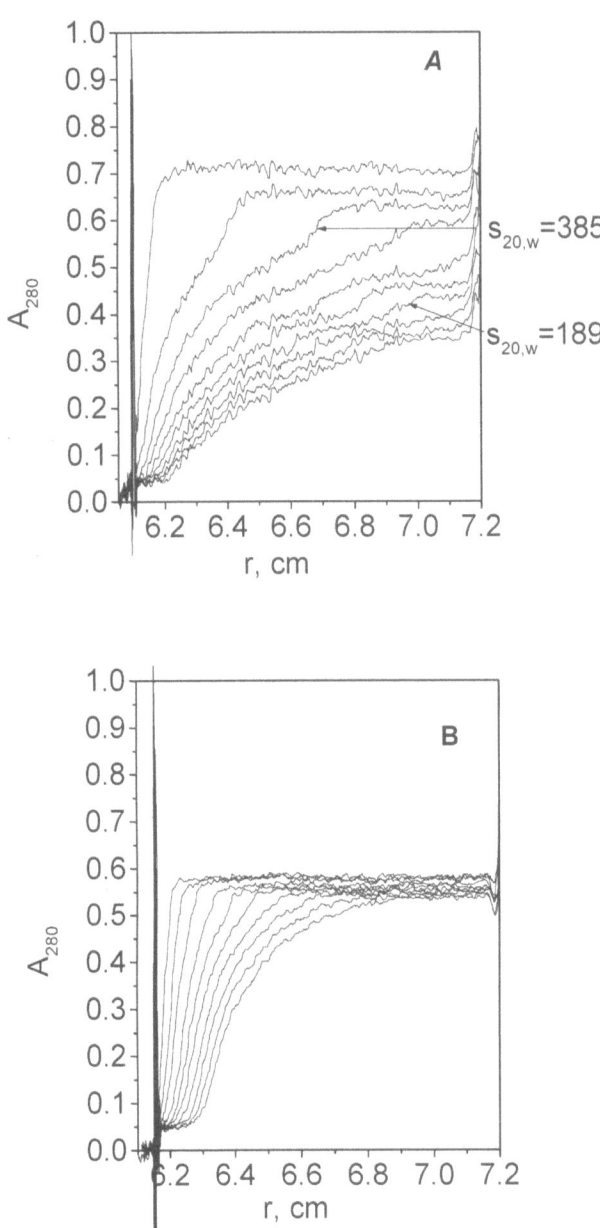

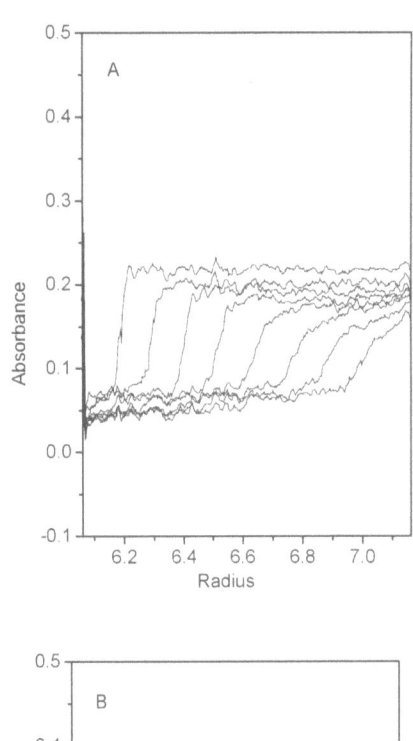

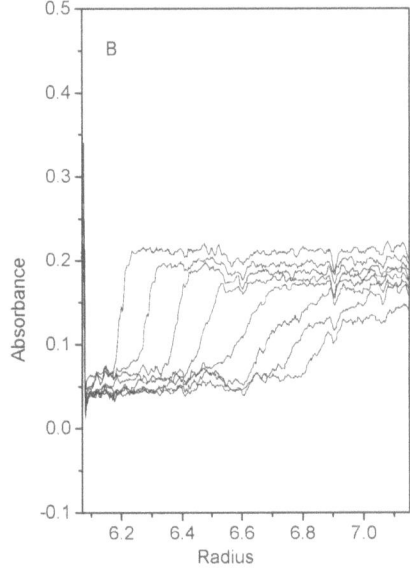

Fig. 5 Sedimentation velocity analysis of self-association of PhK (0.2 g/l) in 40 mM Hepes buffer, pH 6.8, at 20 °C **a** in the absence of TMAO and **b** in the presence of 1 M TMAO. The sedimentation was carried out at 30 000 rpm. The time interval between scans was 3 min

Fig. 4a, b Effect of trimethylamine *N*-oxide (*TMAO*) on self-association of PhK (0.41 g/l) in 40 mM Hepes, pH 6.8, supplemented with 0.1 mM Ca^{2+} and 2 mM Mg^{2+}. The first ten scans of the absorbance at 280 nm as a function of the radius were obtained at 3-min intervals **a** in the presence of 0.6 mM TMAO and **b** in the absence of osmolyte. The sedimentation was carried out at 20,000 rpm at 20 °C

The results indicate clearly that TMAO shifts the self-association equilibrium towards the formation of the dimeric form.

Effect of TMAO on the kinetics of self-association of PhK

suggesting that the interconversion between the monomeric and dimeric forms is rather slow. Therefore, sedimentation velocity is applicable for the determination of the dimerization constant (K_{ass}) for a monomer–dimer equilibrium. The K_{ass} value was found to be 1.8×10^{6} M^{-1} in the absence of TMAO and 3.5×10^{6} M^{-1} in the presence of 1 M TMAO under the same conditions (i.e., in the absence of Ca^{2+} and Mg^{2+}).

We studied the kinetics of PhK association induced by TMAO by following the increase in the optical absorbance of the enzyme solution at 360 nm. The time course of self-association of PhK recorded at various concentrations of the enzyme in 40 mM Hepes buffer, pH 6.8, is shown in Fig. 6a. The values of the initial rate of association, v_0, (expressed in units of the optical absorbance per minute) were plotted as a function of PhK concentration squared (Fig. 6b). A linear dependence was obtained, suggesting that the initial step of the self-association process is dimer formation. It should be noted that the kinetics curves do not flatten out: a monotonic increase in the optical absorbance occurs over a long period of time. The reason for this is evidently that the association of the enzyme is accompanied by the

formation of associates of sizes larger than the dimer. Analogous results were obtained at pH 8.2.

The dependence of the initial rate of PhK association on TMAO concentration obtained at pH 6.8 and pH 8.2 is shown in Fig. 7. The fact that the change in pH from 6.8 to 8.2 results in a marked alteration in the rate of the enzyme association agrees with the view of substantial conformational differences of the PhK molecule at these pH values.

Addition of 1 mM ethylene glycol bis(β-aminoethyl ether)-N,N,N',N'-tetraacetic acid causes a substantial disintegration of the PhK associates formed in the presence of TMAO. These results suggest that self-association of PhK is a reversible process.

Discussion

The results obtained show that TMAO facilitates the ability of Ca^{2+} and Mg^{2+} to induce self-association of PhK. To interpret the effect of TMAO, it is expedient to discuss the results of electron microscopy studies of self-association of PhK from pigeon breast muscles in the presence of 2.5 mM Ca^{2+} and 5 mM Mg^{2+} [30]. This isoform of PhK has practically the same size and shape as the isoform from the rabbit skeletal muscle (the difference in molecular masses does not exceed 1%) and has very high sensitivity to Ca^{2+} regulation of the activity (K_m for Ca is 2 orders lower than that estimated for the rabbit isoform) [29]. Pigeon PhK forms linear associates of different length [30]. The length of these rodlike linear associates varies from 100 nm to 1–5 μm. Aggregates formed by the sticking together of these "rods" may be

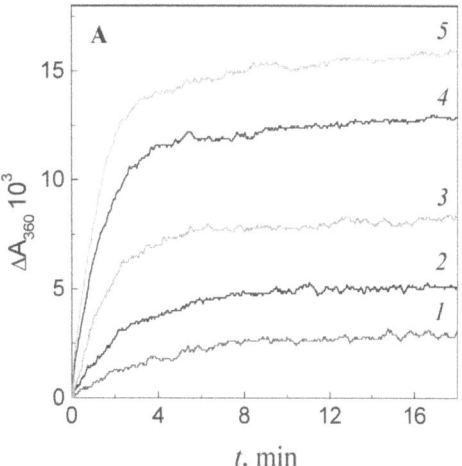

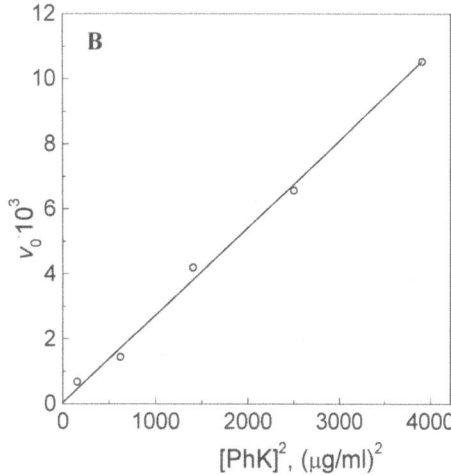

Fig. 6a, b Kinetics of association of PhK induced by TMAO in 40 mM Hepes buffer, pH 6.8 (20 °C). **a** Time course of association at various concentrations of PhK (mg/l): 12.5 (*1*), 25 (*2*), 37.5 (*3*), 50 (*4*), and 62.5 (*5*). ΔA_{360} is the change in the optical absorbance at 360 nm. Association was initiated by the addition of a solution of PhK preincubated with Ca^{2+} and Mg^{2+} at 20 °C for 2 min to a solution containing TMAO, Ca^{2+}, and Mg^{2+}. The final concentrations were 1 M TMAO, 0.1 mM Ca^{2+}, and 10 mM Mg^{2+}. **b** The dependence of the initial rate of association, v_0, (in units of the optical absorbance per minute) on the PhK concentration squared

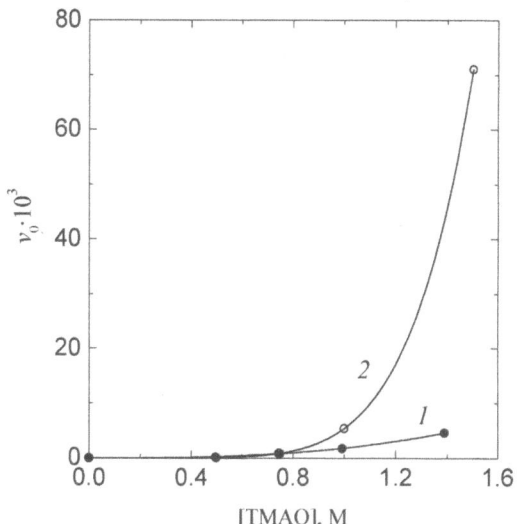

Fig. 7 The dependences of the initial rate of PhK association, v_0, on TMAO concentration in 40 mM Hepes buffer (20 °C) at pH 6.8 (*1*) and 8.2 (*2*)

seen in the electron microscopy images presented in Ref. [30].

The regulatory α subunits form (together with β subunits) lobes on the molecule surface. The ability of α subunits to self-associate has been studied recently by Ayers et al. [31] using a genetic two-hybrid screen. Two regions on each α subunit involved in this interaction were identified: leucine zipper (residues 833–854) and the C-terminal region (residues 1,015–1,237) of the α subunit. The contacts could be realized as three types of interactions:

1. Intrasubunit interactions stabilizing the tertiary structure of the subunit.
2. Intersubunit intramolecular interactions stabilizing the quaternary structure of the $(\alpha\beta\gamma\delta)_4$ enzyme molecule.
3. Intersubunit intermolecular interactions providing contacts between molecules.

It was shown by immunoelectron microscopy studies [32] that an epitope of the α subunit within residues 1,015–1,237 is localized near the distal tips of each lobe of the PhK hexadecamer. These data allow us to suppose that this α subunit area could contribute both to intra- and intermolecular α–α interactions. The leucine zipper identified as a second potential α–α contact region is situated near the previously reported [32] anti-α mAB epitope at the lobe tips and might also be involved in the intermolecular α–α interactions.

The enlargement of the PhK associates is believed to occur not only by linear growth of rods, but by sticking together of rods (side-to-side association due to α–α subunits contacts; Fig. 8). One can assume that the associates of rabbit PhK of relatively small size ($n = 2$–4) are formed as a result of the linear growth of the enzyme oligomers. The distinct large species of PhK ($s_{20,w} = 189$ and 385 S) recorded in the presence of TMAO may be formed by sticking together rodlike associates. The shape of the associates formed by sticking together rather short rods is believed to yield a form not differing markedly from a sphere. It is worth noting that according to molecular crowding theory compact states of macromolecules are favored over asymmetric ones [25]. This allowed us to use Eq. (2) to characterize the dimension of large oligomeric forms (with $s_{20,w} = 189$ and 385 S).

The capability of PhK for forming large- associates in crowded media suggests that as a consequence of the macromolecular crowding effect some of the enzyme molecules in the cell may exist in the form of supramolecular structures. Besides, it is felt that the "sticky" sites responsible for self-association of PhK may provide sites for the interaction of the enzyme with microtubules [33] or certain integral membrane-bound proteins [2]. A similar role of the "sticky" sites for proteins revealing the ability to open association was postulated by Kurganov and coworkers [34, 35].

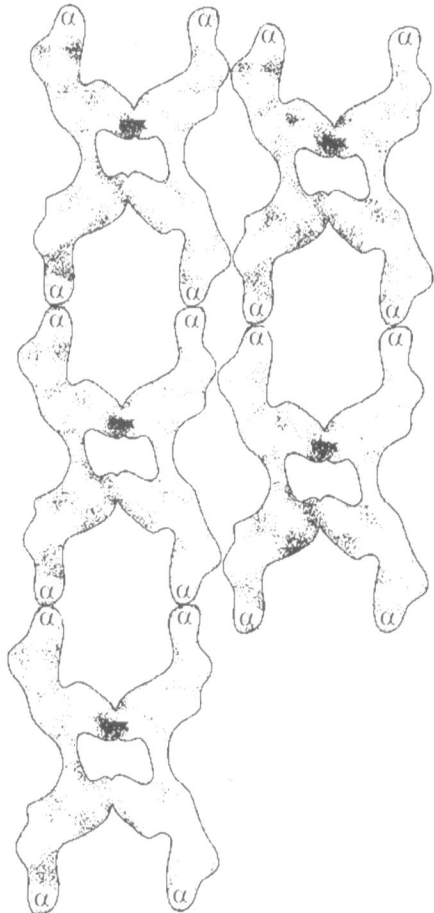

Fig. 8 Scheme illustrating the type of association of PhK

Excluded-volume theory predicts that in crowded media the formation of rodlike or fibrous aggregates is favored [36]. From this theory we can also deduce that the rodlike aggregates so formed may spontaneously align and bundle [14, 37]. We believe that the formation of such rodlike structures with the capacity to stick together takes place for PhK in TMAO solutions. Assuming that condensation reactions in crowded media are transition-state limited, rather than diffusion-limited, excluded-volume theory predicts that the shift in equilibrium towards the aggregated states may be attributed primarily to an increase in the association rate constant [14, 38]. Our data demonstrate the increase in the rate of association of PhK as the fractional volume occupied by TMAO rises, suggesting that the association reaction is transition-state limited.

Acknowledgements This work was supported financially by the Russian Foundation for Basic Research (grants 99-04-48639 and 00-15-97787). Part of the work was performed whilst N.A.C. was an Underwood Fellow of the UK Biotechnology and Biomolecular Sciences Research Council.

References

1. Krebs EG, Graves DJ, Fisher EH (1959) J Biol Chem 234:2867–2873
2. Brusharia RJ, Walsh DA (1999) Frontiers Biosci 4:618–641
3. Cohen P (1973) Eur J Biochem 34:1–14
4. Schramm HJ, Jennissen HP (1985) J Mol Biol 181:503–516
5. Wilkinson DA, Marion TN, Tillman DM, Norcum MT, Hainfeld JF, Seyer JM, Carlson GM (1994) J Mol Biol 235:974–982
6. Norcum MT, Wilkinson DA, Carlson MCh, Hainfeld JF, Carlson GM (1994) J Mol Biol 241:94–102.
7. Ayers NA, Nadeau OW, Carlson GM (2001) J Biol Chem (in press)
8. Nadeaw OW, Traxler KW, Fee LR, Baldwin BA, Carlson GM (1999) Biochemistry 38:2551–2559.
9. Wilkinson DA, Fitzgerald TJ, Marion TN, Carlson GM (1999) J Protein Chem 18:157–164
10. Andreeva IE, Silonova GV, Livanova NB, Eronina TB, Morozov VE, Poglazov BF (1985) Biokhimiya 50:1504–1515 (in Russian)
11. Carlson GM, King MM (1982) FASEB J 41:869
12. Minton AP (2001) J Biol Chem 276:10577–10580
13. Minton AP (1983) Mol Cell Biochem 55:119–140
14. Minton AP (2000) Curr Opinion Struct Biol 10:34–39
15. Tseng H-C, Graves DJ (1998) Biochem Biophys Res Commun 250: 726–730
16. Yang D-S, Yip CM, Huang THJ, Chakrabartty A, Fraser PE (1999) J Biol Chem 274:32970–32974
17. Baskakov I, Bolen DW (1998) Biophys J 74:2658–2665
18. Baskakov I, Wang A, Bolen DW (1998) Biophys J 74:2666–2673
19. Shearwin KE, Winzor DJ (1990) Arch Biochem Biophys 282:297
20. Winzor CL, Winzor DJ, Pagel LG, Jones GP, Naidu BP (1992) Arch Biochem Biophys 296:102–107
21. Hall DR, Jacobsen MP, Winzor DJ (1995) Biophys Chem 57:47–54
22. Mashino T, Fridovich I (1987) Arch Biochem Biophys 258: 356–360
23. Chebotareva NA, Harding SE, Winzor DJ (2001) Eur J Biochem 268:506–513
24. Qu Y, Bolen CL, Bolen DW (1998) Proc Natl Acad Sci USA 95:9268–9273
25. Zimmerman SB, Minton AP (1993) Annu Rev Biophys Biomol Struct 22:27–65
26. Shearwin KE, Winzor DJ (1988) Biophys Chem 31:287–294
27. Timasheff SN (1998) Proc Natl Acad Sci USA 95:7363–7367
28. Cann JR, Coombs RO, Howlett GR, Jacobsen MP, Winzor DJ (1994) Biochemistry 33:10185–10190
29. Morozov VE, Eronina TB, Andreeva IE, Silonova GV, Soloviyova NV, Schors EI, Livanova NB, Poglazov BF (1989) Biokhimiya 54:448–455 (in Russian)
30. Morozov VE (1990) Thesis. Bach Institute of Biochemistry, Moscow
31. Ayers A, Wilkinson, DA, Fitzgerald TJ., Carlson GM (1999) J Biol Chem 274:35583–35590
32. Wilkinson DA, Marion TN, Tillman DM, Norcum MT, Hainfeld JF, Seyer JM, Carlson GM (1994) J Mol Biol 235:974–982
33. Paudel H.K. (1997) J Biol Chem 272:1777–1785
34. Kurganov BI, Sugrobova NP, Mil'man (1986) Molekulyarnaya Biologiya 20:41–51 (in Russian)
35. Lyubarev AE, Kurganov BI (1996) In: Kurganov BI, Lyubarev AE (eds) Organization of biochemical systems: structural and regulatory aspects. Nova, New York, pp 1–81
36. Ross PD, Minton AD (1979) Biochem Biophys Res Commun 88:1308–1314
37. Madden TL, Herzfeld J (1993) Biophys J 65:1147–1154
38. Minton AP (1990) Int J Biochem 22:1063–1067

Progr Colloid Polym Sci (2002) 119:77–83
© Springer-Verlag 2002

Gottfried Mayer
Oliver Anderka
Bernd Ludwig
Dieter Schubert

The state of association of the cytochrome bc₁ complex from *Paracoccus denitrificans* in solutions of dodecyl maltoside

G. Mayer · D. Schubert (✉)
Institut für Biophysik,
Johann Wolfgang Goethe-Universität,
60590 Frankfurt am Main, Germany
e-mail: schubert@biophysik.uni-frankfurt.
de
Tel.: +49-69-63015835
Fax: +49-69-63015838

O. Anderka, B. Ludwig
Institut für Biophysikalische Chemie und
Biochemie, Abt. Molekulare Genetik,
Johann Wolfgang Goethe-Universität,
60439 Frankfurt am Main, Germany

Abstract The cytochrome bc₁ complex is an intrinsic membrane protein of many respiratory chains. We studied the association state of solubilized cytochrome bc₁ from *Paracoccus denitrificans* by sedimentation equilibrium experiments in the analytical ultracentrifuge. Since the stability of the solubilized complex was only maintained in solutions containing low concentrations of the nonionic detergent *n*-dodecyl-β-D-maltoside (DDM), this detergent was also applied in the ultracentrifuge experiments. DDM is an unfavorable detergent in ultracentrifuge studies owing to its high density ($\rho = 1.23$ g/ml), which strongly complicates "density-matching", the standard method for eliminating the contribution of protein-bound detergent to protein molar mass. Use of high sucrose concentrations for matching the DDM density is basically possible but probably induces significant changes in the protein's partial specific volume, $\bar{v}$; therefore we tried to decrease the necessary sucrose concentration by using mixtures of sucrose and 95% D₂O/ 5% H₂O (v/v). The effect of the solvent on $\bar{v}$ was controlled by

studying a related protein of known $\bar{v}$, cytochrome c oxidase, under identical conditions. In sedimentation equilibrium experiments the cytochrome bc₁ complex behaved as an ideal homogeneous compound in the presence of 0.02% (w/v) DDM. Its molar mass was determined to be (240,000 ± 30,000) g/mol (mean and maximum error, respectively). Since the calculated mass of the protein protomer is 117,000 g/mol, the solubilized complex represents a dimer. Measurements in water-containing buffers, in the absence of sucrose, showed that the amount of DDM bound by the complex was (0.86 ± 0.12) g/g protein. A dimeric structure was already established for the much larger mitochondrial cytochrome bc₁ complexes that had been crystallized. In the case of two other bacterial complexes experimental evidence points in the direction of dimers as well.

Key words Cytochrome bc₁ complex · Dodecyl maltoside · Density matching · Sedimentation equilibrium analysis · *Paracoccus denitrificans*

Introduction

The cytochrome bc₁ complex (or its photosynthetic membrane equivalent, b₆f) is a key constituent of many energy-transducing membranes, coupling the free energy of quinol oxidation to the generation of an electrochemical proton gradient [1–3]. In mitochondrial electron transfer, the complex ("complex III") contains up to 11 different subunits, while the bacterial enzyme typically consists only of those three subunits active in the redox reaction [4–7]: a cytochrome b with two heme moieties b_L and b_H, a subunit liganding a [2Fe · 2 S] center (Rieske protein; ISP), and a cytochrome c_1.

Interest in the mitochondrial redox complexes has recently been spurred by the successful crystallization followed by determination of the 3D structure for several representatives of mammalian and avian origin and from yeast [8–11]. These structures proved consistent with the previously established route of internal electron transport according to a quinone cycle acting on cytochrome b and the FeS subunit; however, to allow for subsequent electron transfer to the heme located in the surface-exposed hydrophilic domain of cytochrome c_1, different positions of the FeS domain had to be invoked and were actually observed in several crystal forms obtained under different experimental conditions [9–12]. Interestingly, this domain mobility of the FeS subunit is intimately linked to the dimeric structure of the bc_1 complex: in a given monomer, electron transfer from the b_L heme is mediated by the FeS domain anchored to the membrane sector of the other monomer. Therefore, the dimeric structure observed in all known complex III structures has immediate functional implications.

Present evidence, mostly from inhibitor studies and from extensive sequence alignments [5–7], suggests that the mechanism of electron transfer and energy transduction described previously for the mitochondrial cytochrome bc_1 complex also operates in the bacterial enzyme(s). To this end, a dimeric form would also be a structural prerequisite for the bacterial complex. However, since the prokaryotic complex lacks about 50% of the mass of the mitochondrial protein (largely owing to the absence of the two core subunits), the assumption of a structural intertwining of both monomers exerted by the two Rieske subunits is not necessarily compelling. So far, only two reports have addressed the oligomeric state of a bacterial bc_1 complex [13, 14]. Sone and Takagi [13] studied the state of association of the complex from the thermophilic bacterium PS3 in the nonionic detergent octaethylene glycol n-dodecyl ether by eluting the protein from a gel filtration column and sequentially monitoring the eluate for UV absorption, light scattering, and refractive index. By rigorous analysis they could show that the solubilized enzyme was a dimer which had bound around 1.0 g detergent/g protein. In the second study, Montoya et al. [14] used gel filtration to show that the corresponding complex from *Rhodovulum sulfidophilum* solubilized in n-dodecyl-β-D-maltoside (DDM) is also a dimer. However, the conclusiveness of molar mass determinations by gel filtration is known to be limited [15]. We therefore found it worthwhile to investigate the oligomeric state of another bacterial cytochrome bc_1 complex, isolated from *Paracoccus denitrificans*, the biochemistry of which has already been studied in some detail [4–6, 16, 17]. Its calculated protomer molar mass,

M, is 117,000 g/mol. We applied a method as rigorous as that applied in Ref. [13], namely sedimentation equilibrium analysis in the analytical ultracentrifuge.

In preliminary experiments applying a variety of nonionic detergents, the cytochrome bc_1 complex from *P. denitrificans* was found to be extremely unstable both during isolation and storage, for periods equal to the duration of a sedimentation equilibrium experiment. The only way of handling the protein without inducing either degradation or irreversible aggregation was to perform all steps of solubilization, purification, and storage in buffers containing low concentrations of DDM. On the other hand, DDM is an unfavorable detergent in ultracentrifuge studies since the contribution of the protein-bound detergent to the parameter supplied by the experiment, the effective molar mass, M_{eff}, of the protein-detergent complex, is difficult to determine or to eliminate:

1. Direct determination of the amount of protein-bound detergent by gel filtration, which would allow subtraction of the detergent contribution from the effective molar particle mass determined by sedimentation equilibrium analysis [15, 18], requires the use of radioactively labeled DDM.
2. The high density of the DDM micelles (1.229 g/ml in aqueous buffers [19]) does not allow application of the standard technique for eliminating the detergent contribution to M_{eff}, namely addition of D_2O to the buffer until the buffer density equals that of the detergent ("density-matching") [15, 19, 20]. Instead, high concentrations of sucrose, glycerol or Nycodenz have to be used [21, 22] which, in turn, most probably lead to distinct changes in the protein's partial specific volume, $\bar{v}$, a most critical auxiliary parameter [15, 20].

In the present work, we therefore applied a variant of the density-matching technique to reduce the problems described. We combined the use of D_2O and sucrose to increase the solvent density, thereby lowering the required sucrose concentration. In addition, we established corrections for the influence of the sucrose on the protein's $\bar{v}$ by performing analogous experiments on a monomeric membrane protein with known M and $\bar{v}$, cytochrome c oxidase [21]. By this method, we could unambiguously show that the cytochrome bc_1 complex from *P. denitrificans*, in solutions of DDM, is a stable dimer as well.

Materials and methods

Materials

DDM was purchased from Calbiochem (La Jolla) and 1,6-diphenyl-1,3,5-hexatriene (DPH) from Serva (Heidelberg). All other reagents were obtained from Merck (Darmstadt) or Sigma/Aldrich (Deisenhofen) and were of analytical grade (if available).

Preparation of the cytochrome bc₁ complex

The complex was isolated from a *P. denitrificans* strain overexpressing the enzyme. Cell growth, membrane preparation, subsequent enzyme purification, and determination of enzyme activity were done essentially as described in Ref. [16], with the following modifications:

1. The vector carrying the fbc operon coding for the bc₁ complex was pRI2 [17] instead of pEG 400.
2. Solubilized membranes were diluted to a salt concentration of 350 mM NaCl with 50 mM 2-morpholinoethanesulfonate(MES)/NaOH (pH 6.0) and 0.02% (w/v) DDM before anion-exchange chromatography.
3. A salt gradient between 350 and 600 mM NaCl was applied.

Prior to analytical ultracentrifugation analysis, pooled fractions were concentrated by ultrafiltration (Centrisart, Sartorius), exclusion limit 100,000 g/mol, and further purified by preparative gel filtration (Superose 6 HR 10/30, Pharmacia) with a Pharmacia-LKB FPLC system. Buffer conditions of 50 mM MES/NaOH (pH 6.0), 300 mM NaCl, and 0.02% (w/v) DDM ("buffer A") were used during gel filtration. The same column was applied in studies on the stability of the protein. Solubilization and all subsequent steps were performed at 4 °C.

Isolation of cytochrome c oxidase

The enzyme was isolated from wild-type *P. denitrificans* (strain ATCC 13543) as described earlier [23]. It was transferred into buffer A by gel filtration (see earlier).

Analytical ultracentrifugation

Sedimentation equilibrium experiments were performed using a Beckman Optima XL-A analytical ultracentrifuge, an An-50 Ti rotor, and standard Epon six-channel centerpieces. The volumes of the sucrose-containing and the sucrose-free protein samples were 110 and 120 μl, respectively; in experiments on protein-free micelles 140 μl was used. The rotor temperature was 4 °C.

The determination of the buoyant density of DDM micelles in the buffer used for the studies on the cytochrome bc₁ complex, 50 mM MES/NaOH (pH 6.0), 300 mM NaCl, 0.02% (w/v) DDM, 95% D_2O/5% H_2O (v/v), and X% (w/v) sucrose ("buffer B"), was performed using a procedure analogous to that described earlier [19, 21]. Solutions analogous to buffer B but containing, instead of 0.02% DDM and a fixed sucrose concentration, 0.2% DDM plus 0.5 mg 1,6-diphenyl-1,3,5-hexatriene (DPH)/g DDM (mixed according to the procedure described in Refs. [19, 21]) and different sucrose concentrations, were studied by sedimentation equilibrium runs at 40,000 rpm. The absorbance versus radius distributions, $A(r)$, measured at 357 nm (an absorption maximum of DPH) were evaluated by one-component fits with zero baseline using the computer program DISCREEQ by Schuck [24, 25]. The best-fit figures obtained for the effective molar mass of the micelles, $M_{d,eff} = M_d(1 - \bar{v}_d \rho_0)$, were plotted versus the corresponding sucrose concentration, and the buoyant density of the micelles was taken from the zero crossover ($M_{d,eff} = 0$) of the resulting straight line (M_d is the molar mass of the micelles, $\bar{v}_d$ is the corresponding partial specific volume, ρ_o is the solvent density). The sucrose concentration, X, which resulted in $M_{d,eff} = 0$ was used in all the studies with buffer B.

Part of the sample of the cytochrome bc₁ complex, in buffer A, was directly used for the ultracentrifuge experiments to allow the determination of the amount of protein-bound detergent, by combining the $A(r)$ data with those in buffer B [15, 19, 20]. The remainder, as well as the sample of cytochrome c oxidase, was first brought to X% sucrose by adding an appropriate volume of buffer A plus 50% (w/v) sucrose. The samples were then dialyzed

for 6 h against 19 volumes of a buffer similar to buffer B but containing 100% D_2O. Immediately afterwards, the proteins (in buffer A or B) were studied by sedimentation equilibrium runs at protein concentrations (determined by absorbance measurements [5, 21]) of 0.08–0.13 g/l (cytochrome bc₁ complex) and approximately 0.1 g/l (cytochrome c oxidase) and rotor speeds of 7,000–20,000 rpm. Owing to the high viscosity of buffer B, equilibrium was reached only after 70 h (but within 40 h in buffer A). The $A(r)$ profiles were measured at wavelengths in the proteins' absorbance peaks, 412 nm (cytochrome bc₁ complex) or 425 nm (cytochrome c oxidase), at initial absorbance values (with 1.2-cm optical pathlength) of 0.2–0.4. After completing the data collection the rotor speed was increased to 40,000 rpm for a few hours to determine the baseline position. The evaluations again used the program DISCREEQ [24, 25], applying one-component as well as multi-component fits. $\bar{v}$ of the cytochrome bc₁ complex was calculated from the known amino acid and pigment composition [5] according to the method and the revised data set of Cohn and Edsall [26, 27]. For the $\bar{v}$ value for cytochrome c oxidase see Ref. [21]; in addition, the value was determined from the present equilibrium sedimentation data. The density of the buffers was measured using a Paar DMA 02 digital density meter at 4 °C.

Results

Density-matching of DDM

The sucrose concentration, X, in buffer B, required to match the density of DDM was determined from a series of sedimentation equilibrium experiments on DPH-labeled DDM micelles, at varying sucrose concentrations. The resulting $M_{d,eff}$ versus sucrose concentration plot is shown in Fig. 1. From its zero crossover, X is

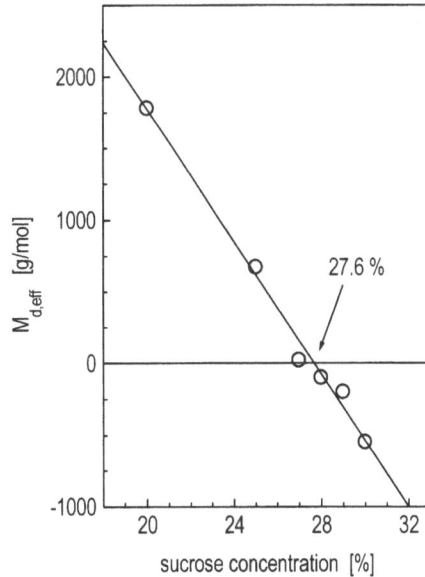

Fig. 1 Determination of the buoyant density of *n*-dodecyl-β-D-maltoside (*DDM*) by sedimentation equilibrium analysis: dependency of the effective molar mass of 6-diphenyl-1,3,5-hexatriene (*DPH*) labeled DDM micelles, $M_{d,eff}$, on the concentration of sucrose (w/v) added to the buffer

obtained as 27.6% (w/v). The density of buffer B is 1.207 g/ml.

The partial specific volume of the cytochrome bc_1 complex

The partial specific volume of the cytochrome bc_1 complex was calculated according to the method described in Refs. [26, 27] from its amino acid and ligand composition [5] and a $\bar{v}$ value for the heme group of 0.82 ml/g (determined by ab initio calculation from volume increments [28]). The result obtained, $\bar{v} = 0.738$ ml/g, refers to the complex at 25 °C and in dilute aqueous buffers. To correct for the influence of temperature and high sucrose concentration we studied, in buffer B and at 4 °C, another heme-containing membrane protein from *P. denitrificans*, cytochrome c oxidase. This protein is known to be homogeneous and monomeric in detergent solutions (also in the presence of sucrose); its molar mass is 128,500 g/mol. Its $\bar{v}$-value calculated, at 25 °C, analogous to that of the cytochrome bc_1 complex is 0.753 ml/g [21]. The analysis of a sedimentation equilibrium experiment of the protein is shown in Fig. 2. In this analysis, one-component fits to the $A(r)$ data were performed using different M_{eff} values for the protein, and the sum of the squared residuals of a fit, σ, is plotted versus M_{eff}. The well-defined minimum corresponds to $M_{\text{eff}} = 8,700$ g/mol (with other samples, the value varied between 8,500 and 8,700 g/mol, with an average of 8,600 g/mol). The fit applying this M_{eff} value

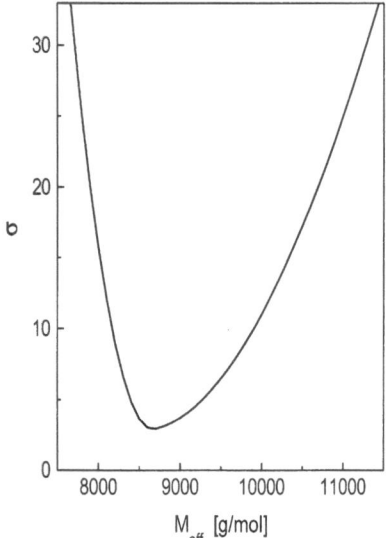

Fig. 2 Determination of the effective molar mass of the cytochrome c oxidase–DDM complex under conditions of density-matching for the detergent (buffer B): sum of the squared residuals, σ, of one-component fits to equilibrium sedimentation data as a function of the assumed M_{eff}. Rotor speed: 20,000 rpm

was of excellent quality (data not shown), which proves the homogeneity of the enzyme also in buffer B. From the M_{eff} value, the known molar mass and solvent density, and a correction for H–D exchange according to the method described in Ref. [29] (but assuming $k = 1.010$ as determined by IR spectroscopy), we obtained $\bar{v} = 0.782$ ml/g, which exceeds the calculated figure given earlier by 0.029 ml/g. By applying this correction to the cytochrome bc_1 complex, its corrected $\bar{v}$ is found to be 0.767 ml/g, with an estimated maximum uncertainty of 0.010 ml/g.

The figure $\bar{v} = 0.782$ ml/g for the cytochrome c oxidase obtained in the present study in solutions of 0.02% DDM/27.6% sucrose/95% D_2O exceeds those determined by analogous procedures in 0.3% nonaethylene glycol *n*-dodecyl ether ($C_{12}E_9$)/46% D_2O, 0.3% $C_{12}E_9$/6.4% sucrose, and 0.3% Triton X-100 (reduced form)/9.7% sucrose [21] by only 0.016, 0.016, and 0.007 ml/g, respectively. Thus, the influence of moderate concentrations of sucrose on the protein's $\bar{v}$ value is surprisingly small. On the other hand, an uncorrected increase in $\bar{v}$ of 0.016 or even 0.029 ml/g, at the given solvent density, would lead to an error in M of 17 or 38%, respectively.

Studies on the cytochrome bc_1 complex in DDM

Purity and stability of the complex

As described previously, gel filtration on Superose 6 is the final step in the purification procedure of the cytochrome bc_1 complex. The elution profile of the column is shown in Fig. 3a; the fractions pooled and used in the ultracentrifuge studies are indicated by a solid line. According to sodium dodecyl sulfate polyacrylamide gel electrophoresis (Fig. 3b) and redox UV–vis spectroscopy (data not shown), the pooled fractions consist of virtually pure cytochrome bc_1 complex [30]. The stability of the protein was monitored by analytical gel filtration and, in addition, by sedimentation velocity experiments (described in detail elsewhere [31]). According to the analyses, the material was essentially stable during the time required for the sedimentation equilibrium studies. However, a low-abundancy component both at lower and at higher elution volumes (larger and smaller molar masses) could be detected (most probably degradation products and aggregates thereof). In most preparations the smaller of the two additional components represented less than 5% of the total absorbance and the larger component represented less than 2%. Some preparations showed more rapid degradation/association; the corresponding sedimentation equilibrium data were discarded.

The stability of the cytochrome bc_1 complex was sufficient only within a narrow range of DDM concentrations. For example, at a DDM concentration of 0.1%

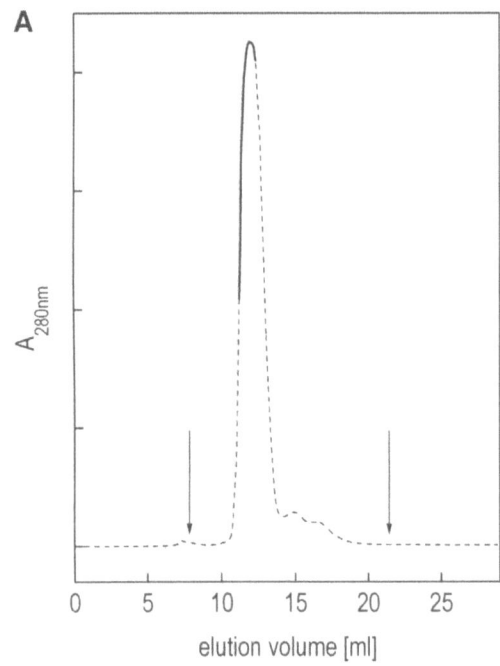

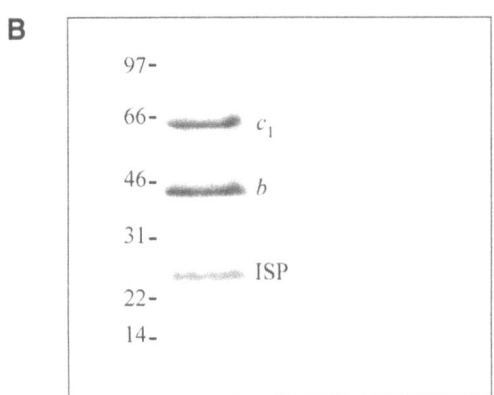

Fig. 3A, B Purity of the isolated cytochrome bc_1 complex. **A** Protein elution profile of the gel filtration in buffer A (absorbance axis not calibrated); fractions pooled and used for ultracentrifuge experiments are indicated by a *solid line*. The *arrows* indicate the void volume and the included volume, respectively. **B** Sodium dodecyl sulfate polyacrylamide gel electrophoresis of the pooled fractions; *left*: position of marker proteins of known molar mass (given in 10^3 g/mol)

(w/v) a considerable amount of the complex was found to be degraded after 24 h. In addition, the use of $C_{12}E_9$ or mixtures of $C_{12}E_9$ and N,N-dimethyldodecylamine N-oxide [21], at detergent concentrations of 0.01–0.1% (w/v), led to predominantly degraded material, whereas $C_{12}E_9$ concentrations of 0.005% or lower yielded irreversibly aggregated material. Thus, the use of DDM, in a narrow concentration range around 0.02% (w/v), seems to be the only possibility for studying the solubilized cytochrome bc_1 complex in its native form. Conservation of the native state of the protein under these conditions is also supported by the finding that the enzymatic activity of the complex did not decrease significantly during

storage of the protein for several days at 4 °C (data not shown).

The state of association of the cytochrome bc_1 complex

The results of a sedimentation equilibrium experiment on the cytochrome bc_1 complex in buffer B, evaluated analogously to that on cytochrome c oxidase (Fig. 2), are shown in Fig. 4. The best-fit figure for M_{eff} from a one-component fit is well-defined at 19,600 g/mol. This M_{eff} value, in combination with the figures for $\bar{v}$ and ρ_o given previously, and with correction for H–D exchange according to the method described in Ref. [29] (but assuming $k = 1.007$ as determined by IR spectroscopy), yields $M = 240,000$ g/mol. The maximum uncertainty of this figure is $\pm$ 30,000 g/mol or $\pm$ 12.5%. The good quality of the fit is demonstrated in Fig. 5: the residuals of the fit are statistically distributed along the r-axis and their absolute values are within the experimental uncertainty of $A(r)$, which suggests that the protein is homogenous with respect to its state of association. Virtually identical results were obtained with other samples of the protein. The fits could not be improved significantly by considering additional terms of $M_{eff} = 10,000$, 30,000 or 40,000 g/mol and of combinations of them; the corresponding concentrations calculated were virtually zero.

The molar mass of the cytochrome bc_1 complex, as determined here, is (2.05 ± 0.26) times that of its monomer mass (per mole). It is thus obvious that the

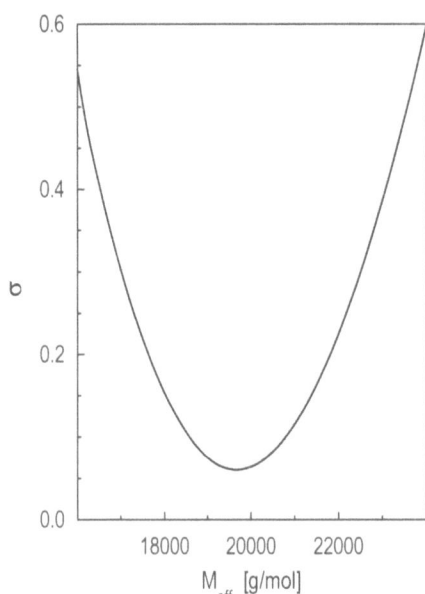

Fig. 4 Effective molar mass of the cytochrome bc_1 complex in buffer B: dependency of σ on the assumed M_{eff}, based on one-component fits to a sedimentation equilibrium profile. Rotor speed: 18,000 rpm

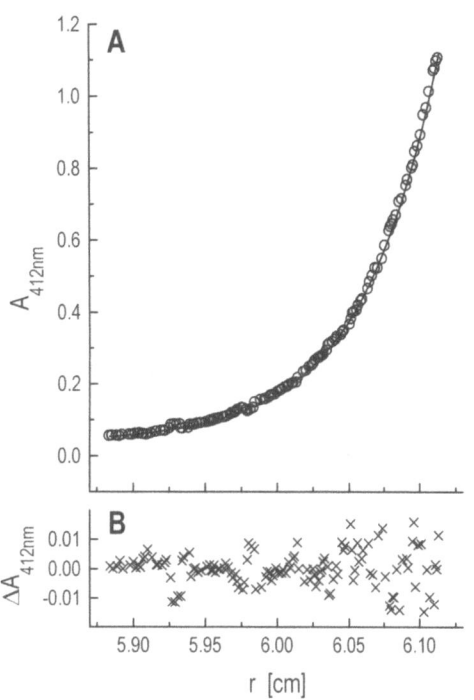

Fig. 5A, B Analysis of sedimentation equilibrium data for the cytochrome bc_1 complex in buffer B. **A** Experimental $A(r)$ data (○) and fit to the data using the best-fit figure for M_{eff} from Fig. 4 (——). **B** Residuals of the fit. The $A(r)$ data are those used in Fig. 4

observed homogeneous protein particle behaves as a stable dimer.

Amount of protein-bound DDM

We also measured sedimentation equilibrium distributions of the cytochrome bc_1 complex in buffer A ($\rho_0 = 1.014$ g/ml). Again, one-component fits to the $A(r)$ data were of good quality; they yielded $M_{eff} = (91{,}000 \pm 5{,}000)$ g/mol (data not shown). By combining the results obtained in the two buffers used, the amount of protein-bound detergent can be determined [15, 19]. According to the evaluation (which applied a value of 0.751 ml/g for the protein's $\bar{v}$, as corrected analogous to the case of buffer B), the cytochrome bc_1 complex binds (0.86 ± 0.12) g DDM/g protein. This figure is within the range found with other intrinsic membrane proteins [18, 19, 32, 33].

Discussion

As shown above, the cytochrome bc_1 complex from *P. denitrificans*, solubilized and studied in buffers containing low concentrations of DDM, is a stable dimer. Since solubilization by suitable nonionic detergents is assumed not to interfere with the native association

behavior of a membrane protein [15, 34], the complex is most probably dimeric in the membrane of *P. denitrificans* as well. It is noteworthy, however, that in the present study among the nonionic detergents tested only DDM seemed to be mild enough to preserve the protein's native structure.

In previous reports, a number of isolated mitochondrial cytochrome bc_1 complexes were shown, by X-ray crystallography, to have a dimeric structure [8–10]. In addition, strong evidence in one case [13] and less conclusive evidence in another [14] indicate that the much smaller bacterial cytochrome bc_1 complexes may possess the same oligomeric state. The present work has added, through rigorous analysis, another example of this type of association.

The mitochondrial complexes contain up to eight additional subunits and their assembly process must be quite distinct from that of the bacterial ones. Nevertheless it is a generally accepted idea that the bacterial enzyme is functionally similar to the eukaryotic complex III and represents a "minimal version" that contains only the three catalytic subunits [4–7]. This work clearly provides further evidence that the dimeric state is an indispensable feature for a functional bc_1 complex. This is rationalized by the observation of an intertwined dimer: the crystal structures of mitochondrial complexes [9–11] show that a given Rieske protein (ISP) interacts with cytochromes b and c_1 from one monomer, but is attached to the other monomer by its N-terminal membrane anchor; thus, its linkage to the adjacent monomer may be a prerequisite to allow the controlled motion of the ISP and provide the necessary geometry. So far, any cooperative effect between the two monomers seems unlikely; therefore, an oligomeric state does not appear to be of any regulatory significance, but is rather of structural and mechanistic significance.

In the present work, emphasis was put not only on determining the state of association of a bacterial cytochrome bc_1 complex but also on a technical aspect, namely the elaboration of a modified version for matching the density of DDM during ultracentrifugal analysis. DDM undoubtedly is an unfavorable detergent for density-matching, owing to its high density requiring extremely high concentrations of additives [21, 22]. This, in turn, causes uncertainties in the partial specific volume of the proteins, which, together with the high solvent density, leads to large uncertainties in the crucial term $(1 - \bar{v}\rho_0)$. We therefore suggested that, if possible, DDM should be replaced, for ultracentrifuge runs, by a detergent of lower density (1.0–1.1 g/ml) [21]. The present study has shown, however, that some membrane proteins are not compatible with such detergents. In this case, reducing the additive concentration necessary for density-matching by the combined use of additive and D_2O (or, expense allowing, by $D_2^{18}O$ [19, 29]) seems to be a useful approach.

Acknowledgements We are grateful to H.-W. Müller for supplying the cytochrome c oxidase, to C. Tziatzios for help with some experiments and for discussion, to P. Hellwig for data on the H-D exchange in the cytochrome bc₁ complex, and to H. Durchschlag for calculating the partial specific volume of the heme group. O.A. thanks the Graduiertenkolleg "Proteinstrukturen, Dynamik und Funktion" and B.L. the Deutsche Forschungsgemeinschaft (SFB 472) for financial support.

References

1. Brandt U, Trumpower BL (1994) CRC Crit Rev Biochem 29:165
2. Trumpower BL, Gennis RB (1994) Annu Rev Biochem 63:675
3. Crofts AR, Berry EA (1998) Curr Opinion Struct Biol 8:501
4. Yang XH, Trumpower BL (1986) J Biol Chem 261:12282
5. Kurowski B, Ludwig B (1987) J Biol Chem 262:13805
6. Trumpower BL (1990) Microbiol Rev 54:101
7. Gennis RB, Barquera B, Hacker B, Van Doren SR, Arnaud S, Crofts AR, Davidson E, Gray KA, Daldal F (1993) J Bioenerg Biomembr 25:195
8. Xia D, Yu C, Kim H, Xia J, Kachurin AM, Zhang L, Yu L, Deisenhofer J (1997) Science 277:60
9. Zhang Z, Huang L, Shulmeister VM, Chi Y-I, Kim KK, Hung L-W, Crofts AR, Berry EA, Kim S-H (1998) Nature 392:677
10. Iwata S, Lee JW, Okada K, Lee JK, Iwata M, Rasmussen B, Link TA, Ramaswamy S, Jap BK (1998) Science 281:64
11. Hunte C, Koepke J, Lange C, Roßmanith T, Michel H (2000) Structure 8:669
12. Darrouzet E, Valkova-Valchanova M, Moser CC, Dutton PL, Daldal F (2000) Proc Natl Acad Sci USA 97:4567
13. Sone N, Takagi T (1990) Biochim Biophys Acta 1020:207
14. Montoya G, te Kaat K, Rodgers S, Nitschke W, Sinning I (1999) Eur J Biochem 259:709
15. Tanford C, Reynolds JA (1976) Biochim Biophys Acta 457:133
16. Schröter T, Hatzfeld OM, Gemeinhardt S, Korn M, Ludwig B, Link TA (1998) Eur J Biochem 255:100
17. Pfitzner U, Odenwald A, Ostermann T, Weingard L, Ludwig B, Richter OM (1998) J Bioenerg Biomembr 30:85
18. Friesen RHE, Knol J, Poolman B (2000) J Biol Chem 275:33527
19. Schubert D, Tziatzios C, van den Broek JA, Schuck P, Germeroth L, Michel H (1994) Prog Colloid Polym Sci 94:14
20. Schubert D, Schuck P (1991) Prog Colloid Polym Sci 86:12
21. Mayer G, Ludwig B, Müller H-W, van den Broek JA, Friesen RHE, Schubert D (1999) Prog Colloid Polym Sci 113:176
22. Lustig A, Engel A, Tsiotis G, Landau EM, Baschong W (2000) Biochim Biophys Acta 1464:199
23. Hendler RW, Pardhasaradhi K, Reynafarje B, Ludwig B (1991) Biophys J 60:415
24. Schuck P (1994) Prog Colloid Polym Sci 94:1
25. Schuck P, Legrum B, Passow H, Schubert D (1995) Eur J Biochem 230:806
26. Cohn EJ, Edsall JT (1943) In: Cohn EJ, Edsall JT (eds) Proteins, amino acids and peptides as ions and dipolar ions. Hafner, New York, pp 370–381
27. Durchschlag H (1986) In: Hinz H-J (ed) Thermodynamic data for biochemistry and biotechnology. Springer, Berlin Heidelberg New York, pp 45–128
28. Durchschlag H, Zipper P (1994) Prog Colloid Polym Sci 94:20
29. Edelstein SJ, Schachman HK (1967) J Biol Chem 242:306
30. Korn M (1999) PhD thesis. Johann Wolfgang Goethe-Universität, Frankfurt
31. Mayer G (2001) PhD thesis. Johann Wolfgang Goethe-Universität, Frankfurt
32. Tziatzios C, Schuck P, Schubert D, Tsiotis G (1994) Z Naturforsch C Biosci 49:220
33. Hendriks J, Warne A, Gohlke U, Haltia T, Ludovici C, Lübben M, Saraste M (1998) Biochemistry 37:13102
34. Helenius A, Simons K (1975) Biochim Biophys Acta 415:29

Progr Colloid Polym Sci (2002) 119:84–91
© Springer-Verlag 2002

Christine Wandrey
Gabriela Grigorescu
David Hunkeler

Study of polyelectrolyte complex formation applying the synthetic boundary technique of analytical ultracentrifugation

C. Wandrey (✉) · G. Grigorescu
D. Hunkeler
Laboratory of Polyelectrolytes
and Biomacromolecules
Department of Chemistry
Swiss Federal Institute of Technology
CH-1015 Lausanne, Switzerland
e-mail: christine.wandrey@epfl.ch
Tel.: +41-21-6933672
Fax: +41-21-6935690

Abstract Polyelectrolyte complex membrane formation has been studied, on-line, using the synthetic boundary technique of an Optima XL-I analytical ultracentrifuge. This recent method, established by the first author, has now been evaluated concerning its sensitivity to monitor differences in the membrane formation if the membrane components and experimental conditions are varied. For this purpose the membrane formation of sodium alginate with three polycations, poly(vinylamine)hydrochloride, poly(L-lysine)hydrochloride, and chitosan oligomers, was examined under various experimental conditions. In addition, two approaches have been developed and applied to evaluate the membrane formation process and the membrane quality on the basis of the experimentally received radial absorption scans. The first approach generalizes the membrane growth and its direction, resulting in a time-dependent description of the membrane geometry. The second one considers the membrane thickness as a function of time and might, therefore, serve to calculate the growth velocity or shrinking of the membrane, but not the directions of growth or shrinking. The investigations revealed a sufficient sensitivity of the method leading to the conclusion that the membranes formed in this study differ in thickness, symmetry, homogeneity, and the kinetics of formation. Consequently, the demonstrated sensitivity of this ultracentrifugation technique offers novel extended possibilities for basic studies of the polyelectrolyte complex membrane formation. Moreover, the goal-directed optimization of such membranes for practical applications can be supported.

Key words Analytical ultracentrifugation ·
Polyelectrolyte complexes ·
Synthetic boundary ·
Membrane formation

Introduction

Polyelectrolyte complexes (PEC) result from Coulombic interaction between oppositely charged polyelectrolytes (PEL) corresponding to anionic PEL + cationic PEL→ PEC + salt. Generally, the spontaneously created PEC precipitate in aqueous solution without any geometrically well defined macroscopic structure formation. Under certain conditions, water-soluble complexes are possible. This is the case if the molar mass of the components is very different and the reaction is terminated before a one-to-one stoichiometry of charge neutralization is reached. On the other hand, a limited number of polyanion/polycation pairs have been identified to form gel-like networks, which do not precipitate. By applying defined reaction conditions, membrane formation can occur.

PEC membranes (PECM) require for their formation that the contact surface of the PEL solutions is well defined, permitting the undisturbed interpenetration of the reaction partners. The contact surface may be flat (layering of two solutions), or spherical (dropping of one solution into a second one). Macromolecular and structural characteristics, as well as the reaction conditions, determine the ultimate properties of the membranes formed. The macromolecular characteristics include the chain length and the chain length distribution of the polyion, as well as the chain architecture of the polyion. The structural characteristics are the chemical structure of the monomer units, the type of the ionic group on the polymer chain, the position of the ionic group (integrated into the polymer backbone or pendent), the charge density, and the charge density distribution along the polymer chain (charge distance). The reaction conditions, which can be varied, are the concentration of the PEL solutions, the ratio of the reacting partners, the temperature, the reaction time, the pH of the solution, or the ionic strength of both PEL solutions.

Membranes obtained by means of electrostatic interaction of oppositely charged polymers have recently attracted increasingly scientific and commercial attention [1, 2]. This is primarily due to the broad potential applications of the resulting permselective "devices". Such materials may be employed in a variety of geometries, including as components of flat sheets or as self-contained hollow fibers, macrocapsules, or microcapsules. Numerous combinations of anionic and cationic PEL have been examined with regard to their ability to form stable membranes having well-defined characteristics; however, only a relatively few PEL pairs have been identified for this purpose [1, 3]. For some applications, multicomponent PEL networks provide favorable properties [4].

The key parameters of the PECM are the permeability, the mechanical stability, elasticity, and durability. Certain applications require, in addition, biocompatibility [5]. Whereas the network density of the complexes and the strength of the electrostatic interaction primarily regulate the former properties, the biocompatibility is influenced by a multitude of parameters, including, in addition to the chemistry, the surface topography and surface charges. Relatively minor changes in the reaction conditions can alter the final membrane properties dramatically; therefore, the adjustment of the desired parameters is time-consuming. Furthermore, PECM are not static materials and all their properties are equilibrium characteristics influenced by the environment's electrochemical characteristics.

Various analytical methods have been employed to study PEC [1, 6]. The complex stoichiometry, the final membrane structure, and surface characteristics have been investigated extensively by scattering methods, microscopic techniques, and surface scanning methods [1, 6–8]. Analytical ultracentrifugation has been employed for the characterization of PEC though not in terms of membrane formation [9].

The prediction of network properties, as a function of macromolecular, physicochemical, structural, and environmental conditions, has not yet been elaborated. A multitude of polymer combinations have been screened to find optimum pairs [10]; however, no approach exists which tries to find general correlation between the physicochemical parameters and the membrane properties. Such screening suffers from the fact that no experimental technique was used which can follow the membrane formation on-line and monitor the various stages of the process, including formation, equilibration, and stabilization/destabilization.

Recently, it has been proposed to study the membrane formation by applying the synthetic boundary experiment and the optical system of an analytical ultracentrifuge (AUC). Both the experimental principle and the reproducibility have been published including a discussion of the advantages and limitations of the method. Sodium alginate (SA) and chitosan were employed as a model system [11–13]. The aim of the studies presented here is to demonstrate the sensitivity of the method, in particular, its suitability to monitor differences in the membrane formation if various polycations are combined with alginate under different reaction conditions. The conclusions of the screening will serve as a basis for further systematic studies to establish a general correlation between the physicochemical characteristics of PEL and the membrane properties under defined reaction conditions.

Experimental

Equipment and method

All experiments were performed using an Optima XL-I AUC (Beckman, Palo Alto, Calif., USA) equipped with two integrated detection systems, scanning UV–vis, and Rayleigh interference optics. To investigate the PECM formation one of the principal techniques of the AUC, the synthetic boundary experiment, was modified. The less viscous polyion solution possessing, in addition, a lower density was filled into the solvent sector and layered over the higher viscous polyion solution in the solution sector. The horizontal projection (view from above) of a synthetic boundary centerpiece and the principle of the experiment are illustrated in Fig. 1. Additionally, the appropriate absorbance plot is presented schematically in this figure. The membrane signal results from the absorption when the light beam passes the 1.2-mm-thick PEL network. It has to be mentioned that the absorbance signal can be caused by the membrane turbidity, the absorption of the polymer components, or a combination of both. A comprehensive description of the experimental details has been published recently [13].

Materials

SA (Keltone HV, lot 54650A, Kelco/NutraSweet, San Diego, Calif. USA) represented the polyanion for all the experiments. The cationic components for the membrane formation were poly (L-lysine)hydrochloride (PLL), poly(vinylamine)hydrochloride (PVAm) and chitosan oligomers (OC). PLL possessing an average molar mass of 18,000 g/mol was purchased from Sigma (Buchs,

86

Switzerland). PVAm was synthesized by a two-steps process [14]. The first step was the radical polymerization of *N*-vinylformamide dissolved in methanol, using 2,2′-azobis(isobutyronitrile) as initiator and 2-mercaptobutanol as a chain-transfer agent to regulate the molar mass. The polymerization was carried out at 60 °C for 24 h. In a second step the stoichiometric base hydrolysis of poly (*N*-vinylformamide) to PVAm took place. During this time the polymer was dissolved in 1.7 M NaOH and heated at 80 °C for 6 h. The polymer solution was acidified by concentrated HCl solution. Precipitated PVAm was washed with methanol and dried in a vacuum. The product was desalted by ultrafiltration. The products of each step were confirmed by NMR and IR analysis (data not shown). The PVAm used in this study had an average molar mass of 17,500 g/mol. Both the controlled radical degradation process of chitosan (radical degradation [15] of raw material

from E-055, Hutchinson/McNeil Int., Philadelphia, Pa. USA) and the characterization of the resulting oligomers (OC) have been described previously [16].

Description of the experiments

A standardized nomenclature will be employed for the chemistry of the membrane forming systems. The chemistry is designated as polyion solution in the solution sector//polyion solution in the solvent sector. If the polyion solution consists of more than one component which participates in the membrane formation the following style is used: A/B//C/D. Here A and B represent the components of the solution placed into the solution sector of the synthetic boundary cell, while C and D are the components in the solvent sector which have to pass the capillary. Low-molar-mass salts or other moieties added to the solutions which do not directly participate in the PECM formation, for example, sodium chloride, are considered as components of the solvent.

Three polycations were selected to study their PECM formation with SA. These were PLL, PVAm, and OC. The solution compositions and the conditions for the screening experiments are summarized in Table 1. The presence of CaCl₂ in some of the polycation solutions caused rapid gelation of sodium alginate. The calcium alginate gel was virtually clear and contributed only minimally to the absorbance signal. The presence of CaCl₂ is of interest for some practical applications.

All the experiments were performed at 5000 rpm after approximately 20 min at 1000 rpm. The scans were taken at wavelengths between 320 and 380 nm with a step width of 10 μm in a radial range where the membrane formation occurred. The filling volumes were 100 μl in the solution sector and 420 μl in the solvent sector. They can be modified if the variation of the polyanion/polycation ratio is intended. To ensure the reproducibility each experiment was carried out at least in duplicate. The layering velocity was controlled by the interference scans and could be reproduced within ±20%.

Results and discussion

The screening revealed differences in the membrane formation by varying the chemical structure of the polycation component. These differences are demonstrated in Fig. 2 for SA//PVAm/CaCl₂ and SA//PLL/CaCl₂ (systems 2 and 3a in Table 1).

By maintaining constant the type and concentration of the polyanion, as well as the experimental conditions, and varying only the polycation solution chemistry different membrane structures were obtained and monitored by the absorption scans. In addition, both PVAm and PLL had comparable average molar masses, 17,500 and 18,000 g/mol, respectively. The thinner SA//PVAm/

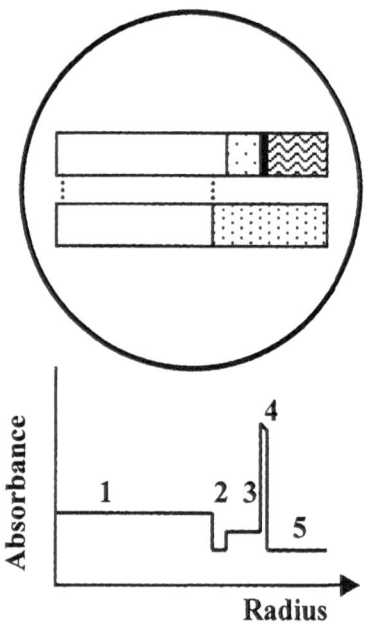

Fig. 1 Principle of the membrane formation experiment in a synthetic boundary cell. The position of the components and the absorption scan are shown for the case when the layering is completed; however, the layered polyelectrolyte (*PEL*) is not yet totally consumed for the membrane formation. The five ranges monitor the different signals of air/air (*1*), air/PEL solution in the solvent sector (*2*), PEL solution which passed the capillary/PEL solution in the solvent sector (*3*), PEL complex membrane/PEL solution in the solvent sector (*4*), and PEL solution in the solution sector/PEL solution in the solvent sector (*5*). The PEL in the solvent sector absorbs, whereas the PEL in the solution sector does not. (This holds for the case of modified chitosan as a polycation. For the other polycations in this study the membrane signal results only from the turbidity.)

Table 1 Composition of the polyanion and polycation solutions: sodium alginate (*SA*), poly(vinylamine)hydrochloride (*PVAm*), poly(L-lysine)hydrochloride (*PLL*), chitosan oligomers (*OC*)

No.	Chemistry	Polyanion solutions	Polycation solutions	Figs.
1	SA//oligochitosan	a) 1% SA in 0.9% NaCl	a) 1% OC in 0.9% NaCl	6a, 7a, 8a
		b) 2% SA in 0.9% NaCl	b) 1% OC in 0.9% NaCl	6b, 7b, 8b
2	SA//PVAm/CaCl₂	2% SA in 0.9% NaCl	0.5% PVAm/0.5% CaCl₂ in 0.45% NaCl	2a, 4a, 5a
3	SA//PLL/CaCl₂	a) 2% SA in 0.9% NaCl	a) 0.5% PLL in 1.22% CaCl₂	2b, 3, 4b, 5b
		b) 1.2% SA in 0.9% NaCl	b) 0.5% PLL in 1.22% CaCl₂	9, 10

$CaCl_2$ membrane was found to be more uniform than that produced with $SA//PLL/CaCl_2$. The two polycations differ in the charge density resulting from the chemical structure, with PVAm possessing the higher charge density. The AUC scans correspond to the microscopic imaging. One of the photomicrographs, which represents $SA//PLL/CaCl_2$, is presented in Fig. 3. It visualizes a well-defined membrane surface (left side of the absorbance scan in Fig. 2b). The heterogeneity, i.e., the less dense structure in the direction to the cell bottom, is clearly visible (right side of the absorbance scan in Fig. 2b).

For a more general evaluation of the membrane formation, two approaches were employed: these are presented in Figs. 4 and 5. Figure 4 shows, as an example for the first approach, that evaluation, where the membrane growth and the growth direction are plotted from the data of Fig. 2a and b. By taking the radial position at various absorbance units and plotting these

data as a function of time, a time-dependent membrane geometry can be estimated. As is evident, such plots provide supplementary information to the absorbance scan plots. They generalize the membrane growth and its direction.

The superposition of the filled symbols in Fig. 4a and b indicates a well-defined surface at the contact interface of the polyanion and polycation solutions for both

Fig. 3 Photomicrograph of the $SA//PLL/CaCl_2$ membrane (Fig. 2b)

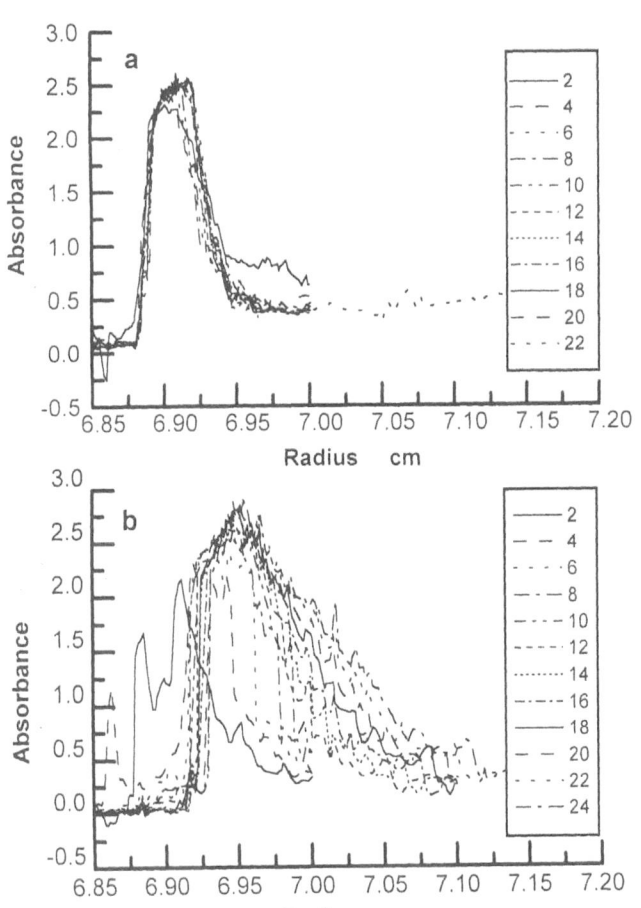

Fig. 2 Analytical ultracentrifuge (*AUC*) absorbance scans (320 nm) for **a** sodium alginate (*SA*)//poly(vinylamine)hydrochloride (*PVAm*)/$CaCl_2$ (2 in Table 1) and **b** SA//poly(L-lysine)hydrochloride (*PLL*)/$CaCl_2$ (3a in Table 1). 5,000 rpm, 20 °C, scan delay 2 min

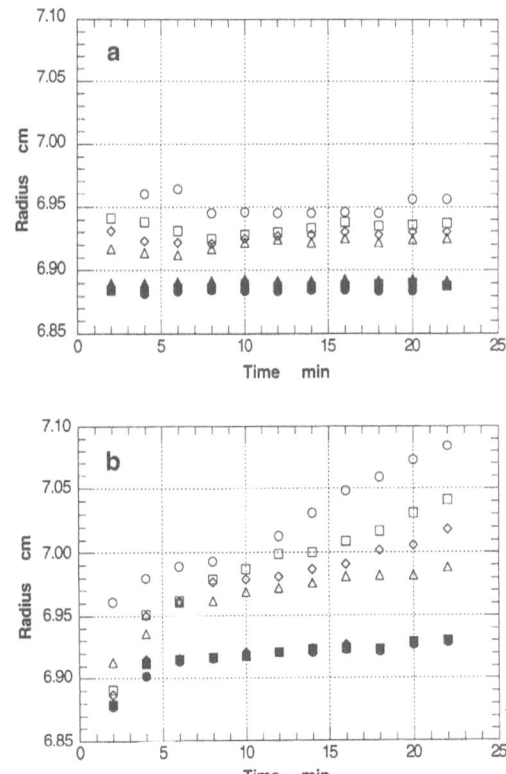

Fig. 4 Evaluation of the network formation. The membrane growth is shown as a function of the reaction time for **a** $SA//PVAm/CaCl_2$ and **b** $SA//PLL/CaCl_2$. The *open symbols* represent the membrane extension in the cell bottom direction, whereas the *filled symbols* represent extension in the cell top direction for the appropriate absorbance units: 0.5 (*circles*), 1.0 (*squares*), 1.5 (*diamonds*), and 2.0 (*triangles*). The data are taken from Fig. 2a and b

chemistries, though a difference is clearly visible in the other (bottom) direction. For SA//PVAm/CaCl$_2$ (system 2 in Table 1) very slow homogeneous membrane growth occurs in the bottom direction, whereas the growth is faster for SA//PLL/CaCl$_2$ (system 3a in Table 1) resulting in a heterogeneous structure (see also Fig. 3). Moreover, a slow shift of the membrane surface in the bottom direction is monitored in Fig. 4b, probably caused by a stronger shrinking which accompanies the network formation and is expected to be more evident for thicker membranes. The limited membrane growth, in the case of Fig. 2a, corresponds to a low permeability (cutoff 70,000 for dextran), which was reported for this PEL combination. The membrane growth is terminated early. Neither the SA nor the PVAm can pass the membrane for further network formation. This results in a stop of the membrane growth.

The second approach of evaluation in Fig. 5 visualizes the increase of the membrane thickness at various selected absorbance units as a function of time. From the slopes a growth velocity may be calculated.

The plots in Fig. 5 indeed quantify the total membrane growth or shrinking, though they do not provide information about the growth direction or the possible shrinking direction as can be concluded from the curves in Fig. 4. Figure 5a monitors general membrane shrinking during the first 6–8 min, indicated by the negative slop for all absorbance intensities. The shrinking is followed by slight membrane growth (approximately between 6 and 16 min) before no further changes are detected. Another formation process can be concluded from Fig. 5b. Here the network extension increases continuously. The exception is the curve for 0.5 absorbance units. By comparison with Fig. 2a some inaccuracy may result from the scan noise close to the baseline.

From another experimental series, Fig. 6 demonstrates the influence of the SA concentration on the membrane formation for SA//OC (systems 1a and 1b in Table 1).

In general, the SA//OC system does not form stable membranes under physiological conditions [16]. The membrane dissolves after a certain time, detected by the membrane broadening in the top direction in Fig. 7. It is faster and more significant for the lower alginate concentration in Fig. 7a. The membrane formation differences monitored should not be the result of the different densities of the 1 and 2% SA solution. In both cases the membrane grows in two directions, not only in

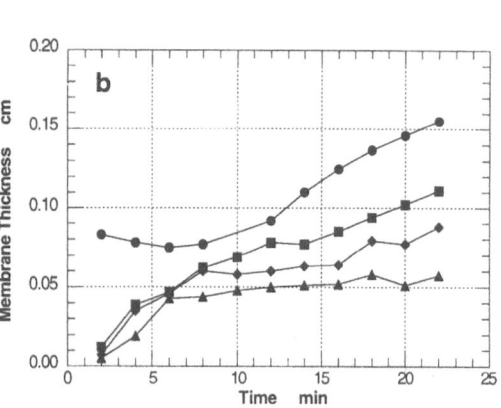

Fig. 5 The membrane thickness as a function of time for **a** SA//PVAm/CaCl$_2$ and **b** SA//PLL/CaCl$_2$. Absorbance units: 0.5 (*circles*), 1.0 (*squares*), 1.5 (*diamonds*), and 2.0 (*triangles*). The data are taken from Fig. 2a and b

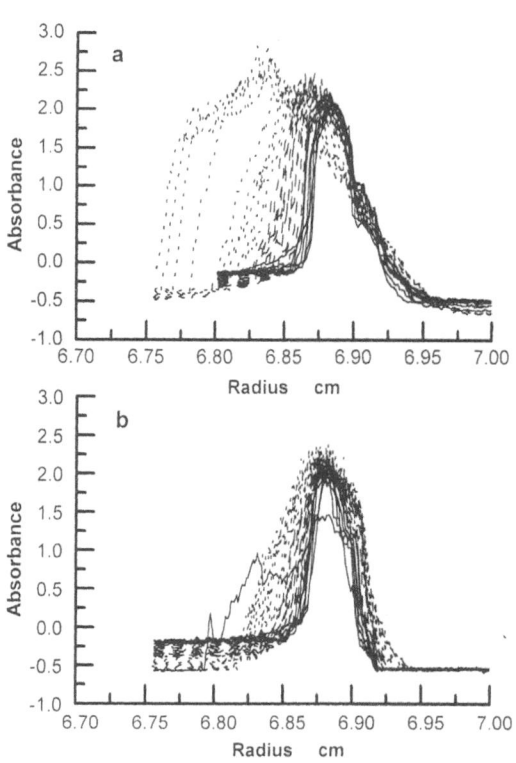

Fig. 6 AUC absorbance scans for SA//chitosan oligomers (*OC*) (380 nm): **a** 1% SA (1a in Table 1) and **b** 2% SA (1b in Table 1). 5,000 rpm, 20 °C, scan delay 2 min (—), 5 min (- - -), 20 min (·····)

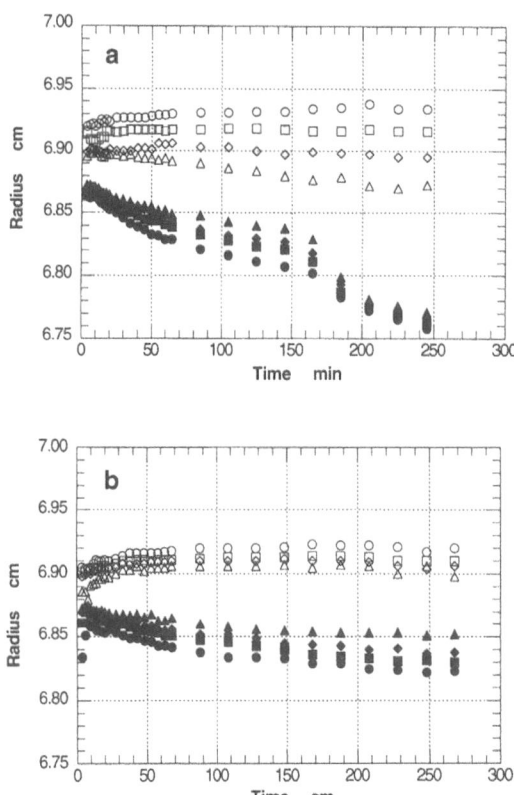

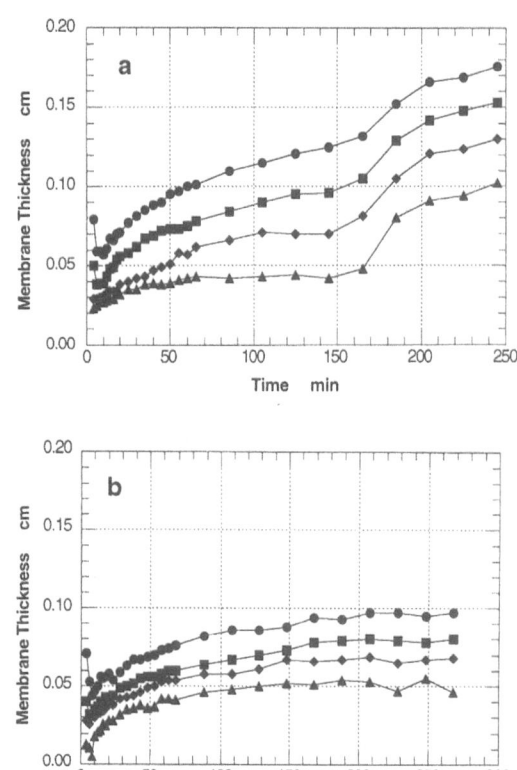

Fig. 7 Evaluation of the network formation. Membrane growth as a function of the reaction time for SA//OC: **a** 1% SA and **b** 2% SA. The *open symbols* represent the membrane extension in the cell bottom direction, whereas the *filled symbols* represent extension in the cell top direction for the appropriate absorbance units: 0.0 (*circles*), 0.5 (*squares*), 1.0 (*diamonds*), and 1.5 (*triangles*). The data are taken from Fig. 6a and b

Fig. 8 The Membrane thickness as a function of time for SA//OC: **a** 1% SA and **b** 2% SA. Absorbance units: 0.0 (*circles*), 0.5 (*squares*), 1.0 (*diamonds*), and 1.5 (*triangles*). The data are taken from Fig. 6a and b

the bottom direction. In general, the dissolution in the top direction was independent of the SA concentration, also for other experiments which are not shown here. By comparison with Fig. 4, a less defined surface at the contact interface of the polyanion and polycation solutions may be concluded, indicated by differences in the radial positions at various absorbance units. The two chemistries in Fig. 4 revealed a superposition.

The network dimensions with the same approach as in Fig. 5 are quantified in Fig. 8. The sudden increase of the expansion velocity after a certain time (during this experiment in Fig. 8a after 160 min) could be reproduced, though it is not, as of yet, understood.

The last experiment presented here gives a selected example for the resolution of the optical detection to monitor the membrane heterogeneity. By varying the experimental conditions for SA//PLL/CaCl$_2$ (system 3b in Table 1) the absorbance scans plotted in Fig. 9 were obtained. The scans monitor a high absorbance in the center of the membrane, with lower absorbance values in the top and bottom directions. A well-defined sharp

surface was not developed on the top or on the bottom of the interface where the polyion solutions came into contact. The absorbance signal corresponds to the photomicrograph in Fig. 10, which shows clearly a more compact zone in the middle of the polymer network, demonstrating the sensitivity of the optical detection by absorbance measurements.

In addition to the absorbance scans, all the membrane formation experiments were scanned by the interference optics. Since their interpretation cannot be unambiguously correlated to the membrane dimensions, they are not discussed here; however, these scans were used to calculate the layering velocity as was described previously [13] to ensure similar reaction conditions for the experiments compared.

By comparing the absorption scans and the evaluation approaches obtained from the various chemical compositions and experimental conditions, it becomes obvious that different network structures result. They differ in dimensions, symmetry, homogeneity, and formation kinetics as summarized in Table 2.

Concentration variation, within a reasonable range, seems to influence the kinetics but not the general network type. Membranes from PLL were always more

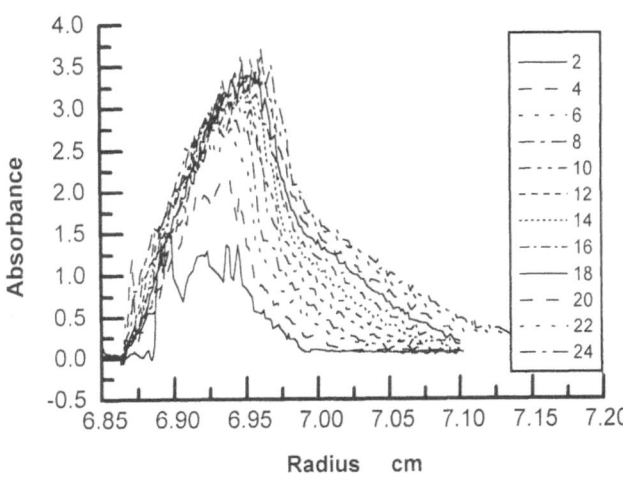

Fig. 9 AUC absorbance scans (320 nm) for SA//PLL/CaCl$_2$ (3b in Table 1). 5,000 rpm, 20°C, scan delay 2 min

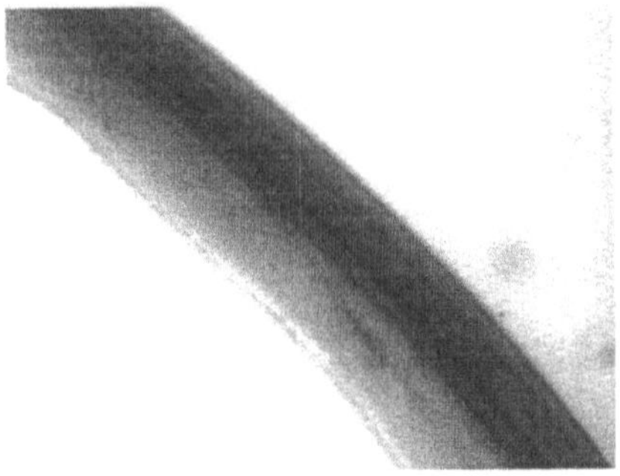

Fig. 10 Photomicrograph of the SA//PLL/CaCl$_2$ membrane (Fig. 9). *Right*: top direction, *left*: bottom direction

Table 2 Comparison of the membranes formed by reacting SA with various polycations. Absorbance units (*AU*)

Chemistry	Membrane dimensions (after 20 min)	Membrane surface (after 20 min)	Membrane symmetry	Formation kinetics
SA//OC	"Thin": a) 0.7 mm; b) 0.6 mm at 0.0 AU	Defined: shift 0.2 mm/1.5 AU	Symmetric, concentration influence	Continuous growth followed by destabilization
SA//PVAm/CaCl$_2$	"Thin": 0.5 mm at 0.5 AU	Well-defined: shift 0.1 mm/1.5 AU	Symmetric	Limited growth
SA//PLL/CaCl$_2$	"Thick": 1.5 mm at 0.5 AU	Well-defined: shift 0.1 mm/1.5 AU, increases with decreasing concentration	Nonsymmetric	Continuous growth

heterogeneous and thick, whereas those from PVAm were thin and compact for the molar masses investigated.

Conclusions and outlook

The screening results presented demonstrate the sensitivity of the optical detection during AUC synthetic boundary experiments for the study of PECM formation at a flat interface of polyanion and polycation solutions. The method monitors significant differences in the membrane formation kinetics, the membrane formation direction, and membrane homogeneity. These differences primarily result from the chemical structure of the polycation employed, its charge density, and the concentration of the participating PEL. The two evaluation approaches which were established allow a more detailed comparison of various experiments and general conclusions concerning the membrane structure and membrane formation kinetics, including growth and shrinking.

To further increase the information value of the synthetic boundary experiments, systematic studies are required to generalize, and exploit, all the information which can be taken from the AUC scans. Such studies have to include the comparison of the scans with information obtained from imaging methods. Furthermore, the detection by the Schlieren optics may provide good resolution for the steep refractive index increment in the membrane range. Other types of centerpieces, such as modified band-forming or one-sector centerpieces, might allow the variation of the layering velocity. The application of centerpieces with a shorter optical path length will also be favorable.

The development of an advanced scan evaluation is intended which will contribute to establishing qualitative and quantitative relationships for the influence of the PEL characteristics and the experimental conditions on the membrane formation process and the membrane properties. Finally, such relationships will be used to optimize PECM.

Acknowledgements The authors thank A. Bartkowiak, Polymer Institute of the Technical University of Szczecin, Poland, who provided the chitosan oligomers. The Swiss Federal Institute of Technology and the Commission for Technology and Innovation are gratefully acknowledged for financial support.

References

1. Dautzenberg H, Jaeger W, Kötz J, Philipp B, Seidel C, Stscherbina D (1994) Polyelectrolytes: formation, characterization and application. Hanser, Munich
2. Hunkeler D (1997) Trends Polym Sci 5:286
3. Gander B, Blanco-Prieto MJ, Thomasin C, Wandrey C, Hunkeler D (2001) In: Swarbrick J, Boylan JC (eds) Encyclopedia of pharmaceutical technology, 2nd edn. Dekker, New York, pp 481
4. Prokop A, Hunkeler D, Powers AC, Whitesell RR, Wang TG (1998) Adv Polym Sci 136:53
5. Angelova N, Hunkeler D (1999) Trends Biotechnol 17:409
6. Philipp B, Dautzenberg H, Linow KJ, Kötz J, Dawydoff W (1989) Prog Polym Sci 14:91
7. Arabi H, Hashemi SA, Fooladi M (1996) J Microencapsul 13:527
8. Lukas J, Richau K, Schwarz HH, Paul D (1997) J Membr Sci 131:39
9. Karibyants N, Dautzenberg H, Cölfen H (1997) Macromolecules 30:7803
10. Prokop A, Hunkeler D, Haralson M, Dimari S, Wang TG (1998) Adv Polym Sci 136:1
11. Wandrey C, Bartkowiak A, Hunkeler D (2000) In: Transactions of the 6th world biomaterials congress, May 15–20, 2000, Hawaii, USA, vol II. Society for Biomaterials,USA, p 893
12. Wandrey C (2000) Polym News 25:299
13. Wandrey C, Bartkowiak A (2001) Colloids Surf A 180:141
14. Van Treslong Bloys CJ, Morra CFH (1975) J R Neth Chem Soc 94:101
15. Mullagaliev IR (1995) Dokl Akad Nauk 345:199
16. Bartkowiak A, Hunkeler D (2000) Chem Mater 12:206

Progr Colloid Polym Sci (2002) 119:92–100
© Springer-Verlag 2002

D. Kisters
A. Straatmann
W. Borchard

The sedimentation behaviour of gels – the generalised Lamm's differential equation

D. Kisters
DGF Stoess AG
Gammelsbacher Strasse 2, 69412 Eberbach,
Germany

A. Straatmann · W. Borchard (✉)
Institut für Physikalische und
Theoretische Chemie der
Gerhard-Mercator-Universität Duisburg
47048 Duisburg, Germany

Abstract Lamm's differential equation for polymer solutions is a well-known tool to describe the time-dependent change of the polymer concentration as a function of diffusion and sedimentation of the polymer component in a centrifugal field. Based on the phenomenological equations, which describe the flux of the polymer component as the sum of the products of the phenomenological coefficients and the generalised specific forces, this was derived. The phenomenological definition of the flux is valid for polymer solutions as well as for gels. It is shown that the phenomenological equation in the case of gels leads to a "generalised Lamm differential equation", which describes the change of the concentration with respect to the time as a function of the diffusion, the sedimentation and a so-called "elastically active coefficient". All changes between the sedimentation behaviour of a polymer in solution and a polymer in a swollen elastic network can be attributed to this coefficient. Ultra-centrifugal measurements of gelatin gels (physical networks) yield the ratio of the diffusion coefficient and the sedimentation coefficient of the polymer at different overall concentrations of the gels. From the literature values of non-cross-linked and cross-linked polystyrene in chlorobenzene the ratio of the mobilities and sedimentation coefficients are calculated and discussed.

Key words Gelatin · Polystyrene · Sedimentation · Lamm's differential equation for gels · Elastically active coefficient

Theory

All formulations of the fluxes or flux densities in the theoretical part of this work refer to a reference frame in the laboratory coordinate system. This means that the plane of the reference frame is fixed at a defined position in the measuring cell perpendicular to the flux and this position does not change during the experiments.

According to Fujita [1, 2] the flux of the component k can be related to this frame, if the partial specific volumes of the components k do not change as a function of the pressure.

General

The continuity equation, applied to the conservation of mass, leads to an equation which describes the time-dependent change of the partial density, ρ_k, of component k as a function of the flux density of component k in the radial direction [1, 2]. This equation is valid for polymer solutions as well as for gels in a centrifugal field.

$$\frac{\partial \rho_k}{\partial t} + \frac{1}{r}\frac{\partial}{\partial r}(rJ_k) = 0 \tag{1}$$

where J_k is the flux density of component k in mass per time (seconds) and cross section centimetres squared,

ρ_k is the partial density (mass concentration) of component k and r is the radial distance from the axis of rotation.

J_k describes the mass of component k which is transported during the time unit through an area of 1 cm^2; therefore, the unit of the flux density is grams per centimetre squared seconds.

The generalised specific force, $\tilde{X}_k$, which acts on the mass of component k in the direction of the centrifugal field, can be described for gels as well as for polymer solutions as the sum of the gradient of the applied centrifugal field and the gradient of the specific chemical potential of component k. This force is positive in the direction of the external field. Disregarding Coriolis forces, this leads to

$$\tilde{X}_k = \omega^2 r - \left(\frac{\partial \tilde{\mu}_k}{\partial r}\right)_T = (1 - \tilde{V}_k \rho)\omega^2 r$$
$$- \sum_{j=2}^{N} \left(\frac{\partial \tilde{\mu}_k}{\partial \rho_j}\right)_{T,P\rho_{m \neq j}} \frac{d\rho_j}{dr}; \qquad (2)$$
$$k = 1, 2, 3 \ldots N; \; m = 2, 3, \ldots N \text{ and } m \neq j$$

where ω is the angular velocity, $\frac{\partial \tilde{\mu}_k}{\partial P} = \tilde{V}_k$ is the partial specific volume of component k and ρ is the density of the solution or the gel.

The component index k extends over all components. In a system with N components there are N–1 independent concentration gradients and fluxes.

The flux density of component k can be described by means of the phenomenological equations and is a homogeneous linear function of the generalised specific forces of all N components.

$$J_k = \sum_{i=2}^{N} L_{ki} \tilde{X}_i; \quad k = 1, 2 \ldots N \qquad (3)$$

The flux densities and the forces are conjugated if the following condition is fulfilled:

$$\Phi_S = \frac{1}{T} \sum_{k=1}^{N} J_k \tilde{X}_k . \qquad (4)$$

Φ_S is defined as a function which describes the entropy production per volume unit in an infinitesimal volume element.

For a set of conjugated fluxes and forces Onsager's reciprocal relation is valid. This means that the matrix which is built up by the phenomenological coefficients L_{ki} is symmetrical.

$$L_{ki} = L_{ik} \quad \text{for} \quad k \neq i \qquad (5)$$

All the fluxes which are derived in the following sections are conjugated.

Polymer solutions

Combination of Eqs. (2) and (3) leads to

$$J_k = \sum_{i=2}^{N} L_{ki} \left[(1 - \tilde{V}_i \rho)\omega^2 r - \sum_{j=2}^{N} \left(\frac{\partial \tilde{\mu}_i}{\partial \rho_j}\right)_{T,P,\rho_{m \neq j}} \frac{d\rho_j}{dr} \right] \qquad (6)$$

or

$$J_k = \sum_{i=2}^{N} (L_{ki})(1 - \tilde{V}_i \rho)\omega^2 r - \sum_{i=2}^{N} (L_{ki}) \sum_{j=2}^{N} \left(\frac{\partial \tilde{\mu}_i}{\partial \rho_j}\right)_{T,P,\rho_{m \neq j}} \frac{d\rho_j}{dr} . \qquad (7)$$

From the definition of the phenomenological sedimentation coefficient and the diffusion coefficient (Eqs. 8, 9),

$$s_k = \frac{1}{\rho_k} \sum_{i=2}^{N} (L_{ki})(1 - \tilde{V}_i \rho) \qquad (8)$$

and

$$D_{kj} = \sum_{i=2}^{N} L_{ki} \left(\frac{\partial \tilde{\mu}_j}{\partial \rho_j}\right)_{T,P,\rho_{m \neq j}} , \qquad (9)$$

the following equation which describes the flux of the component k in a solution can be derived:

$$J_k = s_k \rho_k \omega^2 r - \sum_{j=2}^{N} D_{kj} \frac{d\rho_j}{dr} \quad \text{with } k = 2, 3, \ldots N . \qquad (10)$$

The concentration gradients of both components in a binary system ($N = 2$; index 1 = solvent; index 2 = polymer) cannot change independently. If this and the Gibbs–Duhem relation are taken into consideration, it becomes apparent that for a binary system only a single diffusion coefficient, D, exists whereas two sedimentation coefficients, namely s_1 and s_2, are present.

$$J_2 = s_2 \rho_2 \omega^2 r - D \frac{d\rho_2}{dr} \qquad (11)$$

The combination of the Eqs. (1) and (11) leads to the well-known Lamm's differential equation for polymer solutions:

$$\frac{\partial \rho_2}{\partial t} = \frac{1}{r} \frac{\partial}{\partial r} \left(rD \frac{d\rho_2}{dr} - s_2 \rho_2 \omega^2 r^2 \right) . \qquad (12)$$

The difference of the specific chemical potential of the polymer component, $\Delta \tilde{\mu}_2$ can be written as

$$\Delta \tilde{\mu}_2 = \tilde{\mu}_2 - \tilde{\mu}_2^0 = \frac{RT}{M_2} \ln \gamma_2 \rho_2 , \qquad (13)$$

where $\tilde{\mu}_2^0$ is the specific chemical potential of the pure polymer in a reference state, $\tilde{\mu}_2$ is the specific chemical potential of the polymer in solution, γ_2 is the activity coefficient on the mass concentration scale, ρ_2 is the partial density of the polymer (mass concentration) in grams per cubic centimeter and M_2 is the molar mass of the polymer component.

Insertion of Eq. (13) in Eq. (9) results in an equation for a binary solution:

$$D = L_{22} \frac{RT}{M_2} \left(\frac{\partial(\ln \gamma_2 \rho_2)}{\partial \rho_2} \right)_{T,P,\rho_1}$$
$$= L_{22} \frac{RT}{M_2 \rho_2} \left[1 + \rho_2 \left(\frac{\partial(\ln \gamma_2)}{\partial \rho_2} \right)_{T,P,\rho_1} \right] . \tag{14}$$

If Eq. (8) is taken into consideration, the phenomenological coefficient can be eliminated in Eq. (14), leading to

$$M_2 = \frac{RT s_2^s \left[1 + \rho_2 (\partial \ln \gamma_2 / \partial \rho_2)_{T,P,\rho_1} \right]}{D^s (1 - \tilde{V}_2 \rho)} . \tag{15}$$

In the case of an infinite dilution the so-called Svedberg equation is obtained from Eq. (15). This equation describes the molar mass of the polymer component as a function of the ratio of the sedimentation and diffusion coefficients [1–3].

For $\lim_{\rho_2 \to 0} \gamma_2 = 1$ it results in $M_2 = \frac{s_2^0}{D^0} \frac{RT}{(1 - \tilde{V}_2^0 \rho_{01})}$,

$$\tag{16}$$

where s_2^0 is the sedimentation coefficient of the polymer for infinite dilution, D^0 is the mutual diffusion coefficient for infinite dilution, $\tilde{V}_2^0$ is the the partial specific volume of the polymer component at infinite dilution and ρ_{01} is the density of the pure solvent.

Gels

According to Flory the difference of the specific chemical potential of the polymer component in a network can be described as the sum of two different terms $\tilde{\mu}_{k,\mathrm{mi}}$ and $\tilde{\mu}_{k,\mathrm{el}}$ [4]. The difference between the specific chemical potential of component k and the specific chemical potential of the pure component k is summarised in the term $\tilde{\mu}_{k,\mathrm{mi}}$ [4, 5]; therefore, this term is equal to the statement for the specific chemical potential of the dissolved polymer in Eq. (13).

$$\Delta \tilde{\mu}_k = \tilde{\mu}_k - \tilde{\mu}_k^0 = \tilde{\mu}_{k,\mathrm{mi}} + \tilde{\mu}_{k,\mathrm{el}} . \tag{17}$$

The second term, $\tilde{\mu}_{k_{\mathrm{el}}}$, describes the part of the specific chemical potential which can be assigned to the isotropic deformation of the network, for example, by swelling in a swelling agent. Insertion of Eq. (17) in the equation for the flux density, J_k, leads to

$$J_k = \sum_{i=2}^N L_{kj} \left[(1 - \tilde{V}_i \rho) \omega^2 r \right.$$
$$\left. - \sum_{j=2}^N \left(\frac{\partial(\partial \tilde{\mu}_{k,\mathrm{mi}} + \partial \tilde{\mu}_{k,\mathrm{el}})}{\partial \rho_j} \right)_{T,P,\rho_m} \frac{\mathrm{d}\rho_j}{\mathrm{d}r} \right] \tag{18}$$

and

$$J_k = \sum_{i=2}^N L_{kj} \left[(1 - \tilde{V}_i \rho) \omega^2 r - \sum_{j=2}^N \left(\frac{\partial \tilde{\mu}_{k,\mathrm{mi}}}{\partial \rho_j} \right)_{T,P,\rho_m} \frac{\mathrm{d}\rho_j}{\mathrm{d}r} \right.$$
$$\left. - \sum_{j=2}^N \left(\frac{\partial \tilde{\mu}_{k,\mathrm{el}}}{\partial \rho_j} \right)_{T,P,\rho_m} \frac{\mathrm{d}\rho_j}{\mathrm{d}r} \right] . \tag{19}$$

The first two terms in Eq. (19) resemble the terms which were derived in Eq. (7) for polymer solutions. By using the definitions of the sedimentation coefficient (Eq. 8) and the diffusion coefficient (Eq. 9), Eq. (18) can be written as

$$J_k = s_k^g \rho_k \omega^2 r - \sum_{j=2}^N D_{kj}^g \frac{\mathrm{d}\rho_j}{\mathrm{d}r}$$
$$- \sum_{i=2}^N L_{ki}^g \sum_{j=2}^N \left(\frac{\partial \tilde{\mu}_{k,\mathrm{el}}}{\partial \rho_j} \right)_{T,P,\rho_m} \frac{\mathrm{d}\rho_j}{\mathrm{d}r} , \tag{20}$$

where s_k^g is the sedimentation coefficient of component k in the gel.

A comparison between the resulting equation of the flux J_k in the case of a polymer network (Eq. 20) and the corresponding equation for a polymer solution (Eq. 7) shows clearly that all changes in the flux for a transition from a solution to a gel must be assigned only to the last term in Eq. (20).

In analogy to the diffusion coefficient in Eq. (9), an "elastically active coefficient, E" can be defined. This coefficient describes, together with the corresponding phenomenological coefficient, the part of the flux density which is related to an elastic deformation of the network in the centrifugal field. In analogy to the sedimentation and diffusion velocity the elastically active coefficient E_{kj} describes a deformation velocity in the unit of the diffusion coefficient D (centimetres squared per second).

$$E_{kj} = \sum_{i=2}^N L_{ki} \left(\frac{\partial \tilde{\mu}_{k,\mathrm{el}}}{\partial \rho_j} \right)_{T,P,\rho_m} \tag{21}$$

On the basis of the same considerations that were made for the diffusion coefficient, it follows that for a binary system (index 2 = swollen cross-linked polymer) also only a single elastically active coefficient, E, exists,

$$J_2 = s_2^g \rho_2 \omega^2 r - \frac{\mathrm{d}\rho_2}{\mathrm{d}r} (D + E)^g , \tag{22}$$

where the index g refers to the gel. It can be seen that by the notation $(D+E)^g$ diffusion and elastic deformation are unambiguously coupled in gels; this used to be named the diffusion coefficient of a gel D^g [6–8].

The combination of Eq. (22) and the continuity equation results in the "generalised Lamm differential equation for gels":

$$\frac{\partial \rho_2}{\partial t} = \frac{1}{r}\frac{\partial}{\partial r}\left[r\frac{d\rho_2}{dr}(D+E)^g - s_2^g \rho_2 \omega^2 r^2 \right] \ . \tag{23}$$

For large changes in the composition it is known that both quantities D and E are dependent on concentration, where the concentration dependence of both is different. If the sedimentation–diffusion–deformation, or for short the sedimentation–diffusion, equilibrium is reached, the flux J_k of component k has to vanish, because the polymer concentration is constant with respect to the time at every point in the centrifugal cell. The application of this condition to the equilibrium leads to an equation which can be used to calculate $(D+E)^g$ at equilibrium.

$$(D+E)^g = \frac{s_2^g \rho_2 \omega^2 r}{(d\rho_2/dr)} \tag{24}$$

In this equation the transport quantities are interrelated and belong to the same concentration. If s_2^s and D^s of solutions (index s) are extrapolated to a reference state this can also be done with s_2^g and $(D+E)^g$. By means of Eq. (24) it is possible to determine $(D+E)^g/s_2^g$ of a gel at equilibrium.

For a comparison of the transport coefficients in solutions and gels it is absolutely necessary to measure the values at the same composition of the systems. It is really not possible to compare extrapolated values, for example, at infinite dilution because for gels this state is situated in the unstable phase region.

The diffusion coefficients for systems which show miscibility gaps become very small if the system is very close to the miscibility gap. Rehage [6] showed this for various polymer solutions.

This can be explained by using the thermodynamic description of diffusion, where it is shown that the diffusion coefficient is the product of the mobility u_k of the polymer in a solution or a gel and the thermodynamic factor [5, 7, 9]:

$$D^s = u_k^s \rho_k \left(\frac{\partial \tilde{\mu}_{k,\text{mi}}^s}{\partial \rho_k}\right)_{T,P} \tag{25}$$

and

$$(D+E)^g = u_k^g \rho_k \left(\frac{\partial \tilde{\mu}_k^g}{\partial \rho_k}\right)_{T,P} \ . \tag{26}$$

The mobilities u_k^s and u_k^g describe the mean velocity of component k with respect to the unit of the driving force.

Owing to the fact that the thermodynamic factor, consisting of the product of ρ_k and the derivative of the specific chemical potential $\tilde{\mu}_k^s$ and $\tilde{\mu}_k^g$ with respect to the partial density ρ_k of component k becomes zero at the stability-limiting curve and that the mobility cannot become infinitely large in this case, the diffusion coefficient has to vanish if the system reaches the spinodal curve.

The determination of thermodynamic properties in the gelatin/water system clearly showed that physical gels at 5 °C are very close to a miscibility gap where a highly swollen gel coexists with a slightly swollen one [10, 12]. The latter state corresponds to the collapsed state. In such a system there are at least three stability-limiting curves in different concentration ranges.

The relations between the single fluxes in the centrifugal field

As shown in the last subsection, the flux J_k of component k in a centrifugal field can be described as the sum of three fluxes which are caused by sedimentation, diffusion and elastic deformation of the swollen network:

$$J_k = J_{k,\text{sed}} + J_{k,\text{dif}} + J_{k,\text{el}} \ . \tag{27}$$

The most general case is the result given by Eq. (24), where all contributions have to be considered. In this case $(D+E)^g/s_2^g$ can be determined experimentally. Theoretically, different combinations of vanishing fluxes J_k can be realised in an analytical ultracentrifuge only by the variation of the rotational speed

1. $J_{k,\text{el}} = 0$. This case is equal to an equilibrium run of a polymer solution in which the flux due to the sedimentation is compensated by the flux due to the diffusion. The Svedberg equation (Eq. 16) was derived under this condition for the equilibrium state ($J_k = 0$).
2. $J_{k,\text{sed}} = 0$. After the formation of a concentration gradient in a gel at a defined rotational speed, the flux due to the sedimentation has to vanish if the centrifuge is stopped. In this case the whole flux J_k is described only as the sum of the fluxes $J_{k,\text{el}}$ and $J_{k,\text{dif}}$. This experiment can be performed without the use of a centrifuge, if the rubberelastic cross-linked polymer is brought into contact with pure solvent [6]. It has to be mentioned that the flux $J_{k,\text{el}}$ due to deformation opposes the flux $J_{k,\text{mix}}$, so $(D+E)^g$ is expected to be smaller than D^s in a non-cross-linked system.
3. $J_{k,\text{sed}} = 0$ and $J_{k,\text{el}} = 0$. In analogy to case 2 this experiment can be realised by using a polymer solution instead of a gel [8]. After the formation of a concentration gradient the centrifuge is stopped and the flux due to the sedimentation vanishes. Also the

synthetic boundary experiment at low rotational speeds fulfils these conditions.

4. $J_{k,\text{dif}} = 0$ and $J_{k,\text{el}} = 0$. This case corresponds to a velocity run of a polymer solution for which the diffusion of the polymer molecules can be neglected in comparison to the sedimentation.

5. $J_{k,\text{dif}} = 0$ and $J_{k,\text{sed}} = 0$. If it were possible to deform a dry rubberelastic polymer network in a way that finally a compression equilibrium of the polymer is achieved, in this case the elastically active coefficient describes the rate of compression due to the applied external force.

Experimental

Preparation of the gelatin gels

For all the experiments which were evaluated for this work a type B gelatin from DGF Stoess, Germany, was used. The gelatin gels were prepared in the way that the gelatin was dissolved in water at 45 °C for about 10 h after a preswelling time of 30 min. Afterwards 150 μl Raschit solution (5% 4-chloro-3methylphenol in methanol) was added to prevent bacterial growth in the gels during the measurements.

The hot solutions were filled in the measuring cells with a syringe and the cells were screwed and stored for 48 h at the presumed measuring temperature of 5 °C. Seven gels in a concentration range from 2 to 14.4% were prepared using this procedure.

Gels with concentrations higher than 14.4% could not be investigated because the samples were too turbid to detect the gradient of concentration in the whole measuring cell during the runs.

Equilibrium runs for the gelatin gels

All the equilibrium experiments were carried out at 5 °C and a velocity of 6,000 rpm. The schlieren pictures were taken once or twice a day until the shape of the schlieren curve did not change for at least 4 days. At this time the equilibrium state was reached. These conditions were chosen because the amount of soluble gelatin parts can be minimised at this temperature [13].

Results

The transport coefficients of aqueous gelatin gels

By use of Eq. (24) the ratio $(D+E)^g/s_2^g$ was calculated from the concentration gradient at sedimentation–diffusion equilibrium; this is obtained directly from the schlieren pictures. The concentration distance curve is shown for a gel with an initial polymer weight fraction of $w_{2,0} = 0.076$ in Fig. 1 for a temperature of 5 °C at 6,000 rpm. More details are given in the literature [10].

After 28 days the sedimentation–diffusion equilibrium was established and this was also obtained independently starting from 12,000 rpm, demonstrating the pathway independence. After correction for the dilution effect the partial density–distance curve, $\rho_2(r)$, was obtained, from which the gradient of the osmotically active pressure, π_s^*, was calculated assuming additivity of the volumes [12]. The relation between the thermodynamic factor of a gel and the derivative of the osmotically active pressure with respect to the partial density of gelatin, ρ_2, is given by

$$\left[\left(\rho_2 \frac{\partial \tilde{\mu}_2}{\partial \rho_2}\right)_{T,P}\right]^g = \tilde{V}_{01}\rho_1 \frac{d\pi_s^*}{d\rho_2} \ , \tag{28}$$

where the left-hand side is the thermodynamic factor of a gel (g), ρ_1 is the partial density of the solvent

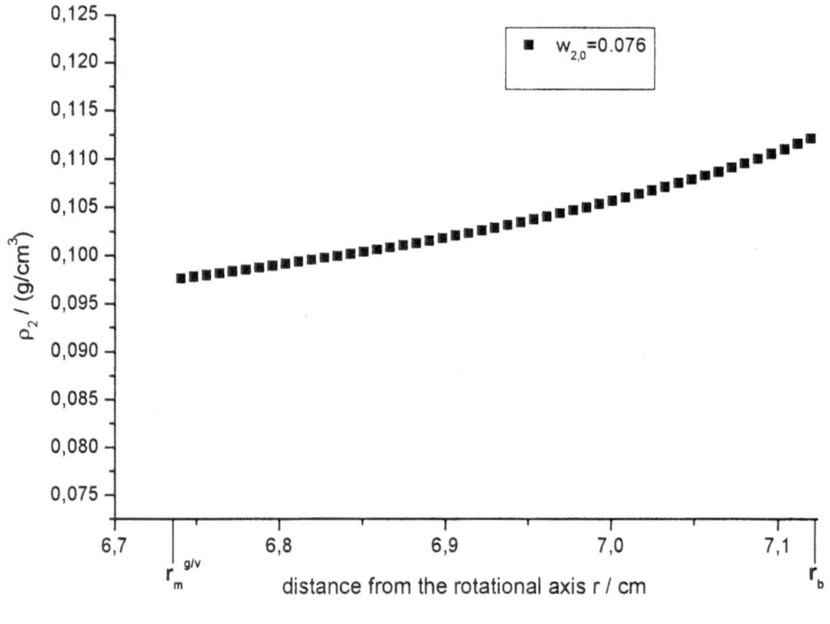

Fig. 1 Partial density, ρ_2, versus radial distance, r, at equilibrium for a gelatin gel with an initial weight fraction of $w_{2,0} = 0.076$ at 5 °C and a rotational speed of 6,000 rpm

and $\tilde{V}_{01}$ is the specific volume of the pure solvent [10, 11].

The thermodynamic factor of a gel with an initial weight fraction of gelatin of $w_2 = 0.076$ is plotted in Fig. 2 versus the partial density of the gelatin in the gel for a temperature of 5 °C. It can be seen that the thermodynamic factor decreases with increasing polymer concentration and reaches very low values in comparison to the polystyrene (PS)/cyclohexane demixing system [14]. This decrease is probably due to the presence of a miscibility gap of the gel at higher concentrations. Such a decrease of the thermodynamic factor between a maximum and a minimum was reported [14] for the non-cross-linked PS/cyclohexane system at 28 and 30 °C, where the upper critical temperature was found at 26.3 °C. In reality an S-shaped concentration dependence of the thermodynamic factor, F_{th}^s, is observed. At low concentrations F_{th}^s increases going through a maximum. After passing a minimum it increases again. As the thermodynamic factor of a gel or a solution has to vanish at the spinodal curve the diffusion coefficient also has to vanish if the mobility remains finite.

The thermodynamic factor of the gel is related by Eq. (26) to the ratio $(D + E)^g/s_2^g$. The mobility, u_2^g, of the gelatin inside the gel is given by the ratio of the sedimentation coefficient of gelatin s_2^g and the buoyancy term of gelatin in parentheses in Eq. (29):

$$\left[s_2^g \left/ (1 - \tilde{V}_2 \rho) \right. \right] = u_2^g \tag{29}$$

Therefore an additional independent measurement of $(D + E)^g$ or of s_2^g is necessary to determine both transport coefficients of a gel.

Comparison of diffusion and sedimentation coefficients of gels and solutions from literature data

Rehage and coworkers [6–8, 14] investigated diffusion in gels and solutions over a large concentration range, thus enabling a comparison of the transport coefficients in an overlapping sector of compositions. Also sedimentation coefficients of the polymer component in solutions have been calculated by use of Eqs. (25) and (29) [7, 14], where the thermodynamic factor was calculated by use of the results of osmotic measurements [15, 17].

In the case of gels the diffusion coefficients of PS have been measured at 30.4 °C in different solvents [6]. Rehage showed that by using the Boltzmann substitution the locus at constant concentration is proportional to the square root of the time. These findings demonstrated, at least for the good solvents chloroform, toluene and chlorobenzene, that the diffusion was normal. In this case the diffusion coefficient is a function of concentration at constant values of pressure and temperature. Later it was shown that the diffusion in the polyethylene/o-xylene system is anormal because the superposition of different processes may occur, leading to non-Fickian diffusion [18, 19].

If the mutual diffusion is normal then the sedimentation of the polymer component in a gel can also be treated by consideration of Eq. (29). We chose the PS/chlorobenzene system at 30 °C because the diffusion coefficients of gels and solutions and also the thermodynamic data could be obtained from the literature. The sedimentation coefficient of the non-cross-linked PS was calculated from the diffusion coefficient [14] and the thermodynamic factor was calculated from the results of osmotic measurements [16]. It has to be mentioned that

Fig. 2 Thermodynamic factor, F_{th}^g, versus ρ_2 for a gelatin gel with an initial weight fraction of $w_{2,0} = 0.076$

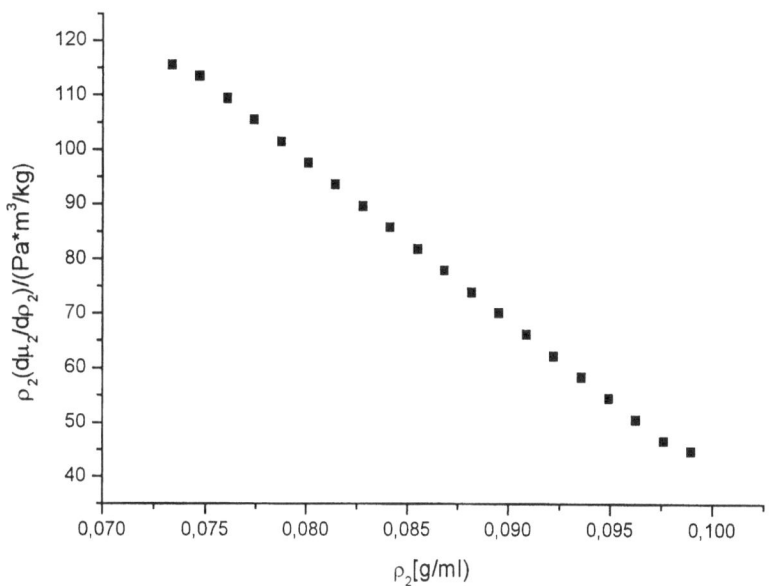

s_2 is negative in the PS/chlorobenzene system, which means that the polymer floats.

The same procedure was applied to the gels by taking the values of the diffusion coefficients from Ref. [6]. The thermodynamic excess values have been reported where swelling and deswelling measurements have been performed [20, 21]. The excess functions of solutions and gels have been evaluated by the Flory–Huggins theories for solutions and gels; therefore, the chemical potentials of the solvent were calculated by use of the interaction parameters for solutions, χ_x^s, and for gels, χ_x^g, on base mole fraction scale (index x).

In this work we calculated the ratio of the sedimentation coefficients of PS in the gels (g) and solutions (s), namely s_2^g/s_2^s, from the ratio of the diffusion coefficients, D^g/D^s, and the ratio of the thermodynamic factors, F_{th}^g/F_{th}^s, for the same compositions of the gels and solutions because all these values and also the ratios are dependent on concentration. Although the diffusion coefficients of the cross-linked PS/chlorobenzene system are known for 30.4 °C we calculated all the other values for 30 °C, neglecting this small temperature difference, which is in the scattering range of the experimental values.

From Eqs. (25), (26) and (29) we get for $\alpha = g, s$

$$\frac{(D+E)^g}{D^s} = \frac{s_2^g \rho_2^g (\partial \tilde{\mu}_2/\partial \rho_2)^g (1 - \tilde{V}_2 \rho)}{s_2^s \rho_2^s (\partial \tilde{\mu}_2/\partial \rho_2)^s (1 - \tilde{V}_2 \rho)} \ , \tag{30}$$

where the thermodynamic factors are given by

$$F_{th}^g = \left[\rho_2 (\partial \tilde{\mu}_2/\partial \rho_2)_{T,P} \right]^g \tag{31}$$

and

$$F_{th}^s = \left[\rho_2 (\partial \tilde{\mu}_2/\partial \rho_2)_{T,P} \right]^s . \tag{32}$$

For the same composition of gel and solution and the same values of the buoyancy terms Eq. (30) reduces to

$$\frac{(D+E)^g}{D^s} = \frac{s_2^g}{s_2^s} \cdot \frac{F_{th}^g}{F_{th}^s} \ . \tag{33}$$

With the Gibbs–Duhem equation (Eq. 34) at constant pressure and temperature

$$w_1 \left(\frac{\partial \tilde{\mu}_1}{\partial w_2} \right)_{T,P} + w_2 \left(\frac{\partial \tilde{\mu}_2}{\partial w_2} \right)_{T,P} = 0 \tag{34}$$

we obtain

$$\left(\frac{\partial \tilde{\mu}_2}{\partial w_2} \right)_{T,P} = -\frac{w_1}{w_2} \left(\frac{\partial \tilde{\mu}_1}{\partial w_2} \right)_{T,P} = \frac{w_1}{w_2} \left(\frac{\partial \tilde{\mu}_1}{\partial w_1} \right)_{T,P} , \tag{35}$$

where $(\partial \tilde{\mu}_1/\partial w_1)_{T,P}$ is related to the derivative of the partial molar chemical potential of the solvent, μ_1, by

$$M_1 \left(\frac{\partial \tilde{\mu}_1}{\partial w_1} \right)_{T,P} = \left(\frac{\partial \mu_1}{\partial w_1} \right)_{T,P} , \tag{36}$$

where M_1 is the molar mass of the solvent or swelling agent. From Eqs. (31), (32) and (35) we get for a gel or solution (index α)

$$\begin{aligned} F_{th}^\alpha &= \rho_2^\alpha \left(\frac{\partial \tilde{\mu}_2}{\partial w_2} \cdot \frac{\partial w_2}{\partial x_2^*} \cdot \frac{\partial x_2^*}{\partial \rho_2} \right)^\alpha \\ &= \left(\rho_2 \frac{w_1}{w_2} \right)^\alpha \left(\frac{\partial \tilde{\mu}_1}{\partial w_1} \cdot \frac{\partial w_1}{\partial x_1^*} \cdot \frac{\partial x_1^*}{\partial \rho_2} \right)^\alpha , \end{aligned} \tag{37}$$

where x_1^* and x_2^* are the base molar fractions of component 1, the solvent, and component 2, the polymer, and are defined by

$$x_1^* \equiv \frac{n_1}{n_1 + n_2^*} \quad \text{and} \quad x_2^* = 1 - x_1^* , \tag{38}$$

where n_2^* is the amount of matter of the base unit of the polymer with the molar mass M_0. Neglecting endgroup effects of the non-cross-linked polymer and the different molar masses of the cross-linker it is absolutely necessary for the calculations of the base molar fractions to start from the relation $n_2^* = m_2/M_0$ as well as $n_1 = m_1/M_1$. If n_2^* for solutions or gels is expressed by rn_2 or zv^*, where r is the degree of polymerization, n_2 the amount of matter of the non-cross-linked polymer, z the degree of polymerisation between cross-links and v^* the amount of matter of the network chains, one term in the Gibbs–Duhem equation is weighted by the factor r [14] or the factor z [29], making a comparison between the properties of gels and solutions impossible.

The Flory–Huggins expression for a polymer solution which these days is named the Flory–Huggins–van Santen–Staverman equation reads using base molar fractions [4, 22–24]

$$\Delta \mu_1^s = RT \left[\ln x_1^* + \left(1 - \frac{1}{r} \right) x_2^* + \chi_x^s x_2^{*2} \right] . \tag{39}$$

$\Delta \mu_1^s$ is the difference between the chemical potential of the solvent in the solution and the pure solvent and depends on concentration, the degree of polymerisation of the non-cross-linked polymer and the interaction parameter in the base molar scale χ_x^s, where R and T have their usual meaning.

For a gel Flory obtained [4]

$$\Delta \mu_1^g = RT \left(\ln x_1^* + x_2^* + \chi_x^g x_2^{*2} + \frac{x_2^{*1/3}}{z} \right) . \tag{40}$$

$\Delta \mu_1^g$ is the difference between the chemical potential of the swelling agent in the gel and the pure solvent. χ_x^g is the interaction parameter of the cross-linked polymer/swelling agent system. By comparing Eqs. (39) and (40) it is seen that Flory considered the molar mass of the cross-linked polymer to be infinitely high. The last term in Eq. (40) describes the part of the chemical potential of the swelling agent due to the isotropic elastic deformation by swelling. If Eqs. (39) and (40) are applied to real systems the quantities r, χ_x^s, z and χ_x^g may be regarded as empirical parameters or functions but describing the systems quantitatively for the conditions given. Therefore a refinement of the theoretical approach will perhaps lead to a better prediction of the behaviour, but, in principle, not to completely different results.

On differentiating Eqs. (39) and (40) with respect to the mass fraction, w_1, and considering Eq. (37) for $\alpha = g$ and $\alpha = s$ the ratio of the thermodynamic factors of a gel and a solution is given by

$$\frac{F_{th}^g}{F_{th}^s} = \frac{\frac{1}{x_1^*} - 1 - 2\chi_x^g x_2^* - \frac{1}{3z}x_2^{*-2/3}}{\frac{1}{x_1^*} - 1 - 2\chi_x^s x_2^*} = F_I \ , \tag{41}$$

where χ_x^g and χ_x^s are assumed to be independent of concentration and the degrees of polymerisation, which means that χ_x^g is independent of z and χ_x^s is independent of r. The expressions RT, M_1, $(\partial x_1^*/\partial w_1)$ and $(\partial x_1^*/\partial \rho_2)$ cancel because they are the same for a gel and a solution at equal composition. It is seen from Eq. (41) that the fourth term in the nominator goes to $-\infty$ for $x_2^* \to 0$. At a certain value between the minimum value for the swelling equilibrium $x_2^* = x_{2,s}^*$ and $x_2^* = 0$ the nominator and with it F_I become zero. This is the stability limit at $x_2^* = x_{2,st}^*$. This is typical for gels. It has been found that to a good approximation the relation $\chi_x^g = \chi_x^s$ holds [21]. The z value has been calculated from the maximum of the partial density, ρ_1, in the swelling equilibrium [6] and is given by $z = 175,46$. The χ_x values at 30 °C are $\chi_x^s = 0.4292$ [16] and $\chi_x^g = 0.4376$ [17, 21]; they differ only by 0.0084, which confirms the statement mentioned.

The ratios of the diffusion coefficients $D^g = (D + E)^g$ taken from Ref. [6] and D^s taken from Ref. [14] are plotted in the overlapping concentration range $0.18 < x_2^* < 0.4$ in Fig. 3. It is seen that this ratio is smaller than 1, indicating that the diffusion process in gels is slower than in solutions. This makes sense and is the same finding as that of Rehage and Ernst [7] for the PS/toluene system; they found that the difference between both coefficients becomes smaller at higher concentrations. At the concentration $\rho_1 = 0.6$ both values seem to be identical. At a very high polymer concentration below the freeze-in concentration the diffusion coefficients of gels and solutions become very similar because the prevailing interaction of non-cross-linked polymer and network chains can be made responsible for this behaviour. At a low solvent concentration the polymer chain segments of the same or different chains are very close together in a highly coiled form. Thus, the fixing of the network chains by the network junction points is unimportant. In the range $x_2^* < 0.4$ it is mainly the contribution of the elastically active coefficient, E, which is negative. Also in this system it leads to lower values of the diffusion coefficients for gels in comparison to those of the solutions. The ratio of the thermodynamic factors or F_I is plotted in Fig. 3 from $x_{2,st}^* = 0.055$ up to $x_2^* = 1$, neglecting the presence of glassy states in PS. It can clearly be seen that the concentration dependence of the ratio D^g/D^s is a little stronger than that of the function F_I. The concentration dependence of the ratio of s_2^g/s_2^s which was calculated by use of Eq. (33) is less steep than that of D^g/D^s. These findings show that the

Fig. 3 Different functions of the polystyrene/chlorobenzene system versus x_2^* at 30 °C (see text)

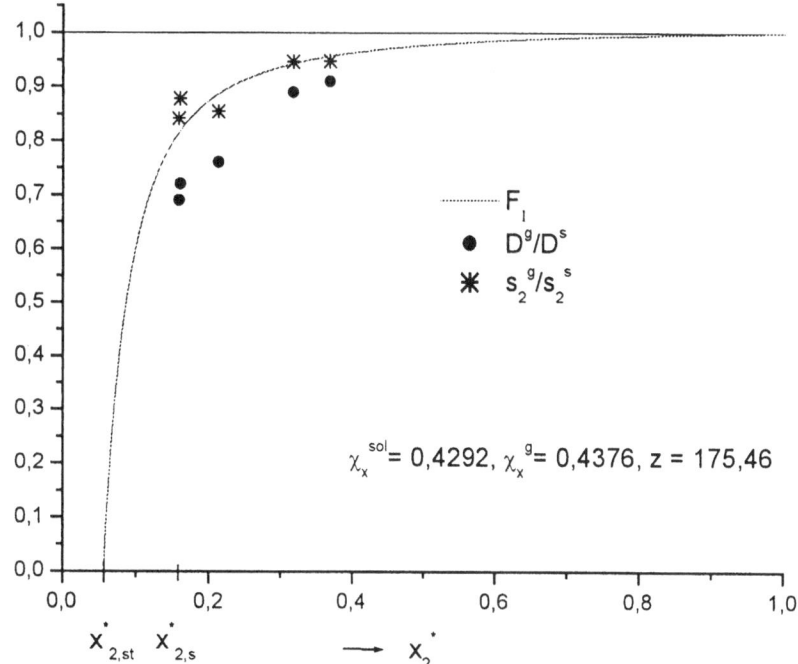

sedimentation (or flotation) of the network component is different from that of the polymer in a solution. The main difference is due to the presence of a network structure leading to the relative movement of a swelling agent in a system of meshes rather than the relative transport in a non-cross-linked system.

Conclusions

- The insertion of the equation for the specific chemical potential of the polymer in a network leads to a so-called "generalised Lamm's differential equation", which represents an extension of the well-known equation for polymer solutions.
- All changes for a transition from a polymer solution to a polymer network can be attributed to an elastically active coefficient, E, which describes a deformation velocity with the same unit as the diffusion coefficient (centimetres squared per second).

- It was shown that the sum of the elastically active coefficient and the diffusion coefficient in the gel, $(D + E)^g$, can be described as the product of the mobility, u_2^g, of the polymer in the centrifugal field and a thermodynamic factor F_{th}^g.
- Literature values of the thermodynamic properties of gels and solutions of the PS/chlorobenzene system made it possible to calculate the thermodynamic factors of the polymer.
- From reported diffusion coefficients of the system the ratio of the sedimentation coefficients of the polymer in gel and solution was calculated for the same compositions. The sedimentation of the cross-linked polymer with respect to that of the non-cross-linked polymer is slower.

Acknowledgements We thank the Deutsche Forschungsgemeinschaft and the Duisburger Universitätsgesellschaft for financial support.

References

1. Fujita H (1962) Mathematical theory of sedimentation analysis. Academic, New York
2. Fujita H (1975) Foundations of ultracentrifugal analysis. Wiley, New York
3. Svedberg T, Pedersen KO (1949) In: Ostwald W (ed) Handbuch der Kolloidwissenschaft. Steinkopff, Dresden, p 5
4. Flory PJ (1953) Principles of polymer chemistry. Cornell University Press, Ithaca
5. Haase R (1956) Thermodynamik der Mischphasen. Springer, Berlin Heidelberg New York
6. Rehage G (1959) Symposium über Makromoleküle, Wiesbaden, II A15
7. Rehage G, Ernst O (1964) DECHEMA Monogr 49:157–179
8. Rehage G, Ernst O (1964) Kolloid Z Z Polym 197:64–70
9. Haase R (1963) Thermodynamik der irreversiblen Prozesse. Steinkopf, Darmstadt
10. Kisters D (2001) Doctoral thesis. Duisburg
11. Borchard W (1991) Prog Colloid Polym Sci 86:84–91
12. Borchard W, Cölfen H, Kisters D, Straatmann A (2001) Prog Colloid Polym Sci 119:101–112
13. Johnson P, Metcalfe JC (1967) Eur Polym J 3:423–447
14. Ernst O (1962) Doctoral thesis. Aachen
15. Rehage G, Meys H (1958): J Polym Sci 30:271
16. Palmen HJ (1960) Diploma thesis. Aachen
17. Rehage G (1960) Habilitation thesis. Aachen
18. Fuhrmann J, Rehage G (1969) Z Phys Chem NF 67:291
19. Fuhrmann J, Driemeyer M, Rehage G (1970) Ber Bunsenges Phys Chem 74:842
20. Rehage G (1964) Kolloid Z Z Polym 194:16
21. Rehage G (1964) Kolloid Z Z Polym 196:97
22. Flory PJ (1942) J Chem Phys 10:51
23. Huggins ML (1943) Ann NY Acad Sci 44:431
24. Staverman AJ, van Santen JM (1941) Recl Trav Chim Pays-Bas 60:76

Progr Colloid Polym Sci (2002) 119:101–112
© Springer-Verlag 2002

W. Borchard
H. Cölfen
D. Kisters
A. Straatmann

Evidence for phase transitions of aqueous gelatin gels in a centrifugal field

W. Borchard (✉) · D. Kisters
A. Straatmann
Institut für Physikalische und Theoretische
Chemie der Gerhard-Mercator-Universität
Duisburg, 47048 Duisburg, Germany

H. Cölfen
Max-Planck-Institut für Kolloid- und
Grenzflächenforschung, 14424 Potsdam
Germany

Abstract Experimental findings of turbid zones in the schlieren patterns of gels in a centrifugal field are explained by a demixing of gels into a highly swollen gel coexisting with a collapsed gel. The thermodynamic analysis of a slightly cross-linked aqueous gel at relatively low centrifugal field, which remains clear, leads to the conclusion that the system is close to its stability limit. A procedure is proposed to show how the stability limits may be extrapolated by means of ultracentrifugal measurements. The phenomena are separated into a transient and a permanent demixing. The permanent demixing has been qualitatively explained by use of state diagrams predicted recently by Khokhlov and coworkers, Ilavsky and also by Moerkerke et al.

Key words Gel · Phase transitions · Demixing · Turbidity · Ultracentrifugation

Introduction

The thermodynamic properties of the gelatin–water system have been studied using methods of static light scattering (LS) [1–3], osmosis [4, 5], mixing calorimetry [6], differential scanning calorimetry [7–12], ultracentrifugation [13–32], swelling [33] and swelling pressure [34]. From all the studies it is known that the second virial coefficients of the osmotic pressure of this system at temperatures close to 20 °C are very low and negative with a Flory Θ temperature probably near 104 °C. At temperatures below 35 °C the system undergoes a coil–helix transition, which is clearly demonstrated by temperature-dependent measurements of LS and optical rotation [35–41]. If the concentration of gelatin is above its critical value for network formation, the solutions (sols) undergo a phase transition to the gel state [42–45]. This change of the system from a viscoelastic liquid to a viscoelastic solid is thermoreversible and is widely used in gelation technology [46, 47].

Long ago it was stated that the process of gelation of macromolecular solutions always occurs very close to miscibility gaps because the tendency of the macromolecules to interact preferably with themselves is a consequence of the negative and small virial coefficients. In non-cross-linked polymer nonelectrolyte solutions the virial coefficient is related to the Flory–Huggins–van Santen–Staverman interaction parameter, χ, [48–50]. If the molar mass of the polymer is infinitely large the system is at its critical point with a value of $\chi = 0.5$.

In aqueous gelatin gels which are physically cross-linked by the aggregation of helices there is evidence that a phase transition occurs inside the gels because the systems become turbid at high enough centrifugal fields; this has been observed by means of schlieren optics [13, 15, 51–56]. Therefore the conclusion was drawn that probably a miscibility gap inside a gel phase might exist at higher polymer concentrations which is responsible for the occurrence of a limiting turbid zone where no concentration gradient can be detected. It will be shown that the phenomenon of demixing of a slightly physically cross-linked gel is preferably detected by applying swelling pressures in centrifugal fields.

Demixing of gels was first envisaged by Dusek and Patterson where a highly swollen gel coexists with a gel of the same cross-linking density at higher polymer concentration [57, 58].

In this contribution we want to remember that terms of higher order of the concentration dependence of the chemical potential of the solvent may be responsible for the phenomenon of demixing in physically or chemically cross-linked gels [59]. An intranetwork phase separation in watery polyelectrolyte gels has been also predicted by Khokhlov and coworkers [60–63] and others [64–66]. Their theory describes a phase diagram similar to that in Ref. [59] but on a different molecular basis. These types of phase diagrams will be used to describe the demixing of poorly cross-linked polymer gels in a qualitative way when submitted to an ultracentrifugal field.

Theoretical considerations

The continuous swelling pressure equilibrium of swollen cross-linked systems in a centrifugal field has been treated theoretically (generalised Svedberg–Pederson equation [19]). From Eq. (9) or Eq. (14b) of Ref. [19] we recall

$$\left(\frac{\partial \tilde{\mu}_2}{\partial \rho_2}\right)_{T,P} = \frac{\omega^2 r (1 - \tilde{V}_2 \rho)}{\mathrm{d}\rho_2/\mathrm{d}r} \quad . \tag{1}$$

Index 2 refers to the network component which is the cross-linked polymer. Thus $\tilde{\mu}_2$ is the specific chemical potential of component 2, ρ_2 its partial density or mass concentration, $\tilde{V}_2$ its partial specific volume, ρ the density of the polymer, ω the angular velocity and r the radial distance. By means of schlieren optics it is possible to determine $(\mathrm{d}\rho_2/\mathrm{d}r)$. $\tilde{V}_2$ and ρ are obtained by density measurements of the gels [27]. The left-hand side of Eq. (1) is an important thermodynamic quantity, which provides, in principle, the stability criterion of a binary phase, here the gel. For stable phases the following relation has to hold [67]

$$\left(\frac{\partial \tilde{\mu}_i}{\partial \rho_i}\right)_{T,P} > 0 \quad \text{for } i = 1, 2 \quad , \tag{2}$$

where index 1 refers to the solvent, here water. At the stability limit one has

$$\left(\frac{\partial \tilde{\mu}_i}{\partial \rho_i}\right)_{T,P} = 0 \quad \text{for } i = 1, 2 \quad . \tag{3}$$

It is well known that stability limits may be extrapolated from LS measurements close to a miscibility gap in non-cross-linked quasibinary polymer solutions [68–70]. Also the sedimentation of a demixed solution in an ultracentrifugal field leading to extremely high concentration gradients at the meniscus between the two demixed phases has been investigated [71]. The turbidity range near this meniscus could be explained by a density inversion where the mechanical equilibrium is not established.

Resolving Eq. (1) with respect to $(\mathrm{d}\rho_2/\mathrm{d}r)$ it is seen that this derivative is inversely proportional to $(\partial \tilde{\mu}_2/\partial \rho_2)_{T,P}$, which means that very high gradients of

the partial density are expected if the system is moving towards the stability limit. Therefore, Eq. (1) may also be considered as the base to determine the stability limits of an isothermal system by means of extrapolation as will be realised later. From the differential form of the generalised Svedberg–Pederson equation the derivative $(\mathrm{d}\pi_s^*/\mathrm{d}r)$ is directly obtained, where π^* is the osmotically active pressure and is identical to the swelling pressure if the polymer concentration at the meniscus is the equilibrium swelling concentration [19].

$$\frac{\mathrm{d}\pi_s^*}{\mathrm{d}r} = \omega^2 r \frac{\rho_2}{\tilde{V}_1 \rho_1} (1 - \tilde{V}_2 \rho) \tag{4}$$

This relation has been used to determine the gradient of the osmotically active pressure $(\mathrm{d}\pi_s^*/\mathrm{d}r)$ from known quantities. It has to be mentioned that for highly swollen gels $\tilde{V}_2 \rho_2 \ll \tilde{V}_1 \rho_1$, so with $\tilde{V}_1 \rho_1 + \tilde{V}_2 \rho_2 = 1$ we have $\tilde{V}_1 \rho_1 \approx 1$ to a good approximation.

It is well known that the partial derivatives of the specific chemical potential, $\tilde{\mu}_i$, for $i = 1$ and $i = 2$ are interrelated by the Gibbs–Duhem equation

$$\rho_1 \left(\frac{\partial \tilde{\mu}_1}{\partial \rho_2}\right)_{T,P} + \rho_2 \left(\frac{\partial \tilde{\mu}_2}{\partial \rho_2}\right)_{T,\rho} = 0 \tag{5}$$

Therefore it is possible to calculate the first term in Eq. (5) having determined the second one by means of Eq. (1) using the relation of the specific chemical potential difference of the solvent and osmotically active pressure under the assumption of an incompressible gel,

$$\Delta \tilde{\mu}_1 = \tilde{\mu}_1 - \tilde{\mu}_{01} = -\tilde{V}_1 \pi_s^* \quad , \tag{6}$$

where $\tilde{\mu}_1$ is the specific chemical potential of the solvent inside an isotropically deformed gel, $\tilde{\mu}_{01}$ that of the pure solvent, $\tilde{V}_1$ the partial specific volume of the solvent water inside the gel and π_s^* the osmotically effective pressure [19] named partial hydrostatic pressure by Svedberg [51]. If $\tilde{V}_1$ is independent of the partial density, ρ_2, which has been found for the gelatin–water system up to 11% by weight of gelatin [27, 56], by means of

$$\frac{\mathrm{d}\pi_s^*}{\mathrm{d}\rho_2} = \frac{\mathrm{d}\pi_s^*}{\mathrm{d}r} \cdot \frac{\mathrm{d}r}{\mathrm{d}\rho_2} = \left(\frac{\mathrm{d}\pi_s^*/\mathrm{d}r}{\mathrm{d}\rho_2/\mathrm{d}r}\right) \tag{7}$$

and Eqs. (1) and (5)

$$\left(\frac{\mathrm{d}\pi_s^*}{\mathrm{d}\rho_2}\right) = \frac{(\mathrm{d}\pi_s^*/\mathrm{d}r)}{\omega^2 r (1 - \tilde{V}_2 \rho)} \cdot \left(\frac{\mathrm{d}\tilde{\mu}_2}{\mathrm{d}\rho_2}\right)_{T,P} \tag{8}$$

is obtained. This equation tells us that the left-hand side of Eq. (8) has to vanish at the stability limit. Equation (8) will be used in the discussion.

By inserting Eq. (4) we get from Eq. (8) and replacing $\tilde{V}_1$ by the speed of volume of water, $\tilde{V}_{01}$

$$\frac{\mathrm{d}\pi_s^*}{\mathrm{d}\rho_2} = \left[\rho_2 \left(\frac{\partial \tilde{\mu}_2}{\partial \rho_2}\right)_{T,P}\right] \cdot \frac{1}{\tilde{V}_{01} \rho_1} \quad . \tag{9}$$

The term in brackets is known as the thermodynamic factor and is part of the Fickian diffusion coefficient [72]. By means of Eq. (5) the dependence of the specific chemical potential of the solvent upon the concentration, ρ_2, can be calculated. For stable binary phases $(\partial \tilde{\mu}_1 / \partial \rho_2)_{T,P}$ is negative.

Experimental

The gelatin under investigation was a bone gelatin of DGF Stoess (Eberbach, Germany) with a water content of 11.95%. After the exchange of ions via an ion exchanger the gelatin solution was neutralised by sulphuric acid. Finally 4‰ calcium hydroxide was added to the solution before drying it. The granules of gelatin were allowed to swell in a defined amount of triply distilled water at nearly 4 °C for 2 days. In order to prevent bacterial attack 0.150 mg 5% solution of 4-chloro-3-methylphenol in methanol was added per gram of dry gelatin. After homogenising the solutions in closed glass vessels with screwed covers at 45 °C for 6–12 h, the solutions were stored at 4 °C until they were heated again to 45 °C before being transferred to the ultracentrifugal cells. The concentration range covered by 14 solutions was in the range 2–14.4% by weight of gelatin. At polymer concentrations higher than 14.4% by weight the turbidity of the gels was so huge that a schlieren pattern during a run of the centrifuge could no longer be detected [56].

To prevent adhesion of the gels at the walls of the centrepieces and the surfaces of the cell windows all contact surfaces were coated by means of Teflon AF 1600 from DuPont Polymers, Wilmington,[1] which was dissolved in fluoroinert 75 of the firm 3M. The centrepieces were coated by adding drops of the viscous solution by means of a syringe to the contact surfaces and removing the solvent by heating the pieces for 2 h at 40 °C.

A special apparatus was built for the spin-coating of the cell windows [56]. A drop of nearly 0.5 ml Teflon AF 1600 in fluoroinert solution was placed on the centre of the window. After rotation around the cylindrical axis at 3,000 rpm by a regulated motor a plane layer of the perfluorinated oligomer of Teflon of thickness of about 360 nm could be deposited. The thickness of the layers was determined with an S 2000 fibre spectrometer (Top-Sensors) at a wave length of 633 nm.

The cells were screwed with a torque of 60 Nm, filled by means of a tempered syringe of 45 °C and placed into the gelling apparatus constructed by Cölfen, where the centrepiece is adjusted in the gravitational field to the same direction as the later applied centrifugal field [54]. The solutions were allowed to gel at 5 °C for 24 h. Before being placed into an eight-hole rotor [54] of the analytical centrifuge (Beckman, model E), the cells were tested for vacuum tightness by measurement of the weight at different time intervals. All runs of the centrifuge were performed at a temperature of 5 °C in order to minimise the amount of soluble parts of the gelatin gels [13, 14, 28, 54]. It was shown that real swelling equilibria are obtained with gelatin–water gels in pure water if the temperature is kept below 10 °C [33]. At higher temperatures the junction zones of the polymer chains are not stable, leading finally to a dissolution of the gels at finite times.

The optical device used and especially the possibility of superimposing the images of the counterbalance cell and any of the seven cells filled with gels have been described elsewhere [24, 25, 73].

To prevent the occurrence of dark zones in the schlieren patterns, only very low rotational speeds were used so that detectable concentration gradients are formed without any move-

[1] We thank DuPont Polymers for placing a small amount of Teflon AF 1600 at our disposal

ment of the meniscus between the gel and the vapour [28, 29]. Furthermore with this method it was possible to prove the existence of real continuous equilibria of κ-carrageenan gels in a centrifugal field. The concentration gradients inside the gels at a selected rotational speed, ω_1, were pathway-independent. This means that they were the same if ω_1 was reached from lower or higher rotational speeds with regard to ω_1 [30].

For this special gelatin all auxiliary measurements, like the refractive index–partial density relationship and the concentration dependence of the gel densities, were determined separately but are not mentioned here. The calibration of the schlieren optics was performed in the usual way [56].

Results and discussion of the investigations of aqueous gelatin gels in an analytical ultracentrifuge

Sedimentation–diffusion equilibria at relatively low centrifugal fields

The sedimentation–diffusion equilibrium applied to gels deserves explanation of the principal. Gels undergo a continuous equilibrium in a centrifugal field as do polymer solutions. The only difference is that a gel is deformed anisotropically in the radial direction. Thus, the gradient of the chemical potential of the solute or solvent does not only depend on the concentration gradient, for example, of the polymer owing to mixing terms but additionally to terms of deformation of the polymer network.

It was shown that diffusion of low-molecular-weight solvents into a chemically cross-linked polymer can be considered to be normal diffusion [74]. In this case the diffusion coefficient only depends on concentration at constant values of temperature and pressure. This has been proven experimentally, because the locus at constant composition of the solvent moves proportionally to the root of the time into the cross-linked polymer [74].

If now the centrifugal field causes a concentration gradient inside the gel, diffusion has to take place, leading finally to a sedimentation–diffusion equilibrium [19, 51], which was also called a continuous swelling–pressure equilibrium [19, 51].

As the deformation of the gel at low external fields remains very small the deformation can be considered to be approximately isotropic. As a consequence of the cross-linking of the gel there is the formation of a new meniscus between the gel and the exuded solvent under high enough external fields. This has been explained by the appearance of the swelling equilibrium at this meniscus caused by mass conservation of the cross-linked polymer in the gel [19, 29–31].

The schlieren patterns of a gelatin gel with an overall polymer weight fraction of $w_{2,0} = 3.6\%$ by weight are shown for different running times in Fig. 1. The distance markings of the counterbalance cell bracket the schlieren picture of the cell. After 15 min the refractive index gradient had already increased at the meniscus gel(g)/

Fig. 1 Schlieren patterns of an aqueous gelatin gel with an initial mass fraction of the polymer of $w_{2,0} = 3.6\%$ for different running times with a rotational speed of $v = 6,000$ rpm at 5 °C [56]

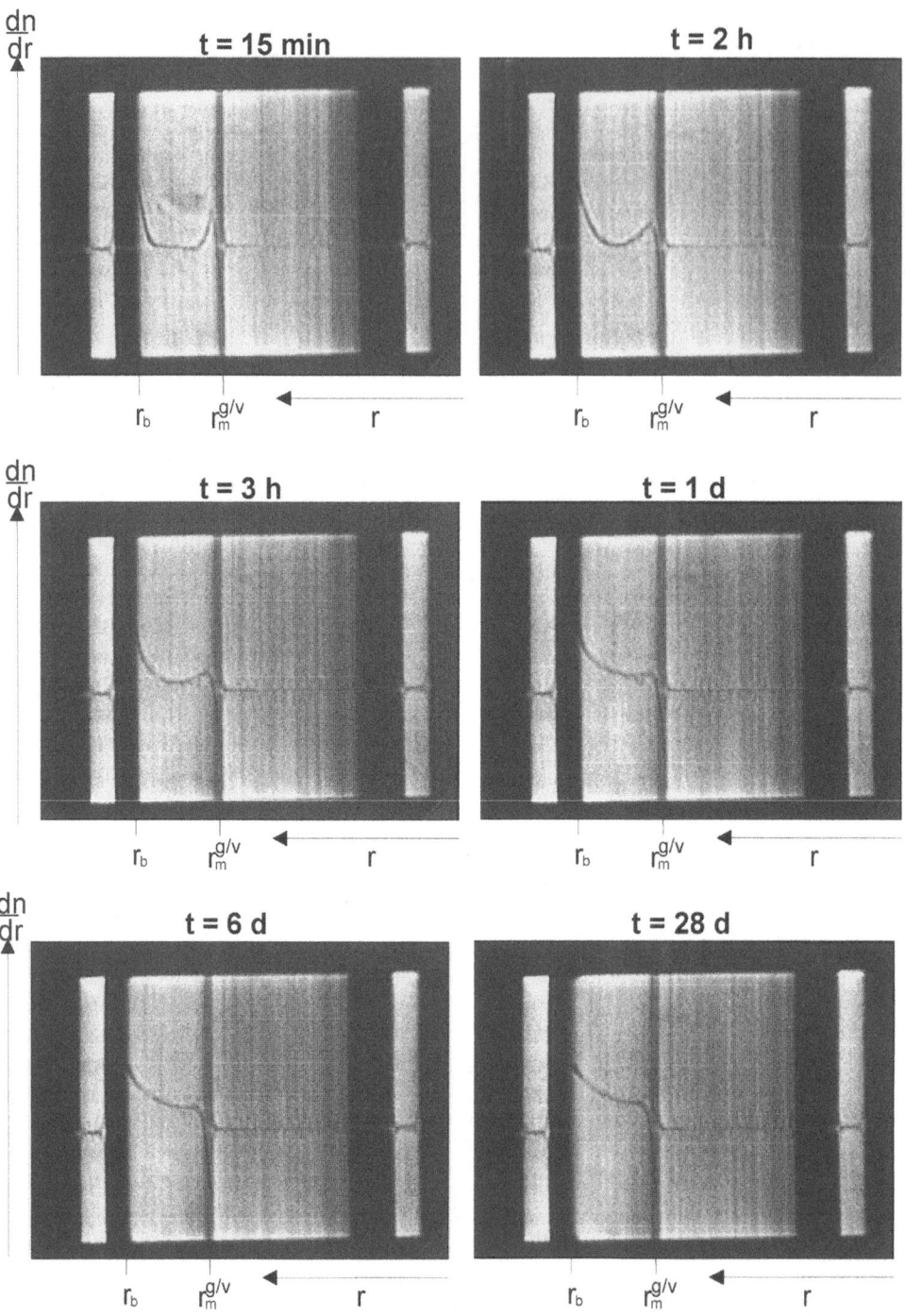

vapour(v) and the bottom (b), corresponding to the distances $r_m^{g/v}$ and r_b from the axis of rotation. At the meniscus an irregular course of the refractive index which stems from reflections at the meniscus of the gel can be seen and this remains unaltered for different running times; therefore, this was not considered in the data evaluation. On the basis of the schlieren patterns of Fig. 1 the partial densities for different running times versus the distance from the axis of rotation at a rotational speed of 6,000 rpm were calculated by integration of the partial density gradient–distance curve under consideration of the mass conservation in the cell. The partial density–distance curves are represented in Fig. 2 at a temperature of 5 °C for different times [56]. Although the height of the gel is about 0.43 cm, the final value of the partial density–distance curve was still not established after 6 days. At a running time of 28 days a typical exponential course of the $\rho_2(r)$ curve can be stated which has also been found for gels with other polymer concentrations. This is at variance with earlier investigations where the concentration gradient could not be detected and a linear course of the $\rho_2(r)$ curve was assumed [54].

Up to now in the evaluation procedure of sedimentation equilibrium experiments of gels the dilution effect occurring in sector-shaped cells has not been considered. As the polymer concentration in an equilibrium run increases from that at the gel–vapour meniscus to that at

the cell bottom there is a dilution of the gel by the movement of the cross-linked polymer to longer distances. The calculations of the osmotically active pressures were performed by previously correcting the $\rho_2(r)$ curves, which is, for example, shown in Fig. 3 for two concentrations [56]. The consideration of the dilution effect leads to changes in ρ_2 at the bottom of the cell in the range 5–6%. This means that this correction has to be considered.

The proof of the presence of real sedimentation–diffusion equilibria in gels was furnished by achieving the

Fig. 2 Partial density of the polymer versus distance to the rotational axis at $v = 6,000$ rpm and 5 °C for different running times [56]

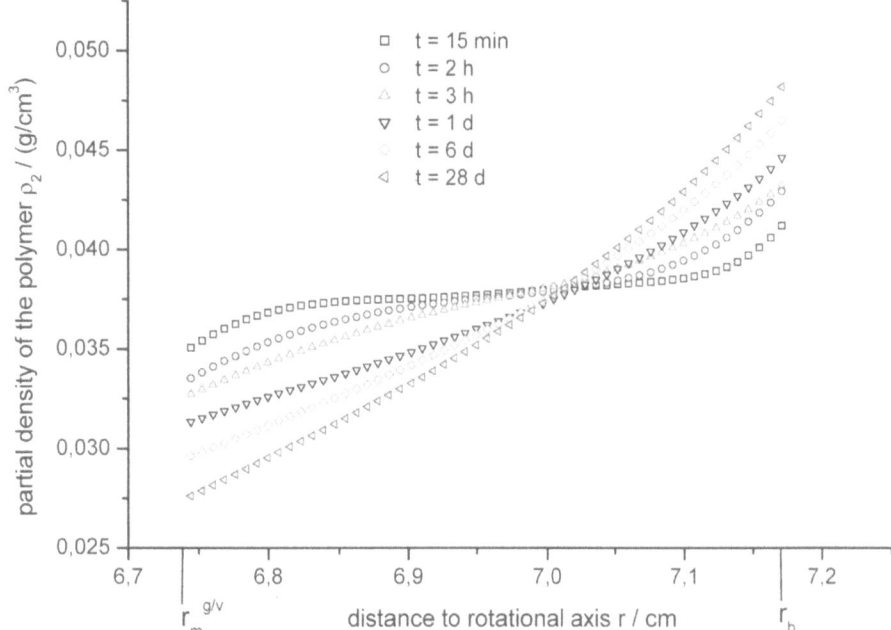

Fig. 3 Correction for the dilution effect. ρ_2 versus r [56]

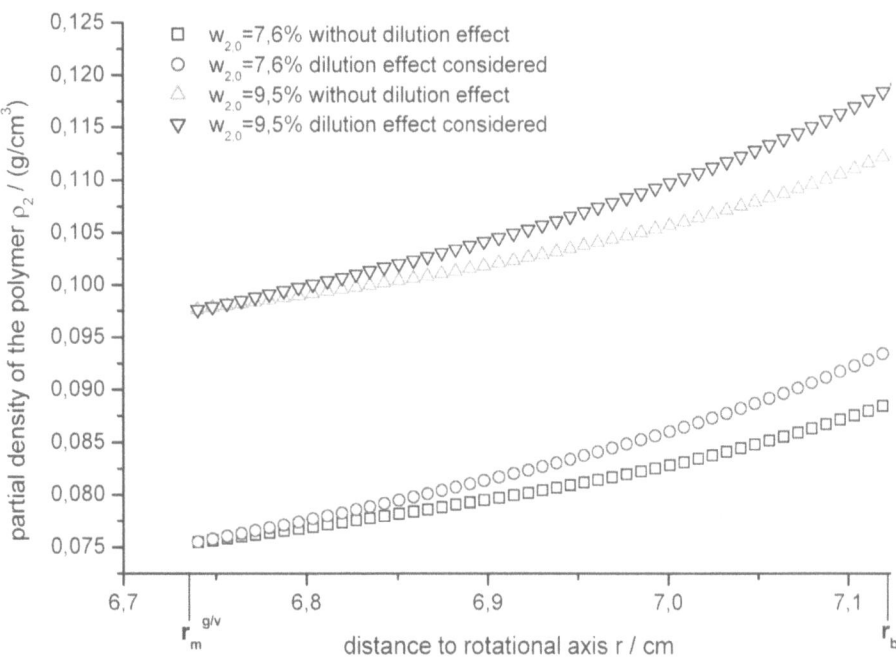

equilibrium at 6,000 rpm from two different speeds, for example, 0 and 12,000 rpm. It can be seen in Fig 4 that the symbols are situated on the lines in the accuracy of the measurements, so for these four gels the proof of the existence of real equilibria is given and this is also true for the other gels under investigation [56].

The gradients of the osmotically active pressures were calculated by use of Eq. (4) under consideration of the dilution effect. The integration of the $(d\pi_s^*/dr)$ curves over the distance r yields the osmotically active pressure π_s^* for every point r in the gel. The π_s^* values are plotted versus r in Fig. 5. As expected, the osmotically active pressures obtained are 1–2 orders of magnitude smaller than those found by Cölfen, who used much higher rotational speeds (14,000–18,000 rpm), where the movement of the gel meniscus was allowed to take place [54]. The $\pi_s^*(r)$ curves for the lowest initial concentrations are extremely flat with respect to those at higher initial concentrations.

As the concentrations ρ_2 as function of the radial distance are known, the $\pi_s^*(\rho_2)$ curves can be constructed (Fig. 6). Comparing the results for different initial

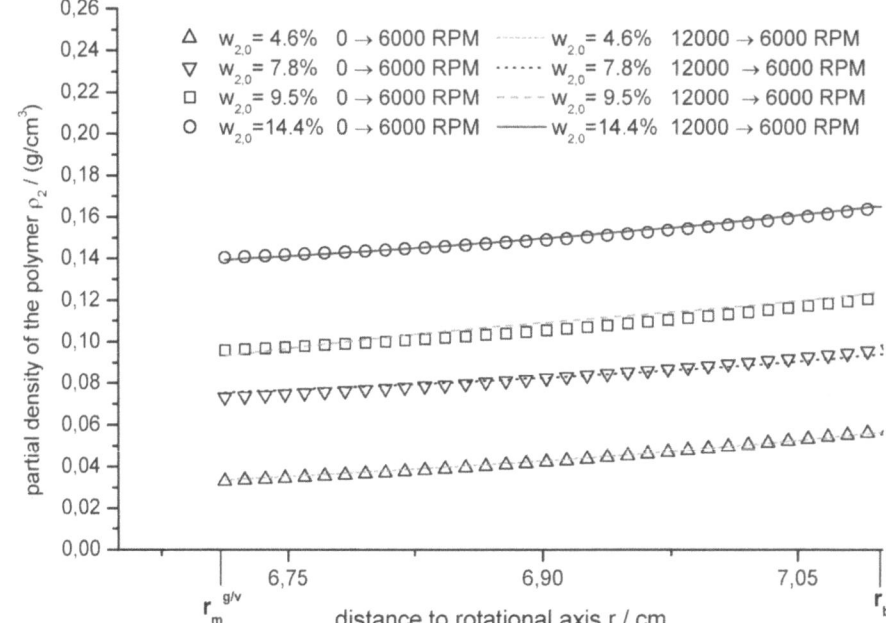

Fig. 4 Partial density of the polymer, ρ_2, versus r for different polymer concentrations at $v = 6,000$ rpm starting from 0 and 12,000 rpm [56]

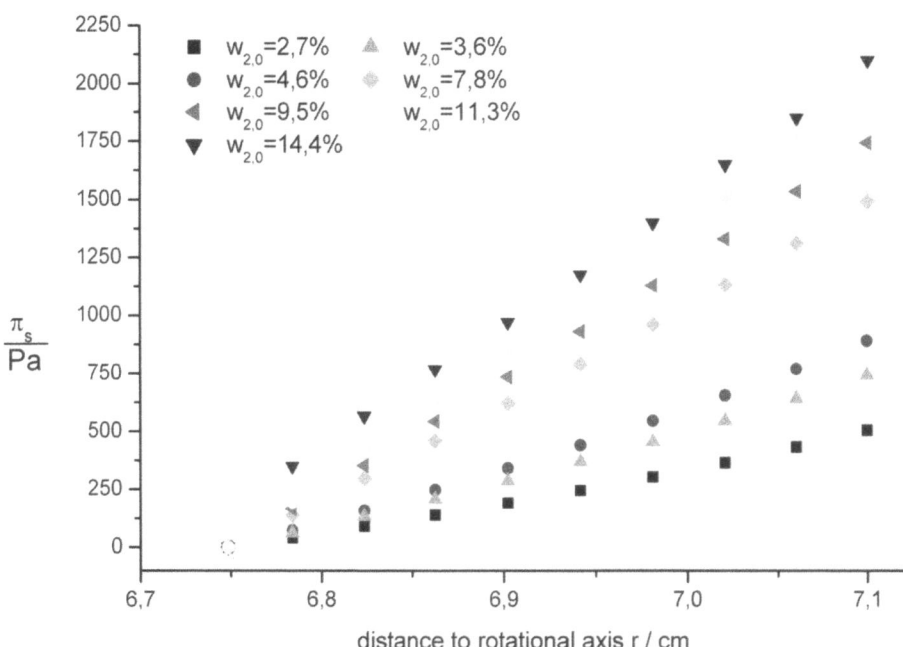

Fig. 5 Osmotically active pressure, π_s^*, versus r for different concentrations at $v = 6,000$ rpm in the gelatin–water system [56]

Fig. 6 π_s^* versus ρ_2 for different initial concentrations [56]

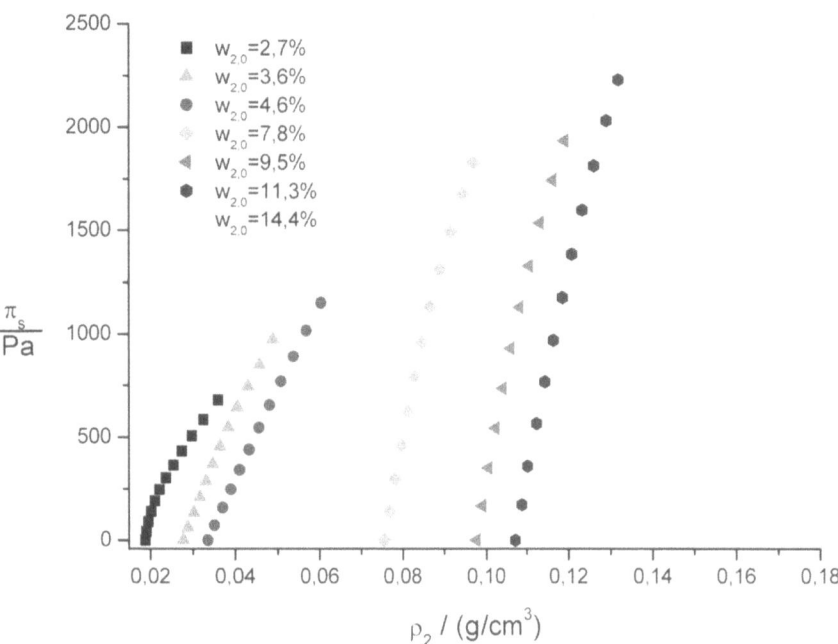

concentrations, it can be seen that the concentration range for the three lowest concentrations and the highest is nearly symmetrical around the initial concentration but not for the initial concentrations between 7 and 11.5%. At $w_{2,0} = 9.5\%$ the initial concentration is close to the value at the meniscus, which is given by $\pi_s^* = 0$. This is only possible when the pure solvent is exuded at the gel–vapour meniscus and a new gel–solvent meniscus is formed. Thereby the polymer concentration in the gel is increased.

In Fig. 6 there is no intersection of the different osmotically active pressure curves versus partial density. This tells us that with increasing initial concentration the network structure also becomes more densely cross-linked, which also leads to thermodynamically more stable gels. This is in agreement with the results of statistical theories for non-polyelectrolyte polymers swollen in solvents. Furthermore the nonintersecting curves in Fig. 6 indicate that the influence of soluble parts, which can lead to additional cros-slinking of the gel, is negligible at the low speed applied [28].

All the $\pi_s^*(\rho_2)$ curves are nonlinear and have a negative curvature; this is more pronounced if the initial concentration is low. These findings are new for the gelatin–water system. It has to be mentioned that the swelling pressures were obtained from the schlieren pattern without an assumption concerning the course of the concentration gradients in the gels.

The course of the curves in Fig. 6 gives a hint that at a given initial concentration, $w_{2,0}$, there might be a maximum in the $\pi_s^*(\rho_2)$ curves, indicating the presence of a stability limit at higher polymer concentrations. At the present time there is no direct experimental proof but there is indirect evidence that at higher rotational speeds

turbidity zones occur; these are expected if a phase transition takes place in the adjacent concentration region.

These considerations led to the calculation of the derivative $(d\pi_s^*/d\rho_2)$ from Eq. (8), which is proportional to the thermodynamic factor introduced by Eq. (9). The result is shown for the gel with an initial concentration $\rho_{2,0} = 7.8\%$ in Fig. 7 with respect to ρ_2 [56]. The function $(d\pi_s^*/d\rho_2)$ decreases linearly with the concentration, ρ_2. If this curve continued linearly it would be zero at 11.54 gcm^{-3}. So we may argue that the system is very close to a stability limit.

Sedimentation–diffusion equilibria at relatively high centrifugal fields

For a better comparison a typical schlieren pattern of a gel, for example, with an initial weight fraction of $w_{2,0} = 2.7\%$ is shown in Fig. 8 for the equilibrium case at a rotational speed of 6,000 rpm [56]. It can be seen that the meniscus has just broadened, indicating the sedimentation of the gel phase, thus establishing the gel–solvent meniscus. The refractive index gradient between the gel–solvent and solvent–air menisci cannot be resolved, but the schlieren curve can be clearly seen in the gel.

If the rotational speed is doubled to 12,000 rpm the series of schlieren patterns of the same gel for different running times shown in Fig. 9 is obtained.

When the experiment is started only a gel–vapour meniscus can be seen. After 24 h solvent is exuded from the gel and two menisci can be obtained: a vapour–solvent meniscus and a gel–vapour meniscus. The gel–solvent meniscus has moved a little farther to the bottom

Fig. 7 Derivative of π_s^* with respect to ρ_2 versus ρ_2 at $v = 6,000$ rpm and 5 °C for $w_{2,0} = 7.8\%$ [56]

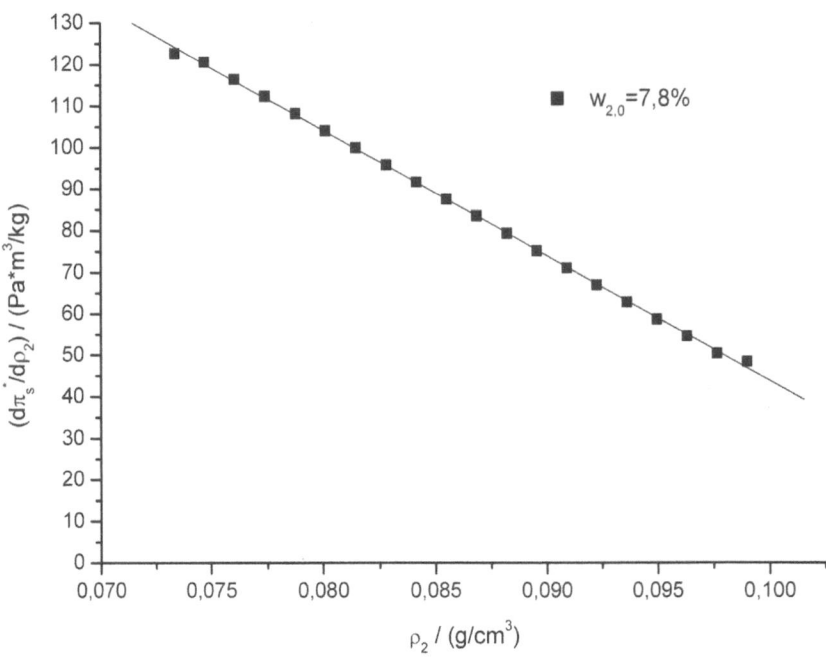

(dn/dr)

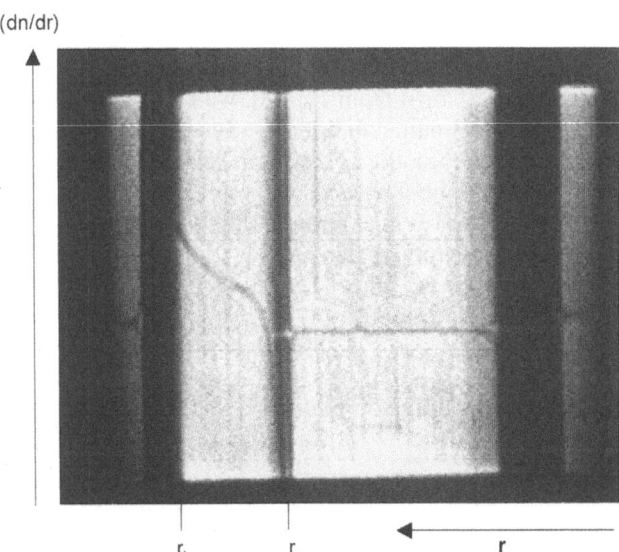

r_b r_m $\leftarrow$ r

Fig. 8 Schlieren pattern of a gel with $w_{2,0} = 2.7\%$ at $v = 6,000$ rpm and 5 °C [56]

compared to 6,000 rpm and dark zones appear at the bottom and at the meniscus. With increasing time the dark zone at the bottom broadens, whereas the dark zone at the meniscus becomes smaller and moves towards the dark zone at the bottom. Finally after 1,200 h two very well contrasted menisci can be seen and a very broad turbid zone extending very close to a small clear zone at the bottom remains with an equilibrium refractive index gradient in that part of the gel situated close to the gel–solvent meniscus. In the dark zone the gradient of the refractive index could not be resolved even if centrepieces

of height 3 mm were used. The transient existence of the dark zone at the meniscus may be explained by the fact that the local polymer concentration does not yet correspond to the equilibrium concentration. The complicated sequence of clear and dark (turbid) zones may be shown schematically by use of the Figs. 10 and 11. If the system with an initial concentration $\rho_{2,0}$ in Fig. 10 is rotated at a speed ω_1 a concentration gradient is built up in the gel, from which the concentration dependence of the osmotically active pressure is calculated (see curve 1). It was assumed that the change in the polymer concentration with respect to $\rho_{2,0}$ is symmetrical. This means that the intersection of curve 1 with the ρ_2-axis gives the concentration at the meniscus, whereas the concentration at the bottom belongs to the end of curve 1. Point D_1 is situated in the middle of curve 1; it corresponds to the concentration $\rho_{2,0}$.

On using a higher speed (ω_2), curve 2 is obtained, where point D_2 in the middle of curve 2 belongs to the initial concentration $\rho_{2,0}$. In both cases the gel remains homogeneous although a concentration gradient is present. If the rotational speed (ω_3) is applied the polymer concentration at the meniscus at $\pi_2 = 0$ equals the lowest possible polymer concentration of the gel, $\rho_{2,S}$. This is the concentration of the swelling equilibrium establishing at the meniscus. We assume that the polymer concentration at the bottom of the cell is given by $\rho_{2,C}$ belonging to point C on the swelling pressure curve. Point D_3 intersects the swelling curve between A and C near the middle and belongs to the initial concentration $\rho_{2,0}$. Every speed higher than ω_3 shifts the gel to the heterogeneous region. In this case, pure solvent has to appear at the gel–vapour meniscus. The gel exudes solvent and a new gel–solvent meniscus is formed which may move towards the bottom of the cell if higher centrifugal fields are acting.

If now $\omega_4 > \omega_3$ is applied to a gel with concentration $\rho_{2,0}$ at every point and the swelling pressure corresponding to D_4 is reached, the system has entered the heterogeneous region. A phase separation has to take place. If no concentration gradient is established the mean concentration $\rho_{2,0}$ has to shift to $\rho_{2,B}$ belonging to point B on the swelling pressure curve. There is coexistence between points B and S, the latter representing the pure solvent. In this situation the solvent is able to

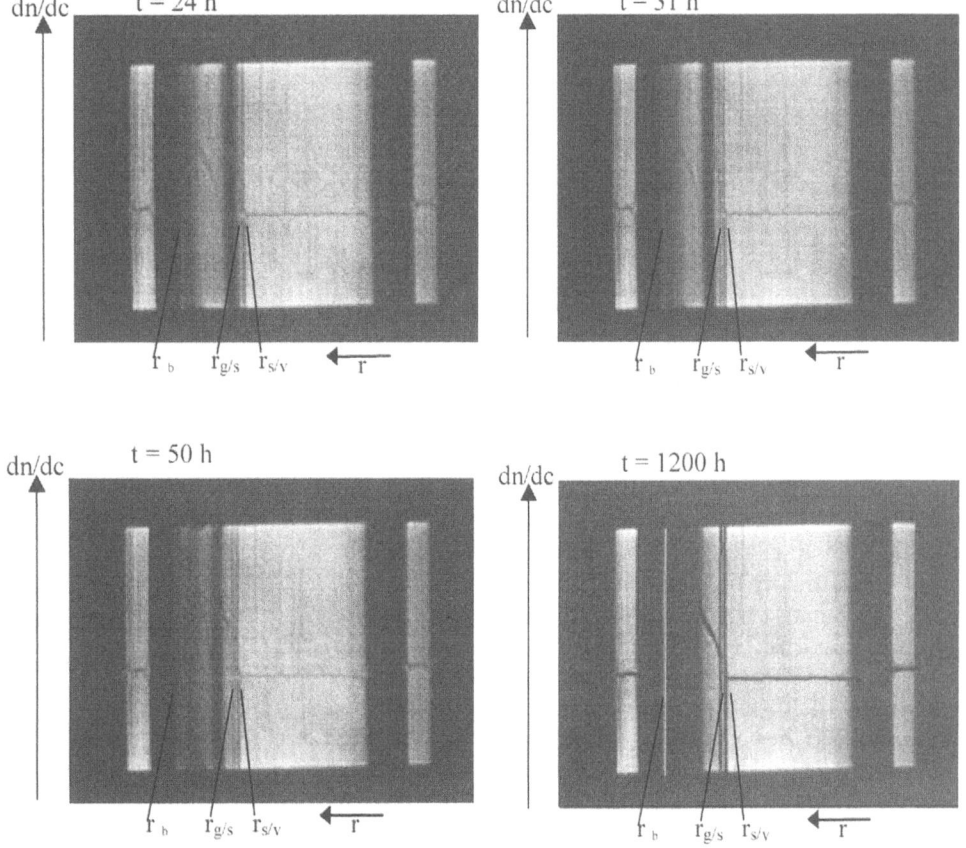

Fig. 9 Series of schlieren patterns of a gel with $w_{2,0} = 2.7\%$ by weight of gelatin after having reached $v = 12,000$ rpm for different running times [56]. For better contrast the pictures have been retouched slightly

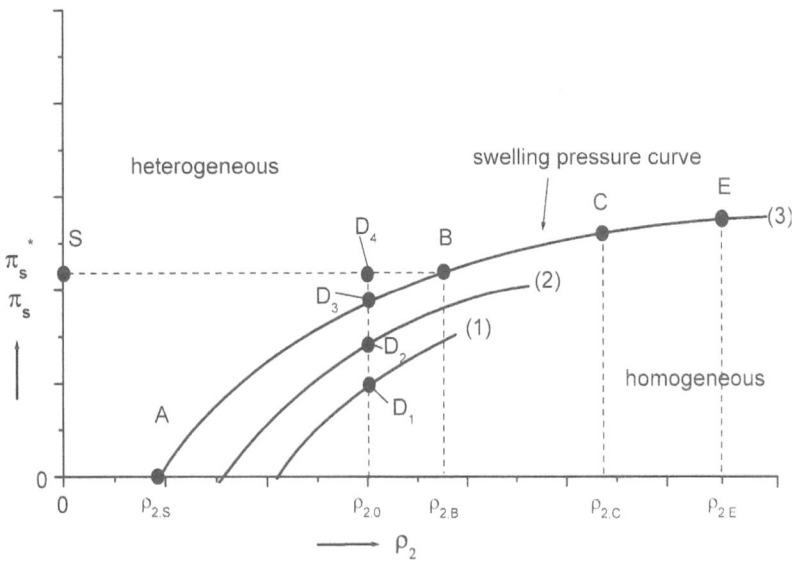

Fig. 10 Schematic representation of π_s^* and π_s versus ρ_2

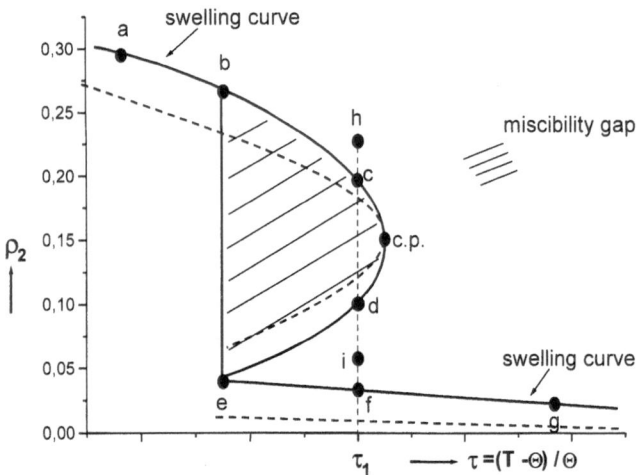

Fig. 11 Schematic phase diagram of a demixed slightly cross-linked gel. T is the temperature, Θ is the Flory Θ temperature and ρ_2 is the partial density of the polymer, taken from Kramarenko and Khokhlov [62]; *dashed line*: branches of the spinodal curve following Moerkerke et al. [59]

build up small droplets of pure solvent inside the gel because at D_4 the gel is in a supersaturated state. This is the explanation for the occurrence of a transient and nonpermanent turbidity; however, because of the de-swelling of the gel the mean concentration belonging to point B will finally be established.

As the gel is now in a centrifugal field a concentration gradient has to be formed around the new mean concentration $\rho_{2,B}$ under the condition that $\rho_{2,S}$ will be reached at the meniscus, point A. Assuming once more symmetry – in the change of concentrations – the swelling pressure curve has to extend up to point E with $\rho_{2,E}$ at the bottom of the cell. For even higher rotational speeds the concentration ρ_2 at the bottom has to increase further. Therefore, in a continuous swelling pressure equilibrium all local concentrations between the meniscus and the bottom are situated on the swelling pressure curve, which is continuous throughout the cell if the system remains thermodynamically stable. If we disregard the very well known stability limit in the concentration region $0 \leq \rho_2 \leq \rho_{2,S}$ of a demixing unstable gel the swelling pressure curve has to have two more points for which $(\partial \pi_s / \partial \rho_2)_{T,P}$ vanishes [57–66]. This corresponds for the gelatin–water system according to Eq. (8) to the condition for the presence of a stability limit. In this case we expect nonvanishing turbidity of a gel in a centrifugal field. This is deduced from a schematic phase diagram as presented in the literature [57–66].

Evidence for the presence of demixed states in slightly cross-linked polyelectrolyte gels

Khokhlov and coworkers and other authors calculated phase diagrams for polyelectrolyte gels starting from the

assumption that the total free energy of a gel can be described by the contributions of four terms [60–63, 66]:

1. An elastic part, F_{el}, due to deformation of the network chains.
2. An interaction term, F_{int}, describing the mixing of network chains.
3. The change in the free energy of electrostatic interactions with swelling.
4. A term considering the presence of additional ions outside and inside the network.

A remarkable result is the prediction of a demixing of the gel phase. Without going further into details, a schematic phase diagram is shown in Fig. 11. τ describes the relative distance from the Θ temperature of Flory, which is given by $\Theta = 377$ K in the present system. At 5 °C τ has a value of $\tau_1 = -0.268$. For a set of parameters and negative τ values, in the concentration range $\rho_2 < 0.3$ gcm^{-3}, a miscibility gap (curve b–c–c.p.–d–e) is predicted, where at τ_1 a highly swollen gel at point d coexists with a gel at point c, which is in a collapsed state; c.p. is the critical point of demixing. Thus going from d to c the gel undergoes a transition from a coiled state to a globular state. Between c and d the gel is heterogeneous, where without a centrifugal field both phases have to settle with a phase mass ratio depending on the overall concentration ρ_2. Curve e–f–g corresponds to the swelling curve of the highly swollen gel. At point e there is a three-phase coexistence between phases b, e and the pure solvent. The swelling curve of the collapsed gel is described by curve a–b. Points i and h belong to the homogeneous phase area of the gel outside the miscibility gap.

If the gel with a concentration belonging to point i is submitted to a centrifugal field a low concentration gradient will keep the system in the homogeneous range. This means that the concentration at the bottom of the centrifugal cell does not exceed that belonging to point d. Also the concentration ρ_2 at the meniscus is higher than $\rho_{2,S}$, which belongs to point f. Discussing only equilibrium states the gel has to expel solvent as soon as the polymer concentration at the meniscus becomes smaller than that belonging to point f. If the centrifugal speed is high enough that the concentration at the bottom is higher than that belonging to d, a demixing of the gel will take place, which was frequently observed by a turbid zone at the cell bottom. A part of the gel is separated with the concentration at point c; however, in a centrifugal field local gradients of the concentration ρ_2 are formed at c and d. This means that the concentration gradient at c produces locally a phase with a concentration belonging to point d, whereas the phase with concentration at d generates again collapsed states belonging to point c. One has to imagine that at every phase boundary between the highly swollen gel belonging to d and the collapsed state at point c there is a finite step in concentration ρ_2 over a very small distance of the phase boundary. Thus, the concentration gradient is extremely high, as it is, for example, at the meniscus between gel and vapour. In the schlieren optics there will be a sharp signal belonging to such a meniscus. In the case of a demixing gel there are, as already explained, many of these layers side by side and therefore many of the menisci which can no longer be resolved. This is the reason for the occurrence of the dark zone at/before the bottom of the centrifugal cell. If the system is brought back to zero rotational speed the concentration gradients and also the turbidity vanish.

The layering of highly swollen and collapsed states is fixed by the polymer network. As both phases coexist locally there is no driving force for the diffusion of solvent or network polymer. This is a persisting density inversion and corresponds to the schlieren patterns shown in Fig. 9. The system remains turbid for the same reason as was found for the superposition of sedimentation and demixing in the polystyrene–cyclohexane system [17]. If the overall polymer concentration is between those belonging to points c and d and the centrifugal field is high enough we expect a clear zone at the gel meniscus and at the bottom, if the collapsed state is transparent again as a homogenous gel. It can be seen in Fig. 9 that after a very long time of 1,200 h a clear zone appears at the bottom, which is too small for a concentration gradient to be resolvable. This seems to correspond to the fact that the concentration at the cell bottom is situated beyond point c in the homogeneous region of the phase diagram in Fig. 11. If, finally, the run is started at an initial overall concentration belonging to point h, a permanent turbidity is expected to occur close to the meniscus if in the sedimentation–diffusion equilibrium the concentration at the meniscus is lower than that belonging to point c, whereas the gel near the bottom of the cell will remain transparent.

In Fig. 11 the branches of the stability limiting curve are also presented schematically. A quantitative prediction of the demixing of gels is formally possible by describing the $\pi_s^*(\rho_2)$ curve at low centrifugal fields and introducing three interaction parameters assuming a non-polyelectrolyte gel, but this is beyond of the scope of this contribution. However, it has been found experimentally that even for non-cross-linked polymer–solvent systems at least three parameters are necessary if demixing occurs [75].

Acknowledgements The authors thank the Deutsche Forschungsgemeinschaft for financial support of the project. H.C. also thanks the Max Planck Society and the Dr. Hermann Schnell Foundation for financial support.

References

1. Boetger H, Doty P (1958) J Phys Chem 58:1246
2. Pezron I, Herning T, Djabourov M, Leblond J (1990) In: Burchard W, Ross-Murphy SB (eds) Physical networks, polymers and gels. Elsevier, London, p 231
3. Bohidar HB (1998) Int J Biol Macromol 23:1
4. Pouradier J, Venet AM (1950) J Chim Phys Phys-Chim Biol 47:391
5. Borchard W, Keese A (1979) 26th IUPAC symposium on macromolecules, Mainz, Germany, vol II, p 888
6. Meerson ST, Lipatov SM (1958) Colloid J USSR 20:336
7. Borchard W, Bergmann K, Rehage G, Cox RJ (eds) (1976) Investigation of gelation phenomena in aqueous gelatin solutions. Photographic gelatin II. Academic, London, p 57
8. Godard P, Biebuyck JJ, Daumerie M, Naveau H, Mercier JP (1978) J Polym Sci Polym Chem Ed 16:1817
9. Borchard W, Bremer W, Keese A (1980) Colloid Polym Sci 258:516
10. Borchard W, Luft B, Reutner P (1984) Ber Bunsenges Phys Chem 88:1010
11. Reutner P, Luft B, Borchard W (1985) Colloid Polym Sci 263:519
12. Borchard W, Luft B, Reutner P (1986) J Photogr Sci 34:132
13. Johnson P (1964) Proc R Soc Lond Ser A 278:527
14. Johnson P, Metcalfe JC (1967) Eur Polym J 3:423
15. Johnson P, King RW (1968) J Photogr Sci 16:82
16. Johnson P (1971) J Photogr Sci 19:49
17. Holtus G, Borchard W (1989) Colloid Polym Sci 267:1133
18. Holtus G, Cölfen H, Borchard W (1991) Prog Colloid Polym Sci 86:92
19. Borchard W (1991) Prog Colloid Polym Sci 86:84
20. Cölfen H, Borchard W (1991) Prog Colloid Polym Sci 86:102
21. Borchard W, Cölfen H (1992) Macromol Chem Macromol Symp 61:143
22. Borchard W (1994) Prog Colloid Polym Sci 94:82
23. Cölfen H, Borchard W (1994) Acta Polym 45:325
24. Cölfen H, Borchard W (1994) Prog Colloid Polymer Sci 94:90
25. Cölfen H, Borchard W (1994) Anal Biochem 219:321
26. Cölfen H, Borchard W (1994) SPIE Proc 2136:307
27. Cölfen H, Borchard W (1994) Macromol Chem Phys 195:1165
28. Cölfen H, Borchard W (1995) Macromol Chem Phys 196:3469
29. Hinsken H, Borchard W (1995) Colloid Polym Sci 273:913–925
30. Hinsken H, Selic E, Borchard W (1995) Prog Colloid Polym Sci 99:154–161
31. Borchard W, Hinsken H (1997) Prog Colloid Polym Sci 107:172–179
32. Borchard W, Selic E (1997) In: Pethrick RA, Dawkins JV (eds) Experimental methods in polymer characterisation, vol 1. Wiley, Chichester, pp 317
33. Hermanns B, Borchard W, Rehage G (1974) Angew Makromol Chem 36:117
34. Borchard W, Emberger A, Schwarz J (1978) Angew Makromol Chem 66:43
35. Smith CR (1919) Am Chem Soc 41:135
36. Flory PJ, Weaver ES (1960) J Am Chem Soc 82:4518
37. Todd A (1961) Nature 191:567
38. Engel J (1962) Arch Biochem Biophys 97:150
39. Djabourov M, Leblond J, Papon P (1988) J Phys 49:319

40. Burg B, Borchard W (1989) Integration of fundamental polymer science and technology, Rolduc meeting III. Elsevier, London, p 100
41. Borchard W, Burg B (1989) In: Baumgärtner, A, Picot C (eds) Molecular basis of polymer networks. Proceedings in physics, vol 42. Springer, Berlin Heidelberg New York, p 162
42. Borchard W, Burg B (1990) Prog Colloid Polym Science 83:200
43. Stauffer D, Coniglio A, Adam M (1982) Adv Polym Sci 44:103
44. de Gennes PG (1979) Scaling concepts in polymer physics. Cornell University Press, Ithaca
45. Borchard W (1998) Ber Bunsenges Phys Chem 102:1580
46. Guenet JM (1992) Thermoreversible gelation of polymers and biopolymers. Academic, New York
47. te Nijenhuis K (1997) Thermoreversible networks. Advances in polymer science, vol 130. Springer, Berlin Heidelberg New York
48. Staverman AJ, van Santen JM (1941) Recl Trav Chim Pays-Bas 60:76
49. Huggins ML (1943) Ann NY Acad Sci 44:431
50. Flory PJ (1953) Principles of polymer chemistry. Cornell University Press, Ithaca
51. Svedberg T, Pedersen KO (1940) The ultracentrifuge. Oxford University Press, Oxford, pp 29–33
52. Metcalfe JC (1965) PhD thesis. Cambridge
53. Holtus G (1990) Doctoral thesis. Duisburg
54. Cölfen H (1993) Doctoral thesis. Duisburg (1994)
55. Cölfen H (1999) Biotechnol Genet Eng Rev 16:133
56. Kisters D (2001) Doctoral thesis. Duisburg
57. Dusek K, Patterson DJ (1968) J Polym Sci A-2 6:1209
58. Dusek K, Prins W (1969) Adv Polym Sci 6:1
59. Moerkerke R, Koningsveld R, Berghmans H, Dusek K, Solè K (1995) Macromolecules 28:1103
60. Khokhlov AR (1980) Polymer 21:376
61. Grosberg A Yu, Khokhlov AR (1994) Statistical physics of macromolecules. Nauka, Moscow (English translation AIP, New York)
62. Kramarenko E Yu, Khokhlov AR (1998) Polym Gels Networks 6:45
63. Khokhlov A, Starodubtzer SG, Vasilevskaya VV (1993) Adv Polym Sci 106:123
64. Sibayamy M, Tanaka T (1993) Adv Polym Sci 109:1
65. Onuki A (1993) Adv Polym Sci 106:63
66. Ilavsky M (1993) Adv Polym Sci 106:173
67. Haase R (1956) Thermodynamik der Mischphasen. Springer, Berlin Heidelberg New York
68. Scholte ThG (1970) J Polym Sci A-2 8:841
69. Scholte ThG (1970) Eur Polym J 6:51
70. Borchard (1972) Ber Bunsenges Phys Chem 76:224
71. Borchard W, Holtus G (1989) Colloid Polym Sci 267:1127
72. Haase R (1963) Thermodynamik der irreversiblen Prozesse. Steinkopff, Dresden
73. Cölfen H (1995) Colloid Polym Sci 273:1101
74. Rehage G (1959) Symposium über Makromoleküle Wiesbaden IIA15
75. Rehage G (1964) Kolloid Z Z Polym 196:97

Progr Colloid Polym Sci (2002) 119:113–120
© Springer-Verlag 2002

THERMODYNAMICS

Peter R. Wills
Donald J. Winzor

Exact theory of sedimentation equilibrium made useful

P. R. Wills
GMD – German National Research Center
for Information Technology
Schloss Birlinghoven, St. Augustin
53754 Germany

P. R. Wills (✉)
Department of Physics
University of Auckland
Private Bag 92 019
Auckland, New Zealand
e-mail: p.wills@auckland.ac.nz
Tel.: +64-9-3737599 ext 8889
Fax: +64-9-3737445

D. J. Winzor
Centre for Protein Structure
Function and Engineering
Department of Bochemistry
University of Queensland, Brisbane
Queensland 4072, Australia
e-mail: winzor@biosci.uq.edu.au
Tel.: +61-7-33652132
Fax: +61-7-33654699

Abstract The exact description of the thermodynamics of solutions has been used to describe, without approximation, the distribution of all the components of an incompressible solution in a centrifuge cell at sedimentation equilibrium. Thermodynamic parameters describing the interactions between solute components of known molar mass can be obtained by direct analysis of the experimental data. Interpretation of the measured thermodynamic parameters in terms of molecular interactions requires that an arbitrary distinction be made between nonassociative forces, like hard-sphere volume-exclusion and mean-field electrostatic repulsion or attraction, and specific short-range forces of association that give rise to the formation of molecular aggregates. Provided the former can be accounted for adequately, the effects of the latter can be elucidated in the form of good estimates of the equilibrium constants for the reactions of aggregation.

Key words Sedimentation equilibrium · Molecular interactions · Thermodynamic nonideality · Equilibrium constants

Introduction

Over the last 15 years or so we have sought to arrive at a formulation of the sedimentation equilibrium equation which gives an account of thermodynamic nonideality that is at the same time both exact and useful. While it is always possible to write down exact thermodynamic expressions describing the relationship between the most basic thermodynamic variables, it is not always possible to preserve that exactness in recasting the equations in a form that relates the commonly measured experimental quantities to one another. In the case of sedimentation equilibrium one normally determines the concentration of solute as a function of radial distance in the ultracentrifuge cell.

The main result of our investigations is that the osmotic activity is the natural variable to be used in describing an incompressible solution at sedimentation equilibrium. The assumption of incompressibility, that is, that the partial specific volumes of both solute and solvent components are independent of solution composition, has been found to hold for a wide variety of macromolecular systems, including solutions of proteins, across a broad range of experimental conditions; however there are very many systems, especially synthetic macromolecules in organic solvents, for which the compressibility of the solution is its defining feature. The analysis in this work is not applicable to any of those cases, but when the assumption of incompressibility is valid a formulation of the thermodynamic theory of sedimentation equilibrium in terms of the variables defined in the usual description of osmotic pressure studies allows the differentials of the experimental variables appearing in the sedimentation equilibrium equation to be separated. The equation can then be integrated without further approximation. When this procedure is followed the buoyancy term contains the density of the solvent as opposed to the density of the solution, variation of which with respect to solute concentration or radial distance has always proved troublesome in conventional data analysis. In fact, when the buoyancy term contains the solution density, its variation with respect to solute concentration is of first order. The need to incorporate this correction into procedures for the analysis of data has obfuscated interpretation of thermodynamic nonideality, whose largest contribution, through the second virial coefficient, is also of first order in the solute concentration.

Formulation of the sedimentation equilibrium equation for incompressible solutions in terms of the osmotic activity has two further advantages. The first is that in a multicomponent solution the single-solute equation is valid for each component independently. Of course the variation of the osmotic activity of each solute varies with radial distance in a way that depends on the concentrations of all solute species and, thus, analysis of thermodynamic nonideality still requires elucidation of the distribution of each component within the centrifuge cell. However, the second advantage of the formulation of the sedimentation equilibrium equation in terms of osmotic activities is that the polynomial coefficients, obtained by using ordinary fitting procedures on appropriately transformed experimental data, have a rigorously defined statistical mechanical meaning. They are osmotic virial coefficients which, through Mayer f functions, depend only on the energy of interaction as a function of the distances of separation within a cluster comprised of a small number of molecules, two or three for the coefficients whose values can usually be reasonably estimated by ordinary data-fitting to traces from sedimentation equilibrium experiments. In the McMillan–Mayer theory of osmosis [1–3] the energy of interaction between solute molecules is determined at constant chemical potential of all dialysable solvent species and thus corresponds to the conditions for sedimentation equilibrium described by Cassasa and Eisenberg [4] for a single nondialysable component in a multicomponent solvent.

It has long been implicit in the analysis of reversible association reactions that they are no more than a special manifestation of thermodynamic nonideality [5]. In fact macroscopic thermodynamic measurements cannot distinguish between the short-range "associative" forces that give rise to the formation of aggregates and the longer-range "nonassociative" forces, whose effect is usually thought of as comprising nonideality. The notional distinction between associative forces and nonassociative forces can usually be made quite unambiguously, especially in view of the fact that the latter are very often exclusively repulsive (excluded-volume and charge–charge forces), but one must be careful to separate the contributions of different forces to the

parameters that are determined from analysis of experimental data. The description of the sedimentation equilibrium of incompressible solutions in terms of osmotic activities allows this to be done with full transparency. It turns out that the osmotic virial coefficients can be separated into contributions due to the different forces, associative or nonassociative, that operate within the cluster of molecules whose interactions define the magnitude of the coefficient in question. In the simplest case the full second virial coefficient for the interaction between two components is the sum of the association constant and the usual integral over the Mayer f function defined for the energy of interaction due to nonassociative forces [6].

Sedimentation equilibrium of a single solute

At sedimentation equilibrium, the variation with radial distance, r, of the chemical potential, μ_i, of a component i of the solution placed in the rotor cell can be expressed as

$$\mathrm{d}\mu_i = M_i \omega^2 r \mathrm{d}r \ , \tag{1}$$

where M_i is the molar mass of the component and ω is the angular velocity at which the rotor is spinning. We begin by restricting consideration to an isothermal two-component solution in which A is the solute and s is the solvent. Following a procedure more familiar in studies of osmotic pressure, we choose the molar concentration of solute, C_A, and the solvent chemical potential, μ_s, as the independent thermodynamic variables in terms of which we express the variation in chemical potential of A:

$$\mathrm{d}\mu_A = (\partial \mu_A / \partial \mu_s)_{T,C_A} \mathrm{d}\mu_s + (\partial \mu_A / \partial C_A)_{T,\mu_s} \mathrm{d}C_A. \tag{2}$$

By combining Eqs. (1) and (2) and applying the sedimentation equilibrium condition to the solvent we obtain

$$\left[M_A - M_s (\partial \mu_A / \partial \mu_s)_{T,C_A} \right] \omega^2 r \mathrm{d}r = (\partial \mu_A / \partial C_A)_{T,\mu_s} \mathrm{d}C_A. \tag{3}$$

Assumption of incompressibility

The factor $(\partial \mu_A / \partial \mu_s)_{T,C_A}$ is the quotient of the partial derivatives of μ_A and μ_s with respect to pressure, P (at constant temperature, T, and molar concentration of solute, C_A). In the case of an incompressible solution, these derivatives can be replaced by appropriate expressions involving only the partial specific volumes of solute and solvent. The definition of the partial specific volume of the solute is

$$\bar{v}_A = (\partial V / \partial n_A)_{T,P,n_s} / M_A \ , \tag{4}$$

where n_A is the number of moles of solute in a solution of volume V. The partial specific volume of the solvent has a similar definition. The assumption of incompressibility renders $\bar{v}_A$ and $\bar{v}_s$ constant, independent of C_A and under this assumption we can make the substitutions $\rho_s \sim 1/\bar{v}_s$,

$$(\partial\mu_A/\partial P)_{T,C_A} \sim M_A\bar{v}_A \tag{5}$$

and

$$(\partial P/\partial\mu_s)_{T,C_A} \sim \rho_s/M_s \tag{6}$$

to obtain the result

$$M_A(1-\bar{v}_A\rho_s)\omega^2 r dr = (\partial\mu_A/\partial C_A)_{T,\mu_s} dC_A. \tag{7}$$

When the solution is so dilute that it can be considered ideal

$$(\partial\mu_A/\partial C_A)_{T,\mu_s} \approx RT/C_A , \tag{8}$$

from which we retrieve the form often used for molecular-weight determination

$$\frac{M_A(1-\bar{v}_A\rho_s)\omega^2}{2RT} \approx \frac{d\ln C_A}{dr^2} , \tag{9}$$

where R is the universal gas constant.

Thermodynamic nonideality

The applicability of Eq. (7) is not restricted to dilute solutions. It is an exact thermodynamic expression for all incompressible solutions and can be used for the study of the thermodynamic nonideality that arises owing to molecular interactions in such solutions. It is convenient to define a thermodynamic activity z_A as a product of the molar concentration and an activity coefficient γ_A,

$$z_A = \gamma_A C_A . \tag{10}$$

The form of the ideal dependence of the chemical potential on solute concentration is then preserved,

$$\mu_A(T,\mu_s,C_A) = \mu_A^0(T,\mu_s) + RT\ln z_A(C_A) , \tag{11}$$

allowing the sedimentation equilibrium equation to be written in the form

$$\frac{M_A(1-\bar{v}_A\rho_s)\omega^2}{2RT} = \frac{d\ln z_A}{dr^2} . \tag{12}$$

The separation of the experimental variables z_A and r means that this equation can be integrated directly to give

$$z_A(r) = z_A^0\psi^0(r) , \tag{13}$$

where the ψ^0 function is defined as

$$\psi^0(r) = \exp[M_A(1-\bar{v}_A\rho_s)\omega^2 r^2/2RT] . \tag{14}$$

The quantity z_A^0 is a constant of integration and was chosen such that it corresponds to the activity of the solute at $r=0$ in a hypothetical column of solution that extends to the centre of rotation. [Use of the usual function $\psi(r)$ defined relative to any other reference point, r_F, requires use of the appropriate constant of integration, $z_A(r_F)$.]

Equations (12), (13) and (14) establish the osmotic activity z_A as the natural variable for the description of sedimentation equilibrium in incompressible solutions and herald the central role of this quantity in the ultracentrifugal study of globular proteins and other polymers under conditions of incompressibility that maintain the constancy of the partial specific volumes $\bar{v}_A$ and $\bar{v}_s$ [7, 8].

Comparison with conventional derivation

In the conventional derivation of the sedimentation equilibrium equation [9] the solute molality, m_A, and pressure, P, are chosen as the independent thermodynamic parameters. The general expression for the variation in the chemical potential of A is then

$$d\mu_A = (\partial\mu_A/\partial P)_{T,m_A} dP + (\partial\mu_A/\partial m_A)_{T,P} dm_A . \tag{15}$$

The variation in pressure, dP, can be expressed in terms of the local solution density, $\rho(r)$, to give

$$M_A[1-\bar{v}_A\rho(r)]\omega^2 r dr = (\partial\mu_A/\partial m_A)_{T,P} dm_A . \tag{16}$$

The appropriate thermodynamic activity, a_A, is a natural product of the molal concentration, m_A, which is n_A/n_s, and an activity coefficient, y_A,

$$a_A = y_A m_A , \tag{17}$$

and is defined through the relation

$$\mu_A(T,P,m_A) = \mu_A^0(T,P) + RT\ln a_A(m_A) , \tag{18}$$

whence

$$\frac{M_A[1-\bar{v}_A\rho(r)]\omega^2}{2RT} = \frac{d\ln a_A}{dr^2} . \tag{19}$$

It might be thought that this equation can easily be integrated to yield an expression useful for the analysis of sedimentation equilibrium data obtained from nonideal solutions, but this is not the case because $\rho(r)$ depends on solute activity, a_A. Another approach might be to transform Eq. (6) from molal concentration, m_A, to molar concentration, C_A. In the case of incompressible solutions this is trivial [9], leading to the relation

$$M_A[1-\bar{v}_A\rho(r)]\omega^2 r dr = (\partial\mu_A/\partial C_A)_{T,P} dC_A , \tag{20}$$

but the partial derivative $(\partial\mu_A/\partial C_A)_{T,P}$ at constant pressure (rather than solvent chemical potential) is not easily reduced to a simple form and the problem of the dependence of $\rho(r)$ on C_A persists, hindering the development of a simple equation that gives an exact account of thermodynamic nonideality. Failure to recognise this has caused enormous confusion and disputation and has, over decades, hindered the proper analysis of sedimentation equilibrium for the study of molecular interactions. Taking into account the relationship for incompressible solutions,

$$[1 - \bar{v}_A \rho(r)] = (1 - \bar{v}_A \rho_s)[1 - M_A \bar{v}_A C_A(r)] \ , \qquad (21)$$

and evaluation of $(\partial \mu_A / \partial C_A)_{T,P}$ allows Eq. (12) to be derived from Eq. (20), but the path is rather tortuous and without the result at hand the simplification that can be achieved is not evident and has remained unhelpfully obscure.

Analysis of experimental data

The reason that Eq. (12) turns out to be so useful is that the relationship between C_A and z_A can be expressed in terms of rigorously defined statistical mechanical parameters. The simplest form of this relationship is [10]

$$C_A = z_A + 2b_{AA} z_A^2 + 3b_{AAA} z_A^3 + \cdots \ , \qquad (22)$$

where the coefficients b come from the expansion of the osmotic pressure, Π, in terms of solute activity:

$$\frac{\Pi}{RT} = z_A + b_{AA} z_A^2 + b_{AAA} z_A^3 + \cdots \ . \qquad (23)$$

The coefficients b are related to the more familiar virial coefficients B found in the expansion of Π in terms of solute molar concentration,

$$\frac{\Pi}{RT} = C_A + B_{AA} C_A^2 + B_{AAA} C_A^3 + \cdots \ , \qquad (24)$$

through the relationships

$$b_{AA} = -B_{AA} \qquad (25)$$

and

$$b_{AAA} = -2B_{AAA}/2 + 2B_{AA}^2 \ . \qquad (26)$$

The experimental trace of a sedimentation equilibrium pattern consists of measurements of solute concentration as a function of radial distance. After making the transformation $r \rightarrow \psi^0(r)$ such data can be analysed by direct fitting to the equation

$$C_A(r) = z_A^0 \psi^0(r) + 2b_{AA} \left[z_A^0 \psi^0(r)\right]^2 + 3b_{AAA} \left[z_A^0 \psi^0(r)\right]^3 + \cdots \qquad (27)$$

to find best-fit values of z_A^0 and a number of coefficients b_{AA}, b_{AAA}, etc. [8]. This method of data analysis has been used to measure the second virial coefficient for ovalbumin under varying electrolytic conditions [11].

It is obviously necessary to exercise wisdom in the use of this procedure, taking into account the number of parameters that can be estimated from data of limited precision as well as correlations between the estimates of different parameters that the fitting procedure returns.

Any procedure used for estimating thermodynamic parameters from sedimentation equilibrium data must take account of such constraints.

Interpretation of parameter estimates

The advantage of using Eq. (27) or derivatives of it is that the osmotic coefficients relate to molecular properties in an exactly known way. The virial coefficients B_{AA}, B_{AAA}, etc. are essentially cluster integrals over Mayer f functions [1, 6] that are specified in terms of the potential of mean force, $u(\mathbf{x})$, between two molecules (whose separation vector is $\mathbf{x}$) and thermal energy kT (in which k is the Boltzmann constant):

$$f(\mathbf{x}) = \exp[-u(\mathbf{x})/kT] - 1 \ . \qquad (28)$$

For a spherically symmetric potential $u(x)$ the second virial coefficient B_{AA} has the form

$$B_{AA} = -2\pi \int_0^\infty f(x) x^2 \mathrm{d}x \qquad (29)$$

and, assuming the triplet potential is decomposable in a pairwise additive fashion, the third virial coefficient B_{AAA} has the more complicated form [12]

$$B_{AAA} = -\frac{8\pi^2}{3} \int_0^\infty f(x_{12}) x_{12} \mathrm{d}x_{12} \int_0^\infty f(x_{13}) x_{13} \mathrm{d}x_{13}$$
$$\times \int_{|r_{12}-r_{13}|}^{r_{12}+r_{13}} f(x_{23}) x_{23} \mathrm{d}x_{23} \ , \qquad (30)$$

reflecting the interactions between pairs i and j within a cluster of three molecules. Even more complicated forms are available for higher virial coefficients, but they are unlikely to be required often or to be useful in the study of macromolecular interactions.

In the case of globular proteins the assumption of effectively spherical geometry allows the coefficients B_{AA} and B_{AAA} to be accounted for at about the level of precision to which they can be determined from analysis of experimental data. The relevant representation of the potential of mean force, $u(x)$, is divided into hard-sphere (excluded-volume) and screened electrostatic contributions. The extent of the former is determined by the molecular diameter, $2R$, and the latter depends on the magnitude of the Debye–Hückel parameter κ, the dielectric constant of the medium, D, and the net charge, Z, on the macromolecules measured in units of electronic charge, e:

$$u(x) = \begin{cases} \infty & x < 2R \\ \dfrac{Z^2 e^2}{D(1+\kappa R)^2} \dfrac{\exp[-\kappa(x-2R)]}{x} & x \geq 2R \end{cases} \qquad (31)$$

Substitution of Eq. (31) into Eq. (28) followed by integration of Eq. (29) yields

$$B_{\mathrm{AA}} = \frac{16}{3}\pi R^3 + \frac{(1+2\kappa R)}{4\kappa^2}\left(\frac{Z}{1+\kappa R}\right)^2\left(\frac{8\pi e^2}{DkT}\right)$$
$$- \frac{1}{128\pi\kappa}\left(\frac{Z}{1+\kappa R}\right)^4\left(\frac{8\pi e^2}{DkT}\right)^2 + \cdots . \tag{32}$$

The first term arises from the hard-sphere potential and the terms that follow arise from consideration of the Mayer f function expanded in orders of $u(x)/kT$:

$$f(x) = -\frac{u(x)}{kT} + \frac{1}{2!}\left(\frac{u(x)}{kT}\right)^2 - \cdots . \tag{33}$$

Because the Debye–Hückel parameter κ is a simple function of the ionic strength, I, of the solvent medium

$$\kappa = \sqrt{\frac{8\pi e^2 I}{DkT}} , \tag{34}$$

equation (32) can be written in terms of I rather than κ. In the low ionic strength limit ($\kappa R \ll 1$) we obtain the very simple result

$$B_{\mathrm{AA}} \approx \frac{16}{3}\pi R^3 + \frac{1}{4}\frac{Z^2}{I} - \frac{\kappa^3}{128\pi}\frac{Z^4}{I^2} + \cdots . \tag{35}$$

The first electrostatic term ($Z^2/4I$) has been used for many years and can provide a reasonable estimate of this contribution even at moderate ionic strength ($\kappa^2 R^2 \ll 1$) provided the charge Z is small enough to maintain the condition $u(2R) \ll 1$ required to justify consideration of only the first term in the expansion of $f(x)$.

Integration of Eq. (30) gives a more complicated result for the third virial coefficient:

$$B_{\mathrm{AAA}} = \frac{160\pi^2 R^6}{9} + 4\pi R^5\left(\frac{Z}{1+\kappa R}\right)^2\left(\frac{8\pi e^2}{DkT}\right)$$
$$\left(\frac{11}{15} - \frac{38}{90}\kappa R + \cdots\right)$$
$$- \frac{R^3\exp(4\kappa R)}{3\kappa}\left(\frac{Z}{1+\kappa R}\right)^4\left(\frac{8\pi e^2}{DkT}\right)^2(1+\cdots)$$
$$+ \frac{\exp(-2\kappa R)}{768\pi\kappa^3}\left(\frac{Z}{1+\kappa R}\right)^6\left(\frac{8\pi e^2}{DkT}\right)^3 + \cdots . \tag{36}$$

Only terms in $(1/kT)^{3/2}$ have been included in Eq. (36) and simplification in the $\kappa R \ll 1$ limit gives

$$B_{\mathrm{AAA}} \approx \frac{160\pi^2 R^6}{9} + \frac{44\pi\kappa^2 R^5}{15}\frac{Z^2}{I} - \frac{\kappa^3 R^3}{3}\frac{Z^4}{I^2}$$
$$+ \frac{\kappa^3}{768\pi}\frac{Z^6}{I^3} + \cdots , \tag{37}$$

which is provided for comparison with Eq. (35).

Multicomponent systems

The entire analysis presented for a single solute can be recapitulated for an incompressible solution comprising

any number of independent "solvent" and "solute" components. Here we present only the main results. We regard small dialysable components, s, as belonging to the solvent and we make the implicit assumption that they do not sediment significantly in the centrifuge cell. This multicomponent "solvent" is the dialysate that is placed in the reference cell of the centrifuge rotor. The basic sedimentation equilibrium relation for a component i of molar mass M_i is

$$\left(M_i - \sum_s M_s(\partial\mu_i/\partial\mu_s)_{T,\{C_j\}}\right)\omega^2 r\,\mathrm{d}r$$
$$= \sum_j (\partial\mu_i/\partial C_j)_{T,\{\mu_s\},\{C_{k\neq j}\}}\,\mathrm{d}C_j , \tag{38}$$

which reduces, in the case of an incompressible solution, to

$$M_i(1 - \bar{v}_i^*\rho_0)\omega^2 r\,\mathrm{d}r = \sum_j (\partial\mu_i/\partial C_j)_{T,\{\mu_s\},\{C_{k\neq j}\}}\,\mathrm{d}C_j , \tag{39}$$

where ρ_0 is the density of the osmotic solvent comprising all its dialysable components at their specified weight concentrations, c_s, that is, $M_s C_s$,

$$\rho_0 = \sum_s c_s , \tag{40}$$

and the buoyancy term corresponds to the density increment determined under osmotic conditions, that is, the chemical potentials of all solvent components held constant [13]

$$(\partial\rho/\partial c_i)_{T,\{C_{j\neq i}\},\{\mu_s\}} = (1 - \bar{v}_i^*\rho_0) . \tag{41}$$

The corresponding partial specific volume quantity is defined as

$$\bar{v}_i^* = (\partial V/\partial n_i)_{T,P,\{n_{j\neq i}\},\{\mu_s\}}/M_i .$$

In an incompressible solution the density increment of each solute component is independent of solution and solvent composition and we can make the substitution $\bar{v}_i : \bar{v}_i^*$. (Complications arise when there are "preferential interactions" between a solute component and different solvent components and the density increment is not independent of solvent composition [13].) The chemical potential of a component, written in terms of osmotic activity z_i measured at constant chemical potential of all solvent components, is a function of the concentrations of all solute components

$$\mu_i(T,\{\mu_s\},\{C_j\}) = \mu_i^0(T,\{\mu_s\}) + RT\ln z_i(\{C_j\}) . \tag{43}$$

In these terms the sedimentation relation for a solute component of an incompressible solution can be written as

$$\frac{M_i(1 - \bar{v}_i\rho_0)\omega^2}{2RT} = \frac{\mathrm{d}\ln z_i}{\mathrm{d}r^2} \tag{44}$$

and then we obtain the integrated form

$$z_i(r) = z_i^0\psi_i^0(r) \ , \tag{45}$$

where

$$\psi_i^0(r) = \exp\left[M_i(1 - \bar{v}_i\rho_0)\omega^2 r^2/2RT\right] \ . \tag{46}$$

In his comments on the corresponding treatment for a simple binary solution (Eqs. 12, 13, 14) Fujita [14] did not rule out such a simple result for incompressible multicomponent solutions. The general expressions for the osmotic pressure (measured at constant chemical potential of all solvent components) in terms of either osmotic activities,

$$\frac{\Pi}{RT} = \sum_i z_i + \sum_{\{i,j\}} b_{ij}z_iz_j + \sum_{\{i,j,k\}} b_{ij}z_iz_jz_k + \ , \tag{47}$$

or molar concentrations,

$$\frac{\Pi}{RT} = \sum_i C_i + \sum_{\{i,j\}} B_{ij}C_iC_j + \sum_{\{i,j,k\}} B_{ij}C_iC_jC_k + \cdots \ , \tag{48}$$

are defined in terms of sets of indices $\{i,j\}$, $\{i,j,k\}$, etc. that run over all combinations rather than permutations of components and the coefficients are related to one another through the equations

$$b_{ii} = -B_{ii} \ , \tag{49}$$

$$b_{ij} = -B_{ij}; \quad j \neq i \ , \tag{50}$$

$$b_{iii} = -B_{iii}/2 + 2B_{ii}^2 \ , \tag{51}$$

$$b_{iij} = -B_{iij}/2 + 2B_{ij}B_{ii} + B_{ij}^2; \quad j \neq i \ , \tag{52}$$

$$b_{ijk} = -B_{ijk}/2 + B_{ij}B_{ik} + B_{ik}B_{jk} + B_{ij}B_{jk}; \quad j \neq i; \\ k \neq j; \quad k \neq i \ . \tag{53}$$

It should be noted that these virial coefficients are defined for particular values of the chemical potentials of all solvent components. The concentration of a solute component C_i is then given in terms of the osmotic activities of all components as

$$C_i = z_i + 2b_{ii}z_i^2 + \sum_{j\neq i} b_{ij}z_iz_j + 3b_{iii}z_i^3 + 2\sum_{j\neq i} b_{iij}z_i^2z_j \\ + \sum_{j\neq i} b_{ijj}z_iz_j^2 + \sum_{j\neq i}\sum_{\substack{k\neq j \\ k\neq i}} b_{ijk}z_iz_jz_k + \cdots \ . \tag{54}$$

Substitution of Eq. (45) into Eq. (54) allows the experimental data, C_i versus r, to be fitted in terms of the fundamental thermodynamic relationship between the experimental variables. It is useful to bear in mind that

$$\psi_i^0(r)^{M_i\phi_i} = \psi_j^0(r)^{M_j\phi_j} \ , \tag{55}$$

where

$$\phi_i = (1 - \bar{v}_i\rho_0)\omega^2/2RT \ , \tag{56}$$

which means that the ψ^0 functions for different components are simple powers of one another and any one can be chosen as the independent variable in data analysis. The result of global fitting to data from any number of centrifuge experiments is the determination of z_i^0 for each component in each experiment and best-fit values of a number of thermodynamic parameters b_{ii}, b_{ij}, etc. which can be interpreted through their statistical mechanical definitions taking into account hard-sphere and screened electrostatic interactions.

Heterogeneous interactions

When dealing with multicomponent systems it is necessary to take into account virial coefficients that arise owing to interactions between dissimilar molecules. In the case of the second virial coefficient, consideration of the interaction between one molecule of each component suffices. In general

$$B_{AB} = -4\pi \int_0^\infty f(x)x^2\mathrm{d}x, \quad A \neq B \ . \tag{57}$$

If both types of molecule, A and B, can be represented as effective spheres of radii R_A and R_B carrying charges Z_A and Z_B on their surfaces, the second virial coefficient, B_{AB}, can be calculated as

$$B_{AB} = \frac{4\pi}{3}(R_A + R_B)^3 + \frac{(1 + \kappa R_A + \kappa R_B)}{2\kappa^2} \\ \times \left(\frac{Z_AZ_B}{(1+\kappa R_A)(1+\kappa R_B)}\right)\left(\frac{8\pi e^2}{DkT}\right) \\ - \frac{1}{64\pi\kappa}\left(\frac{Z_AZ_B}{(1+\kappa R_A)(1+\kappa R_B)}\right)^2\left(\frac{8\pi e^2}{DkT}\right)^2 + \cdots, \quad A \neq B \tag{58}$$

In the case that A is a sphere of radius R_A and B is an uncharged linear random-walk polymer of mean-square end-to-end length l_B the second virial coefficient has the form [15]

$$B_{AB} = \frac{2\pi}{3}R_Al_B^2 + 4\sqrt{\frac{2\pi}{3}}R_A^2l_B^2 + \frac{4\pi}{3}R_A^3 \ . \tag{59}$$

This equation has been used to interpret measurements of thermodynamic nonideality arising owing to interactions between proteins and chain polymers [16].

Treatment of aggregation

Equations (21), (22), (23), (24), (25) and (26) are completely general thermodynamic relations describing the behaviour of a solute in a solvent and as such they say nothing about the molecular forces that give rise to the parameters B_{AA}, B_{AAA}, etc. The same equations apply equally whether the only forces between the solute molecules are of the nonassociative hard-sphere and electrostatic kind or whether there are specific associative forces that give rise to stable molecular aggregates. In the latter case it is usual to identify monomers (A) and dimers ($A_2\equiv C$), trimers ($A_3\equiv D$), etc. as separate components for each of which it is possible to assign a thermodynamic activity whose value depends on the monomer activity z_A through a rigorously defined equilibrium constant

$$z_C = K_C z_A^2 \; , \tag{60}$$

$$z_D = K_D z_A^3 \; . \tag{61}$$

Since C and D are being treated as separate components it is necessary to use the multicomponent description of the solution given by Eqs. (38), (39), (40), (41), (42), (43), (44), (45), (46), (47), (48), (49), (50), (51), (52), (53), (54) and (55). It is impossible to specify equilibrium constants for aggregation reactions without the notional separation of intermolecular forces into those which are "associative" and those which are "nonassociative" because there is no thermodynamic procedure for labelling configurations of molecules which represent an aggregate or "complex" as distinct from configurations in which the molecules still have separate identities. However, having made the notional separation between associative and nonassociative intermolecular forces, the thermodynamic parameters b_{AA}, b_{AAA}, etc. that can be measured directly become interpreted in terms of the equilibrium constants K_C, K_D, etc. and virial coefficients B_{AA}^*, B_{AC}^*, B_{AAA}^*, etc. accounting for nonassociative interactions between molecules of the notional components A, C, D, etc:

$$b_{AA} = -B_{AA}^* + K_C \; , \tag{62}$$

$$b_{AAA} = -B_{AAA}^*/2 + B_{AA}^{*\,2} - B_{AC}^* K_C + K_D \; . \tag{63}$$

The equilibrium constant K_C accounts for the effect of the short-range associative forces between two monomers that give rise to the formation of the dimer C, and the virial coefficient B_{AA}^* accounts for the effect of nonassociative forces, including hard-sphere and longer-range electrostatic interactions between two monomers, according to Eq. (29). Specification of B_{AC}^* requires consideration of the interaction between a monomer and a dimer.

Dumbbell dimers

For spherical monomers, the virial coefficient for the interaction between a monomer and a simple dumbbell-shaped dimer has the explicit form

$$B_{AC} = 18\pi R^3 - 8\pi \int_0^{\pi/2} \sin\phi\,\mathrm{d}\phi \int_{R_\phi}^\infty f_{el}(x,\phi) x^2 \,\mathrm{d}x \; , \tag{64}$$

where $\mathbf{x}$ is the vector from the centre of the dumbbell dimer to the centre of the monomer and ϕ is the azimuthal angle that the vector $\mathbf{x}$ forms with the major axis of the dumbbell. In Eq. (64) we have already integrated over the hard-sphere potential taking into account that the hard-sphere radius is a function of ϕ given by

$$R_\phi = R\left(\cos\phi + \sqrt{3 + \cos^2\phi}\right) \; . \tag{65}$$

The form of the Mayer f function $f_{el}(x, \phi)$ appearing in Eq. (64) and accounting for the electrostatic contribution to B_{AC} is given in terms of the energy

$$u_{el}(x) = \frac{Z^2 e^2}{D(1 + \kappa R)^2} \left(\frac{\exp[-\kappa(x_1 - 2R)]}{x_1} + \frac{\exp[-\kappa(x_2 - 2R)]}{x_2}\right) \quad x \geq R_\phi \; , \tag{66}$$

where x_1 and x_2 are the distances of the monomer from the centre of each of the spheres comprising the dumbbell,

$$\left.\begin{array}{c} x_1 \\ x_2 \end{array}\right\} = \sqrt{x^2 + R^2 \pm 2xR\cos\phi} \; . \tag{67}$$

These formulae are likely to be useful when the actual shape of the protein under consideration is close to spherical. When this is not the case excluded-volume and charge–charge interaction contributions to B_{AC} can still be approximated reasonably by representing the monomer as a sphere of effective radius R_A carrying Z electronic charges and the dimer as a sphere of effective radius $R_C = \sqrt[3]{2}R_A$ carrying $2Z$ electronic charges.

General case

In the most general case, presented here for the sake of completeness, a solution may contain any number of separate components $i,j,k...$ able to form dimeric complexes, either homomolecular (*ii*) or heteromolecular (*ij*), and trimolecular complexes (*iii*), (*iij*), (*ijk*), and so on. The formation of any complex (*ijk...*) is governed by a thermodynamic equilibrium constant, such that

$$z_{ijk..} = K_{ijk..} z_i z_j z_k \cdots \tag{68}$$

With the notional separation of intermolecular forces among clusters of molecules into those which are "associative" and those which are "nonassociative", Eq. (47) becomes, once again to third order,

$$\frac{\Pi}{RT} = \sum_i z_i + \sum_i \left(b_{ii}^* + K_{ii}\right)z_i^2$$
$$+ \sum_i \left(b_{iii}^* + K_{iii} + b_{i(ii)}^* K_{ii}\right)z_i^3 + \sum_{j \neq i}\left(b_{ij}^* + K_{ij}\right)z_i z_j$$
$$+ \sum_i \sum_{j \neq i} \left(b_{iij}^* + K_{iij} + b_{(ii)j}^* K_{ii} + b_{i(ij)}^* K_{ij}\right)z_i^2 z_j$$
$$+ \sum_i \sum_{\substack{j \neq i}} \sum_{\substack{k \neq j \\ k \neq i}} \left(b_{ijk}^* + K_{ijk} + b_{i(jk)}^* K_{jk}\right)z_i z_j z_k + \cdots ,$$

$$(69)$$

which in turn allows Eq. (54) to be written as

$$C_i = z_i + 2\left(b_{ii}^* + K_{ii}\right)z_i^2 + 3\left(b_{iii}^* + K_{iii} + b_{i(ii)}^* K_{ii}\right)z_i^3$$
$$+ \sum_{j \neq i}\left(b_{ij}^* + K_{ij}\right)z_i z_j$$
$$+ 2\sum_{j \neq i}\left(b_{iij}^* + K_{iij} + b_{(ii)j}^* K_{ii} + b_{i(ij)}^* K_{ij}\right)z_i^2 z_j$$
$$+ \sum_{j \neq i}\left(b_{ijj}^* + K_{ijj} + b_{(ij)j}^* K_{ij} + b_{i(jj)}^* K_{jj}\right)z_i z_j^2$$
$$+ \sum_{j \neq i} \sum_{\substack{k \neq j \\ k \neq i}} \left(b_{ijk}^* + K_{ijk} + b_{i(jk)}^* K_{jk}\right)z_i z_j z_k + \cdots .$$

$$(70)$$

Equations (49), (50), (51), (52) and (53) now specify the relationship between the quantities $b_{ijk\ldots}^*$ and the osmotic virial coefficients $B_{ijk\ldots}^*$ that arise from the intermolecular forces, hard-sphere and electrostatic, that are identified as "nonassociative". It is important to note that some coefficients, like $b_{i(ii)}^*$ and $b_{i(jk)}^*$, arise from nonassociative forces between elementary molecules, like i, and aggregates, like (ii) and (jk). The thermodynamic parameters that can be determined in sedimentation equilibrium experiments are expressed in terms of the notional osmotic virial coefficients $B_{ijk\ldots}^*$ and the equilibrium constants $K_{ijk\ldots}$ as

$$b_{ii} = -B_{ii}^* + K_{ii} , \tag{71}$$

$$b_{ij} = -B_{ij}^* + K_{ij}; \quad i \neq j , \tag{72}$$

$$b_{iii} = -B_{iii}^*/2 + 2B_{ii}^{*2} + K_{iii} - B_{i(ii)}^* K_{ii}, \tag{73}$$

$$b_{iij} = -B_{iij}^*/2 + 2B_{ij}^* B_{ii}^* + B_{ij}^{*2} + K_{iij}$$
$$- B_{(ii)j}^* K_{ii} - B_{i(ij)}^* K_{ij}; \quad j \neq i , \tag{74}$$

$$b_{ijk} = -B_{ijk}^*/2 + B_{ij}^* B_{ik}^* + B_{ik}^* B_{jk}^* + B_{ij}^* B_{jk}^* + K_{ijk}$$
$$- B_{i(jk)}^* K_{jk}; \quad j \neq i; \quad k \neq j; \quad k \neq i . \tag{75}$$

The limited case of a system comprising just two components has been treated explicitly and it has been shown that the equilibrium constant for heterogeneous association, K_{AB}, can be determined by direct fitting to data from multiple sedimentation equilibrium experiments conducted using diverse loading concentrations of each component. Readers are referred to a recent study [17] for a detailed demonstration of the procedures for the estimation of equilibrium constants.

Conclusion

The expression of the sedimentation equilibrium equation for an incompressible solution in terms of osmotic activities allows its exact form to be used directly for the analysis and interpretation of experimental data. Procedures based on the use of the osmotic activities as the independent thermodynamic variables should find wide application in biomolecular systems for which the assumption of solution incompressibility is justifiable, but an equivalent treatment applicable to compressible solutions of synthetic macromolecules in organic solvents awaits development.

Acknowledgements P.R.W. thanks the Alexander von Humboldt Stiftung for its generous support by way of a research fellowship and John McCaskill for providing a congenial environment as well as mathematical help.

References

1. McMillan WG, Mayer JE (1945) J Chem Phys 13:276
2. Hill TL (1955) J Chem Phys 23:623
3. Hill TL (1955) J Chem Phys 23:2270
4. Casassa EF, Eisenberg H (1964) Adv Protein Chem 19:287
5. Prigogine I, Defay R (1954) Chemical thermodynamics (Everett DH translator). Longhams, London
6. Mayer JE (1950) J Chem Phys 18:1426
7. Winzor DJ, Wills PR (1994) In: Schuster TM, Laue TM (eds) Modern analytical ultracentrifugation. Birkhäuser, Boston, pp 66–80
8. Wills PR, Jacobsen MP, Winzor DJ (1996) Biopolymers 38:119
9. Williams JW, Van Holde KE, Baldwin RL, Fujita H (1958) Chem Revs 58:715
10. Hill TL, Chen YD (1973) Biopolymers 12:1285
11. Wills PR, Hall DR, Winzor DJ (2000) Biophys Chem 84:217
12. Hill TL (1956) Discuss Faraday Soc 21:31
13. Wills PR, Winzor DJ (1993) Biopolymers 33:1627
14. Fujita, H (1994) In: Schuster TM, Laue TM (eds) Modern analytical ultracentrifugation. Birkhäuser, Boston, pp 3–14
15. Jansons KM, Phillips CG (1990) J Colloid Interface Sci 137:75
16. Wills PR, Georgalis Y, Dijk J, Winzor DJ (1995) Biophys Chem 57:37
17. Wills PR, Jacobsen MP, Winzor DJ (2000) Biophys J 79:2178

Progr Colloid Polym Sci (2002) 119:121–130
© Springer-Verlag 2002

HYDRODYNAMICS

Helmut Durchschlag
Peter Zipper

Correlations between crystallographic, small-angle scattering and hydrodynamic data of biopolymers

H. Durchschlag (✉)
Institute of Biophysics and Physical
Biochemistry, University of Regensburg
Universitätsstrasse 31
93040 Regensburg, Germany
e-mail: helmut.durchschlag@biologie.uni-
regensburg.de
Tel.: +49-941-9433041
Fax: +49-941-9432813

P. Zipper
Physical Chemistry, Institute of Chemistry
University of Graz, Heinrichstrasse 28
8010 Graz, Austria

Abstract Whole-body and multi-body approaches as well as empirical relations allow the establishment of manifold correlations between structural and hydrodynamic parameters of biopolymers. Hydrodynamic properties can be predicted reliably from crystallography, small-angle X-ray scattering or other structural data. The inverse procedure, the prediction of structural parameters typical of X-ray studies, can also be performed starting from the results of hydrodynamics. Predictions based on crystal and some other data, however, are only feasible if reliable values or qualified assumptions for polymer hydration are taken into account. Comparisons between solution scattering and crystallographic data for proteins reveal obvious differences which may be ascribed mainly to differences in hydration and surface roughness.

Key words Biopolymers ·
Parameter predictions ·
Crystallography · Small-angle
X-ray scattering · Hydrodynamics

Introduction

To interpret and exploit the information provided by high-resolution crystallography or NMR techniques, solution scattering (small-angle scattering, light scattering), hydrodynamic (analytical ultracentrifugation, viscometry, etc.) and other physicochemical techniques such as electron microscopy or size-exclusion chromatography, it is necessary to correlate the parameters obtained by these techniques. For combining structural and hydrodynamic parameters, both whole-body (WB) and multibody (MB) approaches can be used. Biopolymers of simple shape can be modeled as spheres or prolate/oblate ellipsoids of revolution (PE, OE) [1–10], or other triaxial bodies with unequal axes [11–13]; for complex structures, however, more sophisticated assumptions [7] or application of MB (bead) modeling [14–31] are required. Examples studied so far include various globular biopolymers (simple and conjugated proteins, viruses, ribonucleic acids) of different mass and shape. Hydrodynamic parameters can be predicted from structural data, but also the "inverse problem", i.e. the

anticipation of structural data from the results of hydrodynamics [9, 10], is feasible. For both types of parameter predictions, knowledge of molar masses, M, preferably from sequence or composition data [32], and partial specific volumes, $\bar{v}$, obtained by densimetric or calculative procedures [33–36], is required. For some approaches, the hydration, δ_1, (or the hydrated partial specific volume, $\bar{v}_h$) must be known additionally.

For parameter predictions, preferably radii of gyration, R_G, volumes, V, surface-to-volume ratios, S/V, axial and frictional ratios, p and f/f_0, sedimentation and diffusion coefficients, s and D, Stokes radii, R_D, and intrinsic viscosities, $[\eta]$, are used (Table 1). Predictions based on hydrodynamic data as input parameters are more problematic than the opposite procedure. If atomic coordinates, electron microscopic structures or data of models are to be compared with solution parameters, appropriate hydration contributions and differences in the surface roughness have to be considered.

A variety of relations concerning structural and/or hydrodynamic parameters (e.g., ratios of radii, surface areas and volumes, volume-to-mass ratios, in addition to

Table 1 Predictions of hydrodynamic and structural parameters of biopolymers, based on different calculation approaches and applied parameters and techniques. Abbreviations: whole body (WB), multibody (MB), small-angle X-ray scattering (SAXS), small-angle scattering (SAS), light scattering (LS), electron microscopy (EM), X-ray diffraction (XD), analytical ultracentrifugation (AUC)

Approaches[a]	Parameters	Techniques	Refs.
Prediction of hydrodynamic parameters from structural data: Input parameters: M, $\bar{v}$, (δ_1); R_G, V, S/V. Output parameters: s, D, $[\eta]$; R_0, f/f_0, p, etc.			
WB[b]	(R_G and V) or (R_G and S/V)	SAXS	[1–9]
WB	R_G, V_{dry} and $\delta_1 \rightarrow R_G$ and V	SAS, LS	[7, 8]
WB	p (model), V_{dry} and $\delta_1 \rightarrow R_G$ and V	SAS, LS, EM, viscometry, etc.	[7, 8]
WB	Coordinates and rescaling (δ_1) $\rightarrow R_G$ and V	XD, EM	[5, 6, 25, 26, 28, 43]
MB[c]	Coordinates and rescaling (δ_1)	XD, EM	[18–22, 25, 26, 28, 44]
MB	Coordinates and water shell (δ_1)	XD	[45]
MB	Coordinates and special hydration algorithm for individual water molecules on the protein surface (δ_1)	XD	[38, 39]
Prediction of structural parameters from hydrodynamic data: Input parameters: M, $\bar{v}$, (δ_1); s, D, $[\eta]$. Output parameters: $(\delta_1, \bar{v}_h, V)$, p, R_G, S/V, S, etc.			
WB[b]	(Two of s, D, M) and (δ_1 or $[\eta]$)	AUC, LS, viscometry	[9, 10]

[a] The table summarizes only MB and WB approaches of relevance to the present study. Details of the approaches applied are described in the references of our work
[b] Simple use of spheres or ellipsoids of revolution for hydrodynamic or solution scattering modeling has a long history and is described in many textbooks (e.g. Refs. [46–48]), while modeling by ellipsoids with three unequal axes has become available only in the last few years [11–13]. By contrast, the WB approach applied in this work deals with relations between hydrodynamics and solution scattering, using mainly the radius of gyration ($R_{G,SAXS}$ and $R_{G,Hydro}$) and the hydrated volume, V, in addition to hydrodynamic parameters such as s, D and $[\eta]$

[c] A variety of MB (bead modeling) approaches have been proposed in the last few years, exploiting quite different strategies (bead per residue or atom models, filling models, shell models, finite elements models, molecular dynamics simulations, etc.) summarized recently by García de la Torre et al. [17]. The MB approach used in this study utilizes the "filling model" strategy; bead coordinates and radii have to be adjusted to account for hydration (e.g., by "rescaling"). In context with scattering aspects, this type of modeling seems to be adequate, since the entire particle contributes to scattering parameters such as R_G and V

more complex expressions) can be used effectively in order to analyze molecular aspects (such as anisometry/anisotropy, particle hydration and rugosity) and to allow comparisons on a broader basis. In this context, empirical equations relating diffraction, scattering and hydrodynamic properties can be exploited. For example, from crystallography of monomeric proteins, relations between molar mass, M, packing volume, V_p, accessible surface area, A_s, and radii derived from V_p and A_s (R_p and R_s) have been obtained [37]. These quantities and expressions derived therefrom (e.g., R_{Gp} and R_{Gs}, radii of gyration related to R_p and R_s) can be correlated to similar quantities derived from small-angle X-ray scattering (SAXS). In this context "reduced parameters" can be used advantageously for comparing crystal and solution data [9].

The present article summarizes the previously mentioned predictive possibilities (Table 1) and illustrates certain aspects in detail.

1. Starting from the results of X-ray diffraction (XD) and SAXS on the enzyme catalase, MB and WB approaches are compared for their ability to predict hydrodynamic data.
2. Using novel approaches and known hydration contributions (δ_1) for some biopolymers, structural parameters typical of SAXS investigations (p, R_G,

S/V, S) are calculated from given hydrodynamic datasets.
3. SAXS parameters and several parameter combinations are chosen to compare solution and crystal data quantitatively; in particular, empirical expressions which have been derived from crystallography of proteins are compared to SAXS data.

Results and discussion

Prediction of hydrodynamic parameters from structural data

Using the tetrameric enzyme catalase in a case study, several steps and results of possible approaches to predict hydrodynamic data on the basis of SAXS and/or XD data are illustrated (Fig. 1, Table 2). For details of MB and WB approaches see Refs. [7–10] and Refs. [25, 26, 28], respectively.

The initial MB model for hydrated catalase was established, with beads being placed at the amino acid centers (Fig. 1). The contribution of individual water molecules on the protein surface was taken into account by application of advanced modeling techniques, exploiting the information obtained from modern surface

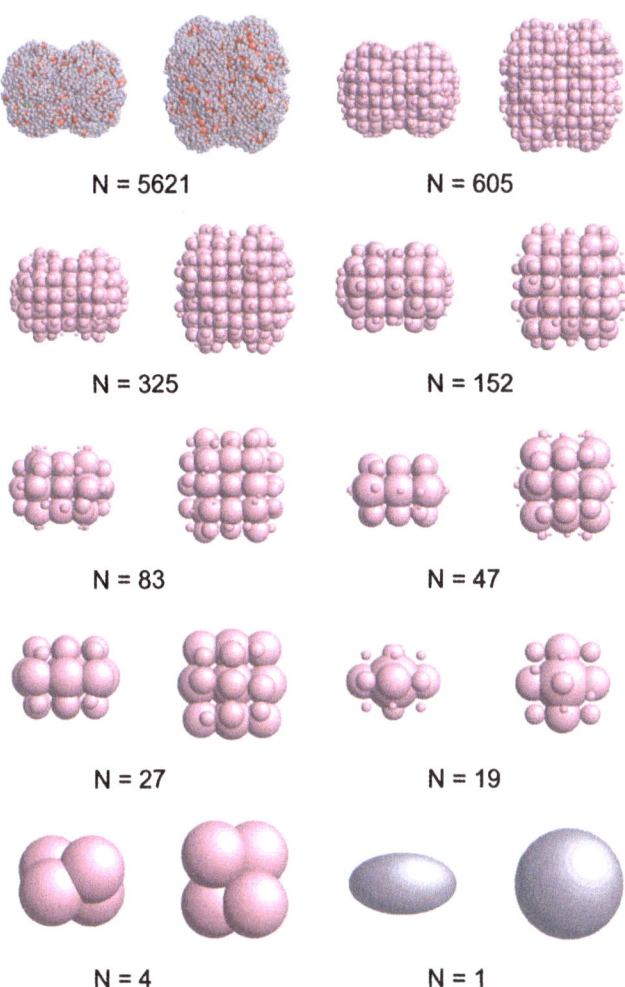

N = 5621   N = 605

N = 325   N = 152

N = 83   N = 47

N = 27   N = 19

N = 4   N = 1

Fig. 1 Side and top views of initial and reduced models for hydrated catalase, consisting of different numbers of beads (*N*). A hydrated model built up from 5,621 beads, which is based on the amino acid and ligand coordinates (2,000 residues), selected SIMS surface points, and a hydration algorithm for individual (3,621) water molecules, served as an initial model (amino acid residues are shown in *red* and bound water molecules in *gray*). For hydrodynamic bead modeling, the number of beads of the model was reduced step by step ($N = 605\rightarrow19$) by means of the cubic grid approach until the structural features of the protein were lost; for $N = 4$, the running mean approach was applied to retain the structural symmetry of the tetrameric protein [25]. For comparison also the whole-body model ($N = 1$; oblate ellipsoid of revolution, *p* about 0.5) is shown. Because of graphics scaling procedures, the models are not exactly to scale if compared to the representation of the initial model. Further details of the respective models are given in Table 2. Graphics were made with the program RasMol [42]

calculation programs together with recently developed hydration strategies [38, 39].

For the application of MB approaches, efficient data reduction algorithms, such as the cubic grid or running mean approaches, have to be applied. In order to test the limits of such reduction procedures, the number of beads was varied over a wide range ($N = 605\rightarrow4$). Using a

sufficient number of beads (more than 100 beads, resolution better than 2 nm), no significant difference between the data reduction approaches can be found (Table 2). Reducing the number of beads considerably leads to obvious disagreements between observed and predicted molecular parameters; the effect is most pronounced when only a few beads ($N = 4$–20) were used. Obviously MB models built up from a few (separated) building blocks simulate the hydrodynamic behavior of proteins insufficiently.

In the case of simple proteins, WB approaches (using ellipsoids of revolution) are more successful, if compared to the MB approaches using only a small number of beads. The finding that proteins with $M < 100$ kg mol^{-1} are preferably of prolate shape, while larger molecules are rather oblate if not roughly spherical [10], facilitates the decision between PE and OE models.

When comparing critically scattering and hydrodynamic data, attention should be paid to the difference between the radii of gyration $R_{G,SAXS}$ (based on the number of excess electrons and the size and coordinates of the beads) and $R_{G,Hydro}$ (calculated only from the size and coordinates of the beads) [38, 39]. As may be followed from the MB examples given in Table 2, this is especially true if a high resolution (large *N*) is used.

Prediction of structural parameters from hydrodynamic data

Modeling homogeneous particles of globular shape by PE or OE of revolution and rearranging the formalism known for the opposite procedure, allows the derivation of typical structural parameters, such as *p*, R_G, S/V and *S* [10].

The combination of two of the three quantities, *s*, *D* and *M*, with the hydration, δ_1, leads to the frictional ratio of the hydrated particle, f/f_0:

1. Combining *s*, *M* and δ_1:

$$\frac{f}{f_0} = \frac{M^{2/3}(1 - \bar{v}\rho)}{\bar{v}_h^{1/3} N_A^{2/3} 6\pi\eta s} \left(\frac{4\pi}{3}\right)^{1/3} . \tag{1}$$

2. Combining *D*, *M* and δ_1:

$$\frac{f}{f_0} = \frac{kT}{6\pi\eta D M^{1/3} \bar{v}_h^{1/3}} \left(\frac{4\pi N_A}{3}\right)^{1/3} . \tag{2}$$

3. Combining *s*, *D* and δ_1:

$$\frac{f}{f_0} = \left(\frac{4\pi k^2 T^2}{3}\right)^{1/3} \frac{1}{6\pi\eta} \left(\frac{(1 - \bar{v}\rho)}{sD^2 \bar{v}_h}\right)^{1/3} . \tag{3}$$

Table 2 Structural and hydrodynamic parameters of tetrameric catalase from bovine liver as revealed from SAXS and hydrodynamic analysis, together with predictions of hydrodynamic parameters based on crystallography and SAXS. Experimental values of the native enzyme were taken from the literature (concerning accuracy of data see Ref. [39]). The atomic coordinates of catalase [49] were obtained from the Brookhaven Protein Data Bank [50], accession code 8CAT. The molar mass of the apoenzyme, as calculated from the sequence deposited in the SWISS-PROT data bank [32], accession number P00432, is 230.342 kg mol⁻¹; the mass of the native enzyme (1 heme b plus 1 $NADPH_2$ per subunit) amounts to 235.776 kg mol⁻¹. The latter value was used as the basis for our predictions, though a few terminal amino acids are missing in the crystal structure. Predictions of hydrodynamic parameters were performed by means of the program HYDRO [14, 17], adapted for the case of overlapping spheres [16, 25, 27]

Method	Approach	Resolution (nm)	N_{beads} or oblate ellipsoid	V (nm³)	$R_{G,SAXS}$ (nm)	$R_{G,Hydro}$ (nm)	$s \times 10^{13}$ (s)	$D \times 10^7$ (cm²s⁻¹)
SAXS, AUC	Experiment			420.0 [1]	3.68 [30], 3.98 [51]		11.3 [52], 11.6 [53]	4.1 [52], 4.5 [54]
XD	MB[a]		5,621	376.1	3.678			
XD	MB[b]	1.0	605	376.1	3.685	3.749	11.67	4.47
XD	MB[b]	1.3	325	376.1	3.676	3.744	11.71	4.48
XD	MB[b]	1.8	152	376.1	3.688	3.738	11.79	4.51
XD	MB[b]	2.4	83	376.1	3.664	3.723	11.90	4.55
XD	MB[b]	3.2	47	376.1	3.658	3.694	12.07	4.62
XD	MB[b]	3.6	27	376.1	3.642	3.676	12.32	4.71
XD	MB[b]	5.5	19	376.1	3.513	3.532	12.28	4.70
XD	MB[c]	≈6	4	235.8		3.221	15.16	5.80
XD	MB[d]	≈6	4	376.1		3.353	14.30	5.47
XD	MB[e]	≈6	4	376.1		3.560	13.72	5.25
XD	MB[f]	≈6	4	376.1		3.680	13.39	5.12
XD, SAXS	WB (R_G, V)		1 ($p = 0.572$)	376.1	3.68		12.23	4.66
XD, SAXS	WB (R_G, V)		1 ($p = 0.406$)	376.1	3.98		11.73	4.47
SAXS	WB (R_G, V)		1 ($p = 0.718$)	420.0	3.68		11.99	4.57
SAXS	WB (R_G, V)		1 ($p = 0.471$)	420.0	3.98		11.54	4.40

[a] Hydrated protein model based on amino acid coordinates, selected SIMS dot surface points [55] and a special hydration algorithm [38, 39]; use of parameter set A as described in Ref. [39]: $r_{probe} = 0.140$ nm, $d_w = 0.280$ nm; $r_w = 0.140$ nm, $V_w = 0.0269$ nm³, $f_K = 2$
[b] Hydrated protein model reduced by the cubic grid approach [25]
[c] Anhydrous protein model reduced by the running mean approach [25]
[d] Running mean model (see footnote c) rescaled by an expansion of the bead radii to fit V
[e] Running mean model (see footnote c) rescaled by a uniform expansion of bead radii and coordinates
[f] Running mean model (see footnote c) rescaled by a nonuniform expansion of bead radii and coordinates to fit V and R_G

N_A symbolizes Avogadro's number, k is the Boltzmann constant, T the temperature, and ρ and η are the density and viscosity of water at 293 K, respectively.

Use of a definite value for δ_1 yields the hydrated partial specific volume, $\bar{v}_h$, and the hydrated particle volume, V:

$$\bar{v}_h = \bar{v}_2 + \delta_1 \bar{v}_1 \quad \text{and} \quad V = \frac{M\bar{v}_h}{N_A} \ , \tag{4}$$

where $\bar{v}_2$ and $\bar{v}_1$ are the partial specific volumes of the macromolecule and water, respectively.

Knowledge of the frictional ratios, f/f_0, allows calculation of the axial ratio, p, by applying Perrin's formulae [40, 41], where the axial ratio $p = a/b$ is defined as the ratio of the semiaxis a of revolution to the equatorial semiaxis b of the ellipsoid.

Since now p and V are known, structural properties such as R_G, S/V and S can be calculated:

$$R_G = \left(\frac{3V}{4\pi p}\right)^{1/3} \left(\frac{p^2 + 2}{5}\right)^{1/2} , \tag{5}$$

$$\frac{S}{V} = \frac{3}{2pR_G} \left(1 + \frac{p^2}{(p^2 - 1)^{1/2}} \sin^{-1} \frac{(p^2 - 1)^{1/2}}{p}\right) \cdot \left(\frac{p^2 + 2}{5}\right)^{1/2} \qquad (p > 1) \tag{6}$$

or

$$\frac{S}{V} = \frac{3}{2pR_G} \left(1 + \frac{p^2}{(1 - p^2)^{1/2}} \tanh^{-1} (1 - p^2)^{1/2}\right) \cdot \left(\frac{p^2 + 2}{5}\right)^{1/2} \qquad (p < 1) , \tag{7}$$

$$S = \left(\frac{S}{V}\right) V . \tag{8}$$

On the basis of known hydrodynamic and hydration data of biopolymers, structural parameters of several biopolymers (p, R_G, S/V, S) were predicted. As may be taken from Table 3, R_G of simple globular proteins, glycoproteins and nucleoproteins, and ribonucleic acids may be predicted reasonably, whereas the accuracy of the

Table 3 Structural and hydrodynamic parameters of biopolymers from hydrodynamic analysis and SAXS, together with predictions of structural parameters based on hydrodynamic techniques and WB approaches. Abbreviations: prolate ellipsoid (PE), oblate ellipsoid (OE), sphere (S)

Biopolymer (source) Input parameters: $M, \bar{v}, s, D, \delta_1$[a]	Reference values[b]		Predicted values[b]					
	Parameters[c]	SAXS	s, M, δ_1		D, M, δ_1		s, D, δ_1	
			PE	OE	PE	OE	PE	OE
1. Globular proteins: Chymotrypsinogen A (bovine pancreas) $M = 25.67, \bar{v} = 0.736,$ $s = 2.58, D = 9.5,$ $\delta_1 = 0.15$	p (PE) R_G S/V S	2.00, 2.12 1.81 1.6 60.5	3.04 2.15 1.71 64.5	0.32 1.98 1.85 69.9	2.59 2.00 1.64 62.0	0.37 1.89 1.73 65.4	2.74 2.05 1.66 62.9	0.35 1.92 1.77 66.8
Albumin (bovine serum) $M = 66.30, \bar{v} = 0.735,$ $s = 4.5, D = 5.9,$ $\delta_1 = 0.55$	p (PE) R_G S/V S	2.49, 3.88 3.06 1.46 207.3	2.35 2.98 1.03 146	0.42 2.86 1.07 152	3.16 3.41 1.11 157	0.30 3.11 1.21 172	2.89 3.26 1.09 154	0.33 3.03 1.17 165
Citrate synthase (pig heart) $M = 97.94, \bar{v} = 0.740,$ $s = 6.2, D = 5.8,$ $\delta_1 = 0.33$	p (PE) R_G S/V S	1.78, 1.65 2.91 0.88 154.2	2.06 3.04 0.94 163	0.48 2.96 0.96 167	2.34 3.19 0.96 168	0.42 3.06 1.00 174	2.25 3.14 0.96 166	0.43 3.03 0.99 172
Cellobiose dehydrogenase (*P. chrysosporium*) $M = 107.2, \bar{v} = 0.741,$ $s = 5.6,$ $\delta_1 = 0.19$	p (PE) R_G S/V S	4.50 4.35	5.56 4.93 1.24 206	0.16 3.98 1.58 262				
Catalase (bovine liver) $M = 235.78, \bar{v} = 0.730,$ $s = 11.6, D = 4.5,$ $\delta_1 = 0.34$	p (OE) R_G S/V S	0.47, 0.43 3.68, 3.98 0.75 315.8	2.01 4.04 0.70 292	0.49 3.94 0.71 298	1.71 3.85 0.68 283	0.58 3.80 0.68 286	1.81 3.91 0.68 286	0.55 3.85 0.69 290
11S seed globulin (rape seed) $M = 300, \bar{v} = 0.729,$ $s = 12.7, D = 3.78,$ $\delta_1 = 0.17$	p (OE) R_G S/V S	0.46, 0.31 4.1 0.91 410	4.13 5.78 0.82 365	0.23 4.99 0.95 425	4.18 5.81 0.82 367	0.22 5.01 0.96 428	4.16 5.80 0.82 366	0.22 5.01 0.95 427
Legumin (*Vicia faba*) $M = 312.48, \bar{v} = 0.729,$ $s = 13.0, D = 3.38,$ $\delta_1 = 0.59$	p (OE) R_G S/V S	0.60, 0.50 4.45 0.62 426	2.13 4.86 0.60 410	0.46 4.71 0.62 421	3.67 6.24 0.68 469	0.26 5.53 0.77 529	3.15 5.76 0.66 450	0.30 5.26 0.72 490
Apoferritin (horse spleen) $M = 466.9, \bar{v} = 0.723,$ $s = 17.60, D = 3.61,$ $\delta_1 = 0.69$	p (S) R_G S/V S	1.00 5.33	1.51 5.16 0.48 528	0.66 5.12 0.48 530				
Artificial top component of bacteriophage fr (*E. coli*) $M = 2,480, \bar{v} = 0.735$ $s = 42.0,$ $\delta_1 = 1.60$	p (S) R_G S/V S	1.00 12.0	2.06 11.6 0.25 2,370	0.48 11.3 0.25 2,430				
Hemoglobin (*Lumbricus terrestris*) $M = 3,500, \bar{v} = 0.733,$ $s = 61.1, D = 1.61,$ $\delta_1 = 0.33$	p (OE) R_G S/V S	0.33 10.71	4.07 13.7 0.34 2,090	0.23 11.9 0.39 2,430	3.81 13.2 0.33 2,050	0.25 11.7 0.38 2,340	3.90 13.4 0.33 2,070	0.24 11.7 0.38 2,370
Pyruvate dehydrogenase core complex (*E. coli*) $M = 5,250, \bar{v} = 0.735,$ $s = 59,$ $\delta_1 = 3.23$	p (S) R_G S/V S	1.00 15.65	1.79 17.0 0.16 5,420	0.55 16.7 0.16 5,490				

Table 3 (Continued)

Biopolymer (source) Input parameters: $M, \bar{v}, s, D, \delta_1$[a]	Reference values[b]		Predicted values[b]					
	Parameters[c]	SAXS	s, M, δ_1		D, M, δ_1		s, D, δ_1	
			PE	OE	PE	OE	PE	OE
2. Glycoproteins:								
Ceruloplasmin (human)	p (PE)	2.00, 2.01	2.52	0.39	1.56	0.64	1.90	0.52
$M = 122.2, \bar{v} = 0.714,$	R_G	3.45	3.77	3.58	3.22	3.20	3.40	3.33
$s = 7.2, D = 5.30,$	S/V	0.82	0.86	0.90	0.78	0.78	0.81	0.82
$\delta_1 = 0.58$	S	213.9	225	236	205	206	212	215
Immunoglobulin IgG1 (human)	p (PE)	5.03	4.44	0.21				
$M = 150, \bar{v} = 0.733,$	R_G	5.84	5.44	4.62				
$s = 6.81,$	S/V		0.92	1.10				
$\delta_1 = 0.59$	S		304	361				
Fibrinogen (bovine)	p (PE)	5.16	5.38	0.17	10.9	0.07	8.77	0.10
$M = 335, \bar{v} = 0.700,$	R_G	12.2	12.5	10.2	19.5	13.3	17.0	12.2
$s = 7.9, D = 1.53,$	S/V		0.48	0.60	0.60	0.99	0.55	0.84
$\delta_1 = 4.45$	S		1,360	1,720	1,710	2,830	1,590	2,400
Fibronectin (human)	p (OE)	0.10	8.54	0.10	8.32	0.10	8.40	0.10
$M = 510, \bar{v} = 0.72,$	R_G	8.75	12.3	8.93	12.1	8.84	12.2	8.87
$s = 13.25, D = 2.27,$	S/V		0.75	1.11	0.74	1.09	0.74	1.10
$\delta_1 = 0.64$	S		858	1,280	851	1,260	854	1,270
Immunoglobulin IgM$_{ser}$ (bovine)	p (OE)	0.07, 0.07	13.8	0.05	11.9	0.07	12.5	0.06
$M = 950, \bar{v} = 0.724,$	R_G	11.5	19.6	12.6	17.7	11.8	18.3	12.1
$s = 17.7, D = 1.73,$	S/V	1.28	0.75	1.39	0.71	1.23	0.73	1.28
$\delta_1 = 0.42$	S	2,300	1,360	2,510	1,290	2,220	1,310	2,320
3. Nucleoproteins, spherical viruses:								
Bacteriophage fr (*E. coli*)	p (S)	1.00	1.33	0.75	3.74	0.25	2.95	0.33
$M = 3,620, \bar{v} = 0.673,$	R_G	10.58	10.4	10.4	15.1	13.4	13.4	12.4
$s = 79, D = 1.4,$	S/V		0.23	0.23	0.29	0.32	0.27	0.29
$\delta_1 = 0.91$	S		2,200	2,200	2,720	3,090	2,560	2,750
Bacteriophage R17 (*E. coli*)	p (S)	1.00	1.45	0.69				
$M = 3,700, \bar{v} = 0.673,$	R_G	10.44	10.5	10.5				
$s = 80,$	S/V		0.23	0.23				
$\delta_1 = 0.89$	S		2,230	2,240				
Brome grass mosaic virus, pH 7	p (S)	1.00	2.45	0.40	1.11	0.90	1.69	0.59
$M = 4,700, \bar{v} = 0.692,$	R_G	12.8	12.9	12.3	10.7	10.7	11.4	11.3
$s = 86.2, D = 1.55,$	S/V		0.24	0.26	0.22	0.22	0.23	0.23
$\delta_1 = 0.72$	S		2,690	2,810	2,400	2,400	2,500	2,530
Turnip yellow mosaic virus	p (S)	1.00	1.07	0.94	2.17	0.45	1.88	0.52
$M = 5,530, \bar{v} = 0.666,$	R_G	11.1	11.1	11.1	12.8	12.4	12.2	12.0
$s = 113.8, D = 1.42,$	S/V		0.21	0.21	0.23	0.24	0.22	0.23
$\delta_1 = 0.67$	S		2,570	2,570	2,810	2,900	2,740	2,780
Wild cucumber mosaic virus	p (S)	1.00	2.31	0.42				
$M = 5,820, \bar{v} = 0.655,$	R_G	11.2	12.8	12.3				
$s = 119,$	S/V		0.24	0.25				
$\delta_1 = 0.53$	S		2,730	2,830				
Southern bean mosaic virus	p (S)	1.00	2.57	0.38	2.27	0.43	2.37	0.41
$M = 6,690, \bar{v} = 0.694,$	R_G	12.3	13.9	13.1	13.2	12.7	13.4	12.9
$s = 115, D = 1.39,$	S/V		0.23	0.25	0.23	0.24	0.23	0.24
$\delta_1 = 0.46$	S		3,010	3,170	2,930	3,030	2,960	3,080

Table 3 (Continued)

Biopolymer (source) Input parameters: $M, \bar{v}, s, D, \delta_1$[a]	Reference values[b]		Predicted values[b]					
	Parameters[c]	SAXS	s, M, δ_1		D, M, δ_1		s, D, δ_1	
			PE	OE	PE	OE	PE	OE
Tomato bushy stunt virus $M = 8{,}700$, $\bar{v} = 0.700$, $s = 132$, $D = 1.15$, $\delta_1 = 0.99$	p (S) R_G S/V S	1.00 13.2			1.87 15.3 0.18 4,330	0.53 15.0 0.18 4,400	1.47 14.4 0.17 4,170	0.68 14.4 0.17 4,190
Rice dwarf virus $M = 65{,}200$, $\bar{v} = 0.696$, $s = 482$, $D = 0.586$, $\delta_1 = 1.11$	p (S) R_G S/V S	1.00 29.8	1.35 28.5 0.085 16,500	0.74 28.4 0.085 16,600	1.51 29.1 0.086 16,800	0.66 28.9 0.086 16,800	1.47 28.9 0.085 16,700	0.68 28.8 0.08-6 16,800
4. Ribonucleic acids: tRNA^{phe} (yeast) $M = 25.0$, $\bar{v} = 0.54$, $D = 7.8$, $\delta_1 = 0.34$	p (OE) R_G S/V S	0.18, 0.17 2.31 2.66 97.5			6.33 3.23 2.14 78.2	0.14 2.52 2.85 104		
5S rRNA (E. coli) $M = 44.0$, $\bar{v} = 0.54$, $s = 5.3$, $D = 6.2$, $\delta_1 = 0.23$	p (OE) R_G S/V S	0.10, 0.10 3.27 2.88 161.0	7.69 4.21 1.97 111	0.11 3.13 2.82 159	8.36 4.44 2.03 114	0.10 3.24 3.00 169	8.13 4.36 2.01 113	0.10 3.20 2.94 165
MS2 RNA (bacteriophage MS2) $M = 1{,}088$, $\bar{v} = 0.457$, $s = 26.6$, $\delta_1 = 11.3$	p (OE) R_G S/V S	0.28 18.1	3.37 18.7 0.21 4,520	0.28 16.9 0.24 5,000				

[a] The experimental input parameters are given in the following units: M: kg mol^{-1}; $\bar{v}$: cm^{3}g^{-1}; s: 10^{-13} s; D: 10^{-7} cm^{2}s^{-1}; δ_1: g g^{-1}. The experimental values were taken or derived (δ_1) from Refs. [7, 8, 10]. In continuation of previous work, $\bar{v}$ of bound water was assumed to be 1 cm^{3}g^{-1}; according to recent findings, $\bar{v}$ of bound water may be lower by 10–20%, finally leading to slightly higher hydration values (for a discussion of this subject see Ref. [38])
[b] The reference values and the predicted structural parameters are given in the following units: R_G: nm; S/V: nm^{-1}; S: nm^2

[c] Calculation approaches: PE ($p > 1$); OE ($p < 1$); S ($p = 1$). The first p value refers to the result from combinations of R_G and V, the second one (if any) to those from R_G and S/V. For hydrodynamic modeling, the particle was assumed to be a solid sphere ($p = 1.00$) in all cases where the particle shape was known to be (i) approximately spherical, whether or not it contained hollow spaces (e.g., spherical viruses) or (ii) could be approximated by a sphere (e.g., pyruvate dehydrogenase core complex)

other structural characteristics (p, S/V, S) is highly influenced by the input data and obvious differences in the surface roughness between models and true particles. A critical analysis of the results, including a critical assessment of errors, discloses that the values applied for δ_1 are crucial for estimating the structural properties of the molecules under consideration. The hydration values to be applied for different classes of biopolymers vary over a wide range. They are relatively low for simple globular proteins; a value of 0.35 g water/g protein may be used as an appropriate default value [10, 38]. Glycoproteins and some other types of biopolymers may exhibit greater hydration.

It should also be borne in mind that contributions caused by cavities and particle inhomogeneities have to be considered in order to model hydrodynamic properties correctly (Table 3; e.g., pyruvate dehydrogenase complex, fibrinogen, artificial top component of bacteriophage fr, many spherical viruses, MS2 RNA). Use of

hydrodynamically effective values for δ_1 leads to significantly improved predictions of hydrodynamic parameters, if compared to predictions based on the default values for simple proteins.

Comparisons between solution and crystal data

Various structural parameters, parameter ratios and combinations can be derived from SAXS and hydrodynamics without any modeling assumptions. They can be used effectively for the characterization of macromolecules in solution (shape, anisometry/anisotropy, hydration, rugosity), for comparing experimental results from different techniques and for probing and establishing empirical relations [7–9].

In the case of simple proteins, empirical expressions have been established on the basis of crystal data [37], or can be derived therefrom. These expressions can be

Table 4 Structural parameters and parameter combinations, as revealed from solution techniques (SAXS or calculated from hydrodynamics) and crystallographic data (based on empirical relations found for a set of globular proteins). R_0 and R_{dry}, radii of spheres corresponding to hydrated volume V and anhydrous volume V_{dry} (surface area of sphere: S_{dry}), respectively; R_η, viscosity radius; R_{Gs} and R_{Gp}, radii of gyration related to R_s and R_p, respectively; $(R_G/R_D)_s$, $(R_G/R_D)_D$, $(R_G/R_\eta)_\eta$, radii ratios calculated from experimental s, D or $[\eta]$ values; names of other parameters are given in the text. For detailed definitions of parameters see Refs. [7–9, 37, 56] and references therein

Parameters	Solution	Crystal	Solution–crystal
Parameters:			
Radius	$R_G, R_0, R_{dry}, R_D, R_\eta$	R_s, R_{Gs}, R_p, R_{Gp}	
Surface	S, S_{dry}	A_s	
Volume	V, V_{dry}	V_p	
Ratios of parameters:			
Radii	$R_G/R_0, R_G/R_{dry}, R_G/R_D, R_G/R_\eta, R_D/R_0,$ $R_\eta/R_0, R_\eta/R_D, (R_G/R_D)_s, (R_G/R_D)_D, (R_G/R_\eta)_\eta$	$R_s/R_p, R_{Gs}/R_{Gp}$[a]	$R_G/R_s, R_G/R_{Gs},$ $R_G/R_p, R_G/R_{Gp}$
Surfaces	S/S_{dry}	A_s/S_{dry}	S/A_s
Volumes	V/V_{dry}	V_p/V_{dry}	V/V_p
Surface–volume	$S/V, S_{dry}/V_{dry}$[b]	A_s/V_p	$(S/V)/(A_s/V_p)$
Parameter combinations:			
Radius–sedimentation	$(R_G s)/M, (R_D s)/M, (R_\eta s)/M$		
Radius–diffusion	$R_G D$		

[a] R_s/R_p or $R_{Gs}/R_{Gp} = 1.4$
[b] $S_{dry}/V_{dry} = 3/R_{dry}$

compared with similar quantities from SAXS or other solution techniques (Table 4). Since "reduced" SAXS parameters turned out to be relatively constant [9], their application is especially useful when comparing the results from solution and crystallographic work. A critical inspection of the values listed in Table 5 reveals significant differences between the results from SAXS and crystallography: $R_{Gp} \ll R_G < R_{Gs}$, $V \gg V_p$, $S < A_s$, and $S/V \ll A_s/V_p$.

Conclusions

Predictions of various molecular properties of biopolymers are possible from correlations between hydrodynamic, X-ray solution and crystal data. Relations are based on modeling the complex molecular structures by simple bodies (WB approaches) or many beads (MB approaches). Empirical relations may be used as an additional approach. All solution techniques have to take hydration contributions into account, either directly or indirectly.

The prediction of hydrodynamic from SAXS data by use of WB approaches exploits the hydrated particle volume (obtainable from SAXS studies), without direct knowledge of δ_1 values. The prediction of structural parameters from the results of hydrodynamics, however, strictly depends on separate information or realistic assumptions concerning the amount of hydration, δ_1. Similarly, definite hydration values (or appropriate rescaling procedures to simulate hydration) are needed for WB predictions based on data from X-ray crystallography or neutron scattering.

The data reduction algorithms for the MB approaches have to conserve the structural characteristics of the biopolymer to be modeled. While equally good results were obtained for bead numbers greater than 100, data reduction to smaller numbers turned out to be less satisfying. This especially holds for models consisting of a few subunits, even if the symmetry of the biopolymer is retained. As is well known, both monomeric and oligomeric proteins form relatively compact entities, different from geometric models built up from a few spherical subunits. On the other hand, MB approaches utilizing simple ellipsoids of revolution may be successful in approximating the overall structure of proteins of simple structure. Of course, simple entities may also be modeled by MB approaches. Complex molecular structures, however, can only be modeled realistically by MB approaches.

For combining solution and crystallographic data, manifold parameter combinations may be derived. In particular, reduced structural parameters represent a high potential of possibilities to compare solution and crystal data. As an essential outcome of the present study, the differences in surface and volume properties of proteins should be mentioned. Obviously, surface and hydration peculiarities of these molecules are responsible for some severe discrepancies between solution and crystal parameters, on the one hand, and true particles and geometric models, on the other. This should be kept in mind before statements on true differences between solution and crystal structure are made.

Acknowledgements The authors are much obliged to J. García de la Torre for making available the program HYDRO and to Y.N. Vorobjev for use of the program SIMS.

Table 5 Reduced structural parameters, as revealed by SAXS and empirical relations based on crystallographic data of globular proteins, together with some parameter combinations obtained from SAXS, crystallography and hydrodynamics

SAXS		Crystallography		SAXS-crystallography	
Parameters	Values[a]	Parameters	Values[b]	Parameters	Values[a]
1. Reduced radii (nm $kg^{-1/3}mol^{1/3}$):				**Ratios of radii (dimensionless):**	
$R_G/M^{1/3}$	0.68 ± 0.08	$R_s/M^{1/3}$	0.94	R_G/R_s	0.72 ± 0.08
$R_0/M^{1/3}$	0.75 ± 0.03	$R_{Gs}/M^{1/3}$	0.73	R_G/R_{Gs}	0.93 ± 0.11
		$R_p/M^{1/3}$	0.67	R_G/R_p	1.01 ± 0.12
		$R_{Gp}/M^{1/3}$	0.52	R_G/R_{Gp}	1.31 ± 0.15
				R_0/R_s	0.80 ± 0.03
				R_0/R_{Gs}	1.03 ± 0.04
				R_0/R_p	1.12 ± 0.04
				R_0/R_{Gp}	1.44 ± 0.06
2. Reduced surfaces ($nm^2 kg^{-2/3}mol^{2/3}$):				**Ratios of surfaces (dimensionless):**	
$S/M^{2/3}$	10.2 ± 2.2	$A_s/M^{2/3}$	11.1	S/A_s	0.92 ± 0.20
Reduced surfaces (dimensionless):					
S/R_G^2	23.2 ± 5.3	A_s/R_s^2	12.6	$(S/R_G^2)/(A_s/R_s^2)$	1.85 ± 0.42
		A_s/R_{Gs}^2	20.9	$(S/R_G^2)/(A_s/R_{Gs}^2)$	1.11 ± 0.25
		A_s/R_p^2	24.6	$(S/R_G^2)/(A_s/R_p^2)$	0.94 ± 0.22
		A_s/R_{Gp}^2	41.0	$(S/R_G^2)/(A_s/R_{Gp}^2)$	0.57 ± 0.13
3. Reduced volumes ($nm^3 kg^{-1}mol$):				**Ratios of volumes (dimensionless):**	
V/M	1.75 ± 0.22	V_p/M	1.27	V/V_p	1.37 ± 0.17
Reduced volumes (dimensionless):					
V/R_G^3	5.9 ± 1.5	V_p/R_s^3	1.53	$(V/R_G^3)/(V_p/R_s^3)$	3.84 ± 1.00
		V_p/R_{Gs}^3	3.29	$(V/R_G^3)/(V_p/R_{Gs}^3)$	1.79 ± 0.46
		V_p/R_p^3	4.19	$(V/R_G^3)/(V_p/R_p^3)$	1.40 ± 0.37
		V_p/R_{Gp}^3	9.01	$(V/R_G^3)/(V_p/R_{Gp}^3)$	0.65 ± 0.17
Ratios of volumes (dimensionless):					
V/V_{dry}	1.44 ± 0.18	V_p/V_{dry}	1.04 ± 0.02		
4. Reduced surface-to-volume ratios ($nm^{-1}kg^{1/3}mol^{-1/3}$):				**Ratios of surface-to-volume ratios (dimensionless):**	
$(S/V)M^{1/3}$	5.8 ± 1.1	$(A_s/V_p)M^{1/3}$	8.8	$(S/V)/(A_s/V_p)$	0.66 ± 0.13
Reduced surface-to-volume ratios (dimensionless):					
$(S/V)R_G$	3.8 ± 0.8	$(A_s/V_p)R_s$	8.2	$(R_G/R_s)(S/V)/(A_s/V_p)$	0.46 ± 0.10
		$(A_s/V_p)R_{Gs}$	6.4	$(R_G/R_{Gs})(S/V)/(A_s/V_p)$	0.59 ± 0.13
		$(A_s/V_p)R_p$	5.9	$(R_G/R_p)(S/V)/(A_s/V_p)$	0.64 ± 0.14
		$(A_s/V_p)R_{Gp}$	4.6	$(R_G/R_{Gp})(S/V)/(A_s/V_p)$	0.83 ± 0.18
5. Parameter combinations:					
Radius–sedimentation (nm s $kg^{-1}mol$):					
$(R_G s)/M$	$(18.9 \pm 1.6) \times 10^{-11}$				
Radius–diffusion (nm $cm^2 s^{-1}$):					
$R_G D$	$(17.5 \pm 1.2) \times 10^{-7}$				

[a] Some values were taken from Refs. [7–9]; experimental SAXS values were derived from a set of 21 simple globular proteins ($M = 14$–220 kg mol^{-1}, $p = 0.2$–5)
[b] Taken or derived from Ref. [37]

References

1. Kumosinski TF, Pessen H (1982) Arch Biochem Biophys 219:89–100
2. Kumosinski TF, Pessen H (1985) Methods Enzymol 117:154–182
3. Pessen H, Kumosinski TF (1993) In: Baianu IC, Pessen H, Kumosinski TF (eds) Physical chemistry of food processes, vol 2. Advanced techniques, structures, and applications. Van Nostrand Reinhold, New York, pp 274–306
4. Müller JJ, Damaschun H, Damaschun G, Gast K, Plietz P, Zirwer D (1984) Stud Biophys 102:171–175
5. Durchschlag H, Zipper P, Purr G, Jaenicke R (1996) Colloid Polym Sci 274:117–137
6. Durchschlag H, Zipper P (1996) J Mol Struct 383:223–229
7. Durchschlag H, Zipper P (1997) J Appl Crystallogr 30:1112–1124
8. Durchschlag H, Zipper P (1997) Prog Colloid Polym Sci 107:43–57
9. Durchschlag H, Zipper P (1998) Biochem Soc Trans 26:731–736
10. Durchschlag H, Zipper P (1999) Prog Colloid Polym Sci 113:87–105
11. Harding SE (1989) In: Harding SE, Rowe AJ (eds) Dynamic properties of biomolecular assemblies. Royal Society of Chemistry, Cambridge UK, pp 32–56

12. Harding SE (1995) Biophys Chem 55:69–93
13. Harding SE, Horton JC, Jones S, Thornton JM, Winzor DJ (1999) Biophys J 76:2432–2438
14. García de la Torre J, Navarro S, López Martínez MC, Díaz FG, López Cascales JJ (1994) Biophys J 67:530–531
15. Carrasco B, García de la Torre J (1999) Biophys J 75:3044–3057
16. Carrasco B, García de la Torre J, Zipper P (1999) Eur Biophys J 28:510–515
17. García de la Torre J, Huertas ML, Carrasco B (2000) Biophys J 78:719–730
18. Beavil AJ, Young RJ, Sutton BJ, Perkins SJ (1995) Biochemistry 34:14449–14461
19. Perkins SJ, Ashton AW, Boehm MK, Chamberlain D (1998) Int J Biol Macromol 22:1–16
20. Byron O (1995) Prog Colloid Polym Sci 99:82–86
21. Byron O (1997) Biophys J 72:408–415
22. Byron O (2000) Methods Enzymol 321:278–304
23. Hellweg T, Eimer W, Krahn E, Schneider K, Müller A (1997) Biochim Biophys Acta 1337:311–318
24. (a) Spotorno B, Piccinini L, Tassara G, Ruggiero C, Nardini M, Molina F, Rocco M (1997) Eur Biophys J 25:373–384; (b) (1977) Eur Biophys J erratum 26:417
25. Zipper P, Durchschlag H (1997) Prog Colloid Polym Sci 107:58–71
26. Zipper P, Durchschlag H (1998) Biochem Soc Trans 26:726–731
27. Zipper P, Durchschlag H (1999) Prog Colloid Polym Sci 113:106–113
28. Zipper P, Durchschlag H (2000) J Appl Crystallogr 33:788–792
29. Chacón P, Morán F, Díaz JF, Pantos E, Andreu JM (1998) Biophys J 74:2760–2775
30. Chacón P, Díaz JF, Morán F, Andreu JM (2000) J Mol Biol 299:1289–1302
31. Banachowicz E, Gapiński J, Patkowski A (2000) Biophys J 78:70–78
32. Bairoch A, Apweiler R (2000) Nucleic Acids Res 28:45–48
33. Durchschlag H (1986) In: Hinz H-J (ed) Thermodynamic data for biochemistry and biotechnology. Springer, Berlin Heidelberg New York, pp 45–128
34. Durchschlag H (2001) In: Hinz H-J (ed) Landolt–Börnstein new series VII/2A. Springer, Berlin Heidelberg New York (in press)
35. Durchschlag H, Zipper P (1994) Prog Colloid Polym Sci 94:20–39
36. Durchschlag H, Zipper P (1997) J Appl Crystallogr 30:803–807
37. Teller DC (1976) Nature 260:729–731
38. Durchschlag H, Zipper P (2001) Biophys Chem (in press)
39. Durchschlag H, Zipper P (2001) Prog Colloid Polym Sci 119:131–140
40. Perrin F (1934) J Phys Radium Série VII 5:497–511
41. Perrin F (1936) J Phys Radium Série VII 7:1–11
42. Sayle RA, Milner-White EJ (1995) Trends Biochem Sci 20:374–376
43. Durchschlag H, Zipper P, Wilfing R, Purr G (1991) J Appl Crystallogr 24:822–831
44. Zipper P, Krebs A, Durchschlag H (2001) Prog Colloid Polym Sci 119:141–148
45. Ashton AW, Boehm MK, Gallimore JR, Pepys MB, Perkins SJ (1997) J Mol Biol 272:408–422
46. Tanford C (1961) Physical chemistry of macromolecules. Wiley, New York
47. Cantor CR, Schimmel PR (1980) Biophysical chemistry. Freeman, San Francisco
48. Van Holde KE, Johnson WC, Ho PS (1998) Principles of physical biochemistry. Prentice Hall, Upper Saddle River, NJ
49. Fita I, Murthy MRN, Rossmann MG, Silva AM (1986) Acta Crystallogr Sect B 42:497–515
50. Sussman JL, Lin D, Jiang J, Manning NO, Prilusky J, Ritter O, Abola EE (1998) Acta Crystallogr Sect D 54:1078–1084
51. Malmon AG (1957) Biochim Biophys Acta 26:233–240
52. Sumner JB, Gralén N (1938) J Biol Chem 125:33–36
53. Tanford C, Lovrien R (1962) J Am Chem Soc 84:1892–1896
54. Tyn MT, Gusek TW (1990) Biotechnol Bioeng 35:327–338
55. Vorobjev YN, Herman J (1997) Biophys J 73:722–732
56. García de la Torre J, Carrasco B (1999) Prog Colloid Polym Sci 113:81–86

Progr Colloid Polym Sci (2002) 119:131–140
© Springer-Verlag 2002

Helmut Durchschlag
Peter Zipper

Modeling of protein hydration with respect to X-ray scattering and hydrodynamics

H. Durchschlag (⊠)
Institute of Biophysics and Physical
Biochemistry, University of Regensburg
Universitätsstrasse 31, 93040 Regensburg
Germany
e-mail: helmut.durchschlag@biologie.uni-
regensburg.de
Tel.: +49-941-943-3041
Fax: +49-941-943-2813

P. Zipper
Physical Chemistry
Institute of Chemistry
University of Graz, Heinrichstrasse 28
8010 Graz, Austria

Abstract Hydration contributions of proteins can be taken into account by advanced modeling techniques starting from the atomic-level structure. Bead modeling may be used both for individual amino acid residues and individual water molecules placed at preferred positions on the protein surface. The exact calculation of the molecular volumes and surfaces of the proteins under analysis is of special importance. Among the approaches tested, the programs MSRoll and SIMS turned out to be particularly effective for calculating "dot surfaces". The dot surface points and the normal vectors to these points were used for placing water molecules in definite positions and under special constraints on the protein surface (program HYDMODEL). After data reduction of the hydrated protein models to appropriate numbers of beads, hydrodynamic parameters were predicted by means of the program HYDRO. In context with the establishment of hydration models, a variety of input parameters were critically tested: calculation approaches, number of surface dot points, probe radius, volume/density, as well as position and number of bound water molecules, distance selection for water molecules. X-ray scattering properties were calculated on the basis of the number of excess electrons and the radii and coordinates of the beads, hydrodynamic quantities only from the bead radii and coordinates. The examples studied comprise the enzymes citrate synthase and catalase. The approaches applied may be used to predict structural and hydrodynamic properties of hydrated proteins more realistically.

Key words Protein hydration · 3D structure · Bead modeling · Hydrodynamics · Small-angle X-ray scattering

Introduction

Hydrodynamic modeling approaches may be used advantageously for various purposes: prediction of hydrodynamic properties of biopolymers from structural parameters, anticipation of structural properties from hydrodynamic quantities, differentiation between different structural models proposed by various solution techniques, comparison of solution, crystal and electron microscopic data, elucidation of surface roughness differences, estimation of hydration/solvation contributions, predictions in connection with size/volume estimates for drug design projects, etc. For modeling purposes, the macromolecules under analysis may be approximated either by simple whole bodies (spheres, ellipsoids of revolution, triaxial bodies, etc.) or by an assembly of spheres ("beads") [1–6]. Of course, the investigation of complex structures and structural details requires "bead modeling" strategies. In the case of calculations starting from the atomic coordinates, theoretical problems (e.g., in context with overlapping of beads) and an efficient handling of enormous amounts of data (e.g., by development of appropriate data reduction algorithms) make the problem rather complicated [5–14]. Proteins are preferred objects of bead modelers, since the atomic coordinates stored in several databases can be used advantageously for modeling attempts. Usually, atoms, amino acid (AA) residues, or small groups of AA residues are approximated by spheres of appropriate size.

The presence of ligands on the surface or in the interior of macromolecules turns out to be a highly delicate problem. In connection with proteins, the binding of cofactors, substrates, products and analogs comes to mind, and the importance of the numerous water molecules "bound" to the macromolecule surface should also be rationalized, from both the structural and the functional points of view. Though many efforts have been made to shed some light on the elucidation of the hydration/solvation problem in solution biophysics [15], many problems, in particular related to the application of biomodeling approaches, have been left. Anyhow, appropriate hydration contributions have to be considered. In bead modeling this demand can be realized by the assumption of reasonable amounts of hydration [5–14], by a uniform or nonuniform expansion of the molecule [5, 7, 8, 10–14]), by construction of water shells around the protein molecules [7] or by positioning individual water molecules at definite protein sites, i.e., at certain

Table 1 Programs and sets of parameters used in simulating the hydration of citrate synthase. r_{probe}: probe radius for determining the molecular surface; N_{dot}: number of dot surface points; r_w: radius of a water molecule (distance of its center from the molecular surface of the protein); d_w: minimum distance between neighboring water molecules; V_w: volume of a water molecule on the protein surface; $N_{w,1}$: the number of water molecules that can be placed on the surface of the protein for $f_K = 1$ (values in the first line); $\delta_{1,1}$: the corresponding hydration; $N_{w,max}$: maximum number of water molecules that can be placed on the surface of the protein for $f_K = 10$ (values in the second line); $\delta_{1,max}$: the corresponding maximum hydration

Parameter sets	SIMS, MSRoll			HYDMODEL						
	r_{probe} (nm)	SIMS N_{dot}[a]	MSRoll N_{dot}	d_w (nm)	r_w (nm)	V_w (nm^3)	SIMS surface		MSRoll surface	
							$N_{w,1}$, $N_{w,max}$	$\delta_{1,1}$, $\delta_{1,max}$ (g/g)	$N_{w,1}$, $N_{w,max}$	$\delta_{1,1}$, $\delta_{1,max}$ (g/g)
A (standard set)	0.140	24,524	43,574	0.280	0.140	0.0245	1,377, 2,136	0.253, 0.393	1,334, 2,072	0.245, 0.381
						0.0257	1,377, 2,136	0.253, 0.393	1,334, 2,072	0.245, 0.381
						0.0269	1,377, 2,136	0.253, 0.393	1,334, 2,072	0.245, 0.381
						0.0284	1,377, 2,136	0.253, 0.393	1,334, 2,072	0.245, 0.381
B	0.140	24,524	43,574	0.280	0.145	0.0245	1,369, 2,130	0.252, 0.392	1,336, 2,073	0.246, 0.381
					0.1475	0.0257	1,370, 2,119	0.252, 0.390	1,339, 2,077	0.246, 0.382
					0.150	0.0269	1,376, 2,127	0.253, 0.391	1,332, 2,073	0.245, 0.381
					0.1525	0.0284	1,375, 2,134	0.253, 0.393	1,338, 2,081	0.246, 0.383
C	0.140	24,524	43,574	0.290	0.145	0.0245	1,341, 2,015	0.247, 0.371	1,299, 1,974	0.239, 0.363
				0.295	0.1475	0.0257	1,322, 1,948	0.243, 0.358	1,291, 1,930	0.237, 0.355
				0.300	0.150	0.0269	1,316, 1,892	0.242, 0.348	1,270, 1,864	0.234, 0.343
				0.305	0.1525	0.0284	1,306, 1,853	0.240, 0.341	1,259, 1,806	0.232, 0.332
D	0.145	23,433	41,852	0.290	0.145	0.0245	1,322, 1,988	0.243, 0.366	1,273, 1,932	0.234, 0.355
	0.1475	22,712	41,050	0.295	0.1475	0.0257	1,284, 1,897	0.236, 0.349	1,250, 1,843	0.230, 0.339
	0.150	22,118	40,478	0.300	0.150	0.0269	1,263, 1,827	0.232, 0.336	1,210, 1,788	0.223, 0.329
	0.1525	21,942	39,852	0.305	0.1525	0.0284	1,237, 1,757	0.228, 0.323	1,196, 1,711	0.220, 0.315

[a] Based on a dot density of 0.1 1/Å^2 as input parameter

AA residues on the protein surface [16]. Application of the latter approach necessitates the calculation of exact protein envelopes and, in addition, the development of special "hydration algorithms" to account for the hydration values of individual AA residues on the surface.

The present study is concerned with the influence of hydration contributions in the bead modeling of two important proteins (citrate synthase, catalase), in particular, with a critical assessment of the parameters to be applied for modeling the hydration of individual AA residues in connection with surface calculation programs. To test the results obtained, the predictions of structural and hydrodynamic parameters were checked by the experimental findings as revealed from small-angle X-ray scattering (SAXS) and hydrodynamics (analytical ultracentrifugation).

Methods

Proteins

For the following modeling purposes, two well-characterized proteins, dimeric pig heart citrate synthase and tetrameric bovine liver catalase, were chosen. Their X-ray crystallographic, SAXS and hydrodynamic properties are well documented, particularly in the case of citrate synthase: atomic coordinates, scattering intensity, $I(h)$, pair-distance distribution function, $p(r)$, molar mass, M, partial specific volume, $\bar{v}$, radius of gyration, R_G, hydrated volume, V, amount of hydration, δ_1, sedimentation coefficient, s, diffusion

coefficient, D, and intrinsic viscosity, $[\eta]$. The AA sequence, stored in databases such as the SWISS-PROT data bank [17], can be used for the calculation of precise M values, and, combined with the hydration values for individual amino acids according to Kuntz [18], for the estimation of the overall protein hydration. Both enzymes have been subjects of various modeling attempts with respect to their overall and fine structures [6, 10–13, 19–26]. In particular, they have been used for the current development of modeling strategies for hydration algorithms [16].

Experimental values were taken from the literature. In particular, the following data were used. Citrate synthase: atomic coordinates [27, 28] as given in Ref. [29], $I(h)$ and $p(r)$ as given in Ref. [30], $M = 97.838$ kg/mol [17], $\bar{v} = 0.740$ cm^3/g [23], $R_G = 2.91 \pm 0.02$ nm [23, 30], $V = 174.4 \pm 2.0$ nm^3 [23], $\delta_1 = 0.339 \pm 0.011$ g/g [23], $s = 6.0 \times 10^{-13}$ s [31] and 6.2×10^{-13} s [32], $D = 5.8 \times 10^{-7}$ cm^2/s [31] and $[\eta] = 3.95$ cm^3/g [31]. Catalase: atomic coordinates [33] as given in Ref. [29], $I(h)$ as given in Refs. [24, 34], $M = 235.776$ kg/mol (mass of the apoenzyme [17] plus masses of the ligands, heme b and NADPH$_2$), $\bar{v} = 0.730$ cm^3/g [35], $R_G = 3.68$ nm [24] and 3.98 nm [34], $V = 420.0$ nm^3 [25], $\delta_1 = 0.34$ g/g [13], $s = 11.3 \times 10^{-13}$ s [36] and 11.6×10^{-13} s [37], $D = 4.1 \times 10^{-7}$ cm^2/s [36] and 4.5×10^{-7} cm^2/s [38] and $[\eta] = 3.9$ cm^3/g [37]. The values of the sedimentation and diffusion coefficients, s and D, had been obtained from runs in the analytical ultracentrifuge [31, 32, 36–38]: s preferably by use of Schlieren optics, and D by runs in synthetic boundary cells, followed by correction of data to standard conditions (20 °C, water as a solvent, extrapolation to protein concentration zero).

Modeling

The crystal data of citrate synthase and catalase were taken from the Protein Data Bank [29] (accession codes: 1CTS and 8CAT). The atomic coordinates stored in the Protein Data Bank files were converted to SAXS and hydrodynamic models step by step.

Table 2 Programs and sets of parameters used in simulating the hydration of catalase. For an explanation of the parameters see Table 1

Parameter sets	SIMS, MSRoll			HYDMODEL						
	r_{probe} (nm)	SIMS N_{dot}[a]	MSRoll N_{dot}	d_w (nm)	r_w (nm)	V_w (nm^3)	SIMS surface		MSRoll surface	
							$N_{w,1}$, $N_{w,max}$	$\delta_{1,1}$, $\delta_{1,max}$ (g/g)	$N_{w,1}$, $N_{w,max}$	$\delta_{1,1}$, $\delta_{1,max}$ (g/g)
A (standard set)	0.140	56,531	98,116	0.280	0.140	0.0245	3,026, 4,048	0.231, 0.309	2,930, 3,897	0.224, 0.298
						0.0269	3,026, 4,048	0.231, 0.309	2,930, 3,897	0.224, 0.298
						0.0284	3,026, 4,048	0.231, 0.309	2,930, 3,897	0.224, 0.298
C	0.140		98,116	0.290	0.145	0.0245			2,810, 3,657	0.215, 0.279
				0.300	0.150	0.0269			2,730, 3,493	0.209, 0.267
				0.305	0.1525	0.0284			2,678, 3,390	0.205, 0.259
D	0.145		94,282	0.290	0.145	0.0245			2,746, 3,577	0.210, 0.273
	0.150		89,564	0.300	0.150	0.0269			2,586, 3,342	0.198, 0.255
	0.1525		87,466	0.305	0.1525	0.0284			2,539, 3,217	0.194, 0.246

[a] Based on a dot density of 0.1 1/Å^2 as input parameter

Fig. 1 Selected hydrodynamic
models for hydrated citrate
synthase, based on MSRoll
and SIMS surfaces together
with hydration program
HYDMODEL (parameter
set A; $V_w = 0.0245$ nm^3; $f_K = 1$,
2 and 10). Amino acid residues
are shown in *red* and bound
water molecules in *gray*;
further details are given in
Table 1. Graphics were made
with the program RasMol [52]

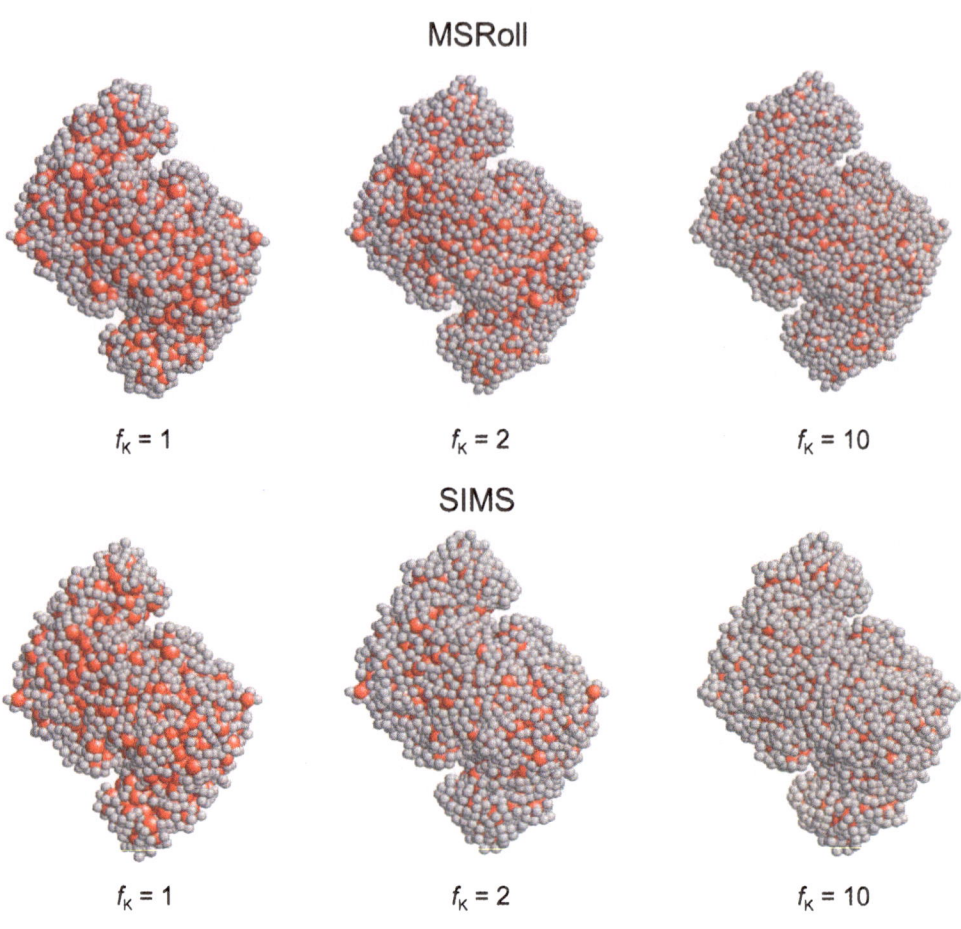

MSRoll

$f_K = 1$ $f_K = 2$ $f_K = 10$

SIMS

$f_K = 1$ $f_K = 2$ $f_K = 10$

Initially, the atoms of each AA residue of the crystal structure were combined in a single sphere. This sphere had a volume equal to the sum of the atomic volumes of the AA residues; it was placed at the mass center of gravity of the AA residues, thereby leading to AA coordinates. The resultant initial "dry" model (of a volume corresponding to the molecular volume of the anhydrous particle) served as a starting point for modifications by hydration and/or reduction procedures.

Pilot tests revealed [16] that, among the surface calculation programs tested, the programs MSRoll [39–41] and SIMS [42] are particularly useful for creating a large number of dot surface points and normal vectors from the original Protein Data Bank file (N_{dot} up to about 50,000–100,000 for the two enzymes under investigation). The probe radius, r_{probe}, was used as an adjustable parameter in these calculations. Potential positions of water molecules on the protein surface were calculated from the dot surface points and normal vectors, taking into account the radius, r_w, of water molecules. To select individual water molecules for hydration, a special hydration algorithm (program HYDMODEL) was applied [16]. To realize an appropriate hydration for each AA residue, the original values given by Kuntz [18] were used as starting values for assigning a definite number of water molecules to each accessible residue. As the minimum distance between neighboring water molecules, the value d_w was used. Selected water molecules were represented by spheres of volume V_w. Since the exact volume of the bound water molecules is still uncertain, the volumes were varied within realistic limits, including also the values preferred in the literature (0.0245 nm^3 [43, 44] and 0.0269 nm^3 [45, 46]). The calculations were performed using the programs MSRoll and SIMS and different sets of the parameters: r_{probe} (0.14–0.1525 nm), r_w

(0.14–0.1525 nm), d_w (0.28–0.305 nm) and V_w (0.0245–0.0284 nm^3); details are given in Tables 1 and 2.

A uniform variation of the original Kuntz hydration values was performed by varying the factor f_K between 1 and 10. A value of $f_K = 1$ means that each accessible AA residue on the protein surface obtains the hydration according to the value given originally by Kuntz [18] or, for steric constraints, a smaller hydration. These water molecules determine $N_{w,1}$. For $f_K \geq 8$ the number of water molecules which can be placed on the surface always reaches an upper level $N_{w,max}$, corresponding to a monolayer of water molecules around the protein. Additional hydration of the accessible Phe residues (which according to Kuntz [18] are devoid of any hydration) may lead to a further increase in the number of water molecules. The selected water molecules were added to the initial dry model. To enable hydrodynamic modeling, the number of beads for the resultant hydrated models (about 3,000 or 6,000 in the case of the two proteins under analysis) had to be decreased to a practicable number (e.g., about 300 or 600), applying the cubic grid approach [10].

Parameter predictions

The hydrodynamic parameters s, D and $[\eta]$ of the reduced models were calculated by means of the program HYDRO [6, 47] which had been adapted for handling overlapping nonequal spheres [10, 48, 49]. Though the data reduction step is not necessary for the calculation of $I(h)$ and $p(r)$, the prediction of the SAXS functions was also based on the reduced models to reach full compatibility with the hydrodynamic predictions. The scattering curves were

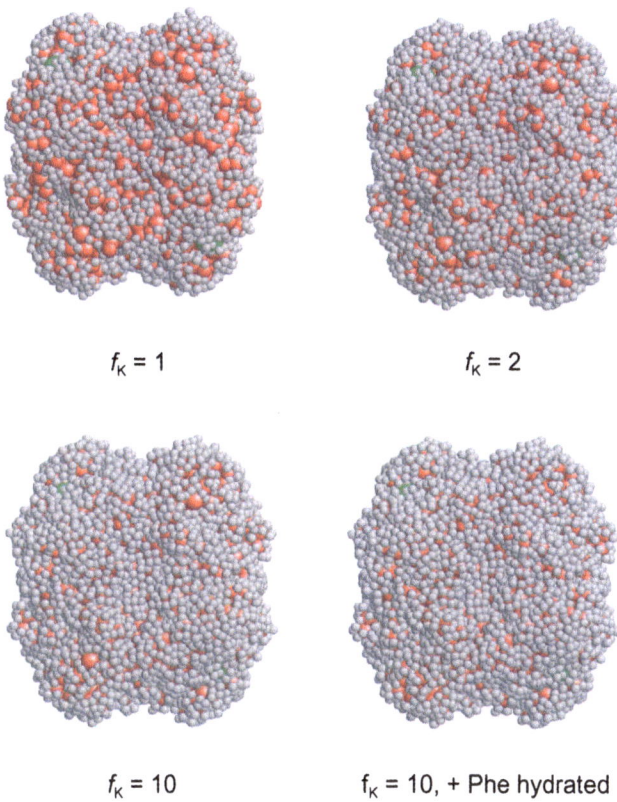

$f_K = 1$ $f_K = 2$

$f_K = 10$ $f_K = 10$, + Phe hydrated

Fig. 2 Selected hydrodynamic models for hydrated catalase, based on MSRoll points and application of HYDMODEL (parameter set A; $V_w = 0.0245$ nm³; $f_K = 1$, 2 and 10, and additional hydration of accessible Phe residues). Amino acid residues are shown in *red*, bound water molecules in *gray*, and NADPH$_2$ ligands in *green*; further details are given in Table 2

calculated by Debye's formula and then converted to $p(r)$ functions by Fourier transformation [50]. In agreement with the theory of SAXS, the spheres were weighted according to the calculated number of excess electrons to predict SAXS functions and parameters.[1] For the prediction of hydrodynamic quantities, however, only the radii (volumes) of the spheres are used in HYDRO. Therefore, two types of radii of gyration of the models were calculated, either from the excess electrons, radii and coordinates of the beads ($R_{G,SAXS}$) or only from bead radii and coordinates ($R_{G,Hydro}$).

Results

As a continuation of preceding studies [16, 51], a variety of models for citrate synthase and catalase were constructed. The modeling attempts aimed at studying the influence of various input parameters on hydrated protein models and their molecular parameters. It was already proven that the data reduction process, required for the

---

[1] The number of excess electrons of a bead was obtained as the product of the bead volume and the calculated electron density difference between the AA residues and/or bound water molecules, which are represented by the bead, and the surrounding bulk water

performance of hydrodynamic computations, does not alter the essential structural features of the original models [10–13, 16, 51]. Predicted structural and hydrodynamic parameters (V, R_G, s, D) were compared with existing experimental data to test the models under consideration. In agreement with previous results [22], the comparison of [η] values turned out to be rather unreliable.

Hydrated protein models

As outlined in Tables 1 and 2, the hydration of the enzymes citrate synthase and catalase was simulated in a systematic fashion, using parameter sets A–D with various values for r_{probe}, d_w, r_w and V_w. In addition, the hydration values given by Kuntz were modulated by variation of f_K, leading to a protein hydration between $\delta_{1,1}$ and $\delta_{1,max}$.

The application of surface calculation and hydration algorithms allows the construction of realistic models for hydrated proteins [16, 51]. The creation of dot surface points by the programs MSRoll and SIMS allows selective positioning of water molecules on the protein surface. This is shown explicitly for selected models for citrate synthase (Fig. 1) and catalase (Fig. 2). The models clearly demonstrate increasing numbers of water molecules on the protein surface as a function of f_K modulating the Kuntz hydration values. While the additional hydration of accessible Phe residues only leads to an insignificant increase of two water molecules in the case of citrate synthase (not shown), an increase of nearly 60 molecules has to be encountered in the case of catalase (Fig. 2). A visual inspection of Fig. 1 suggests that the hydration based on the SIMS surface is slightly higher than that derived from the MSRoll surface.

Though the number of calculated surface points (N_{dot}) is quite different for the surface calculation programs SIMS and MSRoll, the application of our program HYDMODEL finally leads to comparable results concerning N_w and δ_1 (Tables 1, 2) and also to similar parameter predictions. For convenience, we present in the following only the results obtained from application of MSRoll.

To illustrate the differences in the number and positions of the water molecules in the models based on parameter sets A–D, we superimposed the models in pairs. This is shown for citrate synthase for the pairs A and B (Fig. 3a), and A and D (Fig. 3b). Water molecules differing in their coordinates by less than 0.01 nm were considered as being common to both models of the respective pair. On the basis of this criterion, the water positions differ considerably for the pair A and D, but less markedly for the pair A and B. This fact is obviously caused by the far-reaching variation of parameters in the case of set D, if compared to set B (Table 1).

(a) A and B

(b) A and D

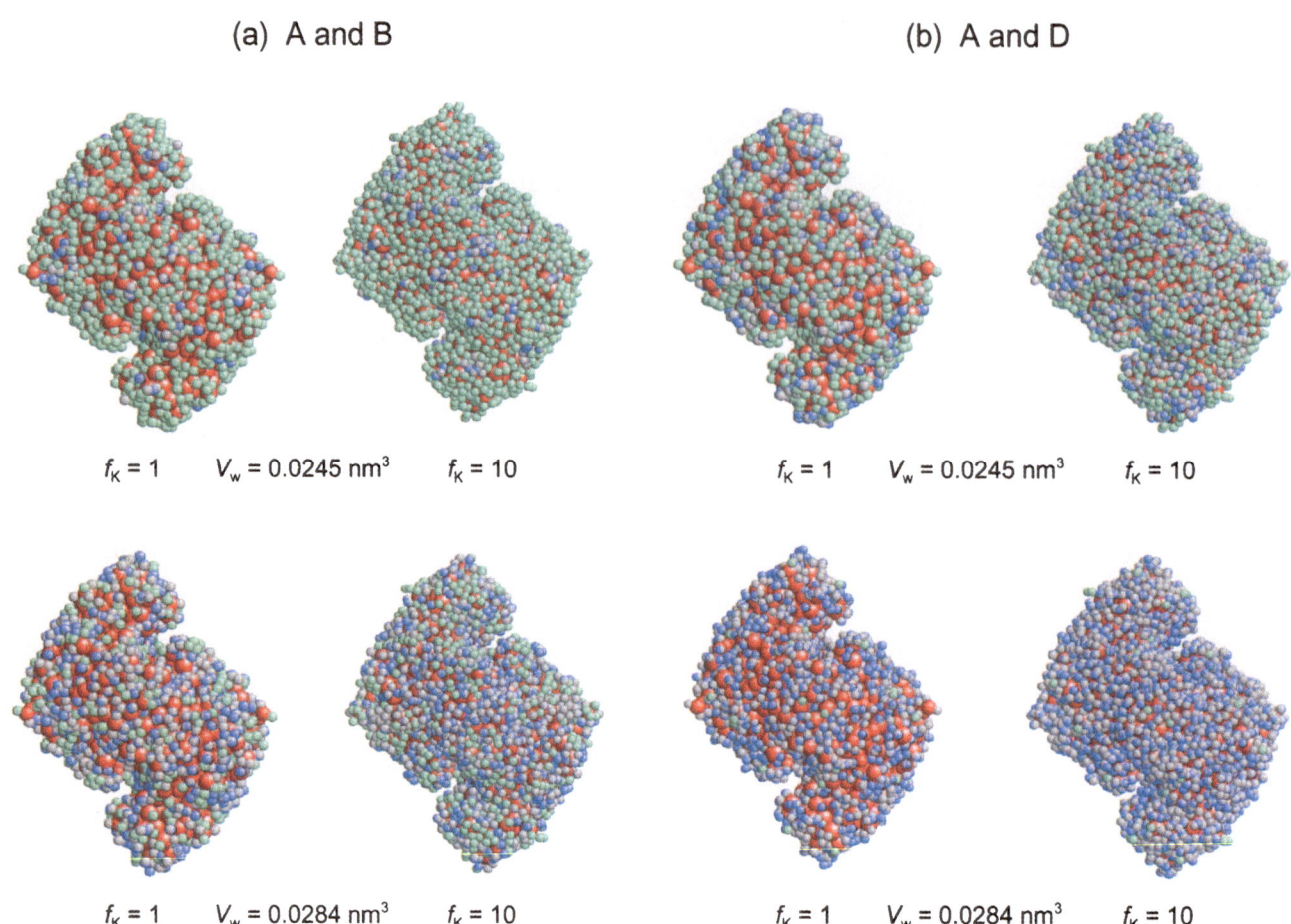

Fig. 3 Comparison of hydrated citrate synthase models, based on MSRoll points and HYDMODEL for parameter sets A and B (**a**) and A and D (**b**): $V_w = 0.0245$ and 0.0284 nm^3; $f_K = 1$ and 10. Amino acid residues are given in *red*. Bound water molecules are shown in *gray* (variant A) or *blue* (variant B or D) if not common to both models; in the latter case they are highlighted in *green*

Parameter predictions

From a comparison of the detailed results obtained for citrate synthase and for catalase with the different programs (MSRoll, SIMS) and calculation variants (parameter sets A–D, combined with different V_w values) the following generalizations can be made.

The use of the dot surface obtained from the program SIMS always leads to slightly higher hydration values compared to the hydration values derived from the MSRoll dot surface when using the same factor f_K (Table 1). The differences are most expressed when parameter set A is used. Differences in the number and positions of water molecules cause slight deviations in the radii of gyration, $R_{G,SAXS}$ and $R_{G,Hydro}$, and the hydrodynamic parameters of the models in both directions. On the whole, however, the differences due to the use of different surface calculation programs are quite

marginal and are usually lower than 0.1% for the radii of gyration, 0.4% for s and D and 1% for $[\eta]$.

With a given surface calculation program and f_K value, the use of different parameter sets may lead to larger differences in the hydration and the radii of gyration of the models, whereas the hydrodynamic parameters s, D and $[\eta]$ appear to be less affected. With MSRoll, for instance, applying parameter sets C and particularly D resulted in significantly lower hydration ($\Delta\delta_1$ up to -0.07 g/g) and radii of gyration (ΔR_G up to -0.6%) if compared to the results obtained with parameter set A at the same f_K value; the differences in s, D and $[\eta]$ were still of the size mentioned before. In contrast, parameter set B yielded almost the same hydration and radii of gyration as set A, and the deviations of the hydrodynamic parameters were lower than with sets C and D.

The influence of the parameter sets and V_w values on the molecular parameters of the hydrated models can be followed from Fig. 4, which shows calculated values for V, $R_{G,SAXS}$, s and D of citrate synthase with respect to the hydration, δ_1. In order to make a decision in favor of a definite set of input parameters, predicted and observed parameters have to be compared critically. All interpre-

Fig. 4 Calculated parameters of hydrated models of citrate synthase versus the hydration δ_1: **a** volume, V, **b** radius of gyration, $R_{G,SAXS}$, **c** sedimentation coefficient, s, and **d** diffusion coefficient, D. The curves were calculated from the MSRoll dot surface, using parameter sets A (——, *black*), B (– · – · –, *blue*), C (– ·· –, *green*) and D (– – – –, *red*), in combination with water volumes of 0.0245 nm³ (*circles*), 0.0257 nm³ (*squares*), 0.0269 nm³ (*up triangles*) and 0.0284 nm³ (*down triangles*); for details see Table 1. For clarity, the representations for B have been omitted in most cases, because they were similar to A. The *horizontal dashed lines* indicate the experimental values

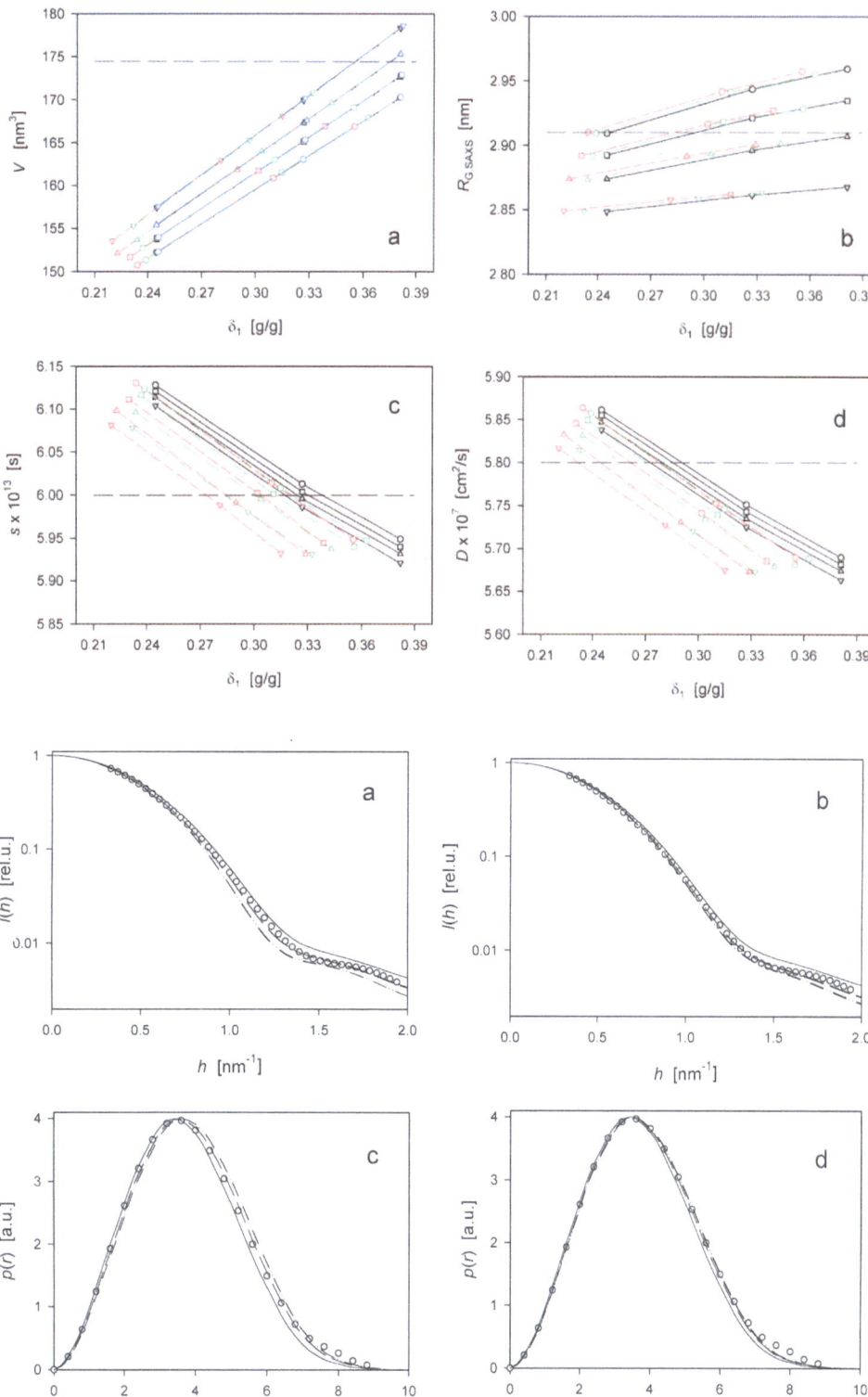

Fig. 5 Small-angle X-ray scattering functions, $I(h)$ (**a**, **b**) and $p(r)$ (**c**, **d**), of hydrated models (MSRoll points, parameter sets A and D) of citrate synthase and comparison with the experimental functions (*circles*), where $h = 4\pi\sin\theta/\lambda$ (2θ = scattering angle, λ = wavelength) and r is the distance between two arbitrarily chosen excess electrons within the molecule. For better comparison the $p(r)$ functions have been normalized to the same height of the maximum. **a, c** $V_w = 0.0245$ nm³; **b, d** $V_w = 0.0284$ nm³. Anhydrous model: (——): $\delta_{1,1}$ (– – –); $\delta_{1,max}$ (– ·· –). The curves for sets A and D coincide within the accuracy of the drawing

tations, of course, depend on the accuracy of the experimental parameters used for this comparison.

An inspection of Fig. 4a discloses that the experimental volume V can only be simulated with parameter sets A and B, using $V_w \geq 0.0269$ nm³ (at $\delta_1 \geq 0.36$ g/g). The experimental value of R_G (Fig. 4b) can be obtained with parameter sets A–D, preferably with V_w values of about 0.0257 nm³ (at $\delta_1 \approx 0.28 - 0.32$ g/g). An experi-

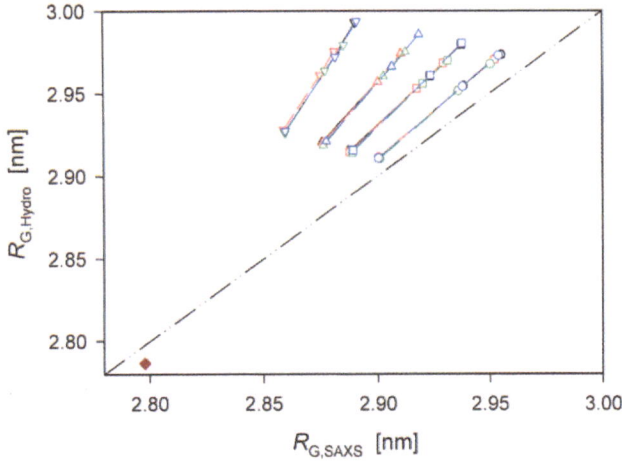

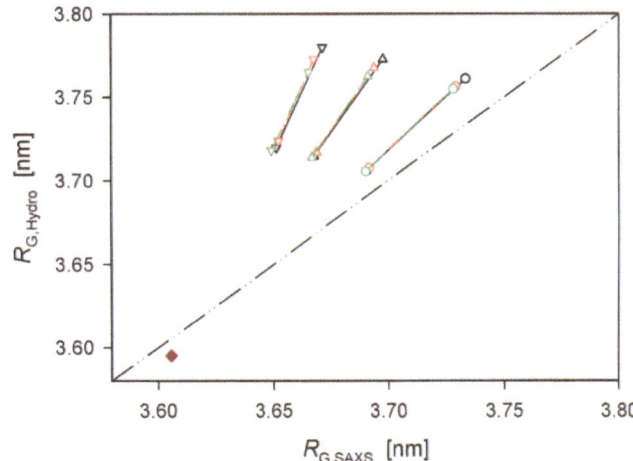

Fig. 6 Plot of the calculated radius of gyration, $R_{G,SAXS}$, of citrate synthase, obtained from excess electrons and bead radii and coordinates, versus the radius of gyration, $R_{G,Hydro}$, derived from bead radii and coordinates. The calculations were based on MSRoll points. The usage of *lines* and *symbols* is the same as in Fig. 4; the *diamond* indicates the anhydrous model; – ·· – symbolizes the median

Fig. 8 Plot of the calculated radius of gyration, $R_{G,SAXS}$, of catalase, obtained from excess electrons and bead radii and coordinates, versus the radius of gyration, $R_{G,Hydro}$, derived from bead radii and coordinates. The calculations were based on MSRoll points. The usage of *lines* and *symbols* is the same as in Fig. 4; the *diamond* indicates the anhydrous model; – ·· – symbolizes the median

mental *s* value of 6.0×10^{-13} s can be reached with all parameter sets (Fig. 4c), irrespective of the value used for V_w (at $\delta_1 \approx 0.27 - 0.34$ g/g); the alternative experimental sedimentation coefficient of $s = 6.2 \times 10^{-13}$ s, however, cannot be simulated within a range of acceptable hydration values. The simulation of the mean of the mentioned *s* values, on the other hand, is feasible. Fig. 4d reveals that also the experimental *D* value can be

mimicked with all parameter sets and V_w values, however, at a slightly lower hydration ($\delta_1 \approx 0.23$–0.29 g/g). The inspection of Fig. 4c and d shows that application of parameter set D enlarges the hydration range. Summarizing these results, it can be concluded that the experimental values for R_G, *s* and *D* can be fitted at a hydration of about 0.29 g/g, although with different sets of parameters.

Fig. 7 Plot of calculated parameters for hydrated models for catalase versus the hydration δ_1: **a** volume, V, **b** radius of gyration, $R_{G,SAXS}$, **c** sedimentation coefficient, *s*, and **d** diffusion coefficient, *D*. The curves were calculated from the MSRoll dot surface, using parameter sets A, C and D; for details see Table 2. The *symbols* are explained in the legend to Fig. 4

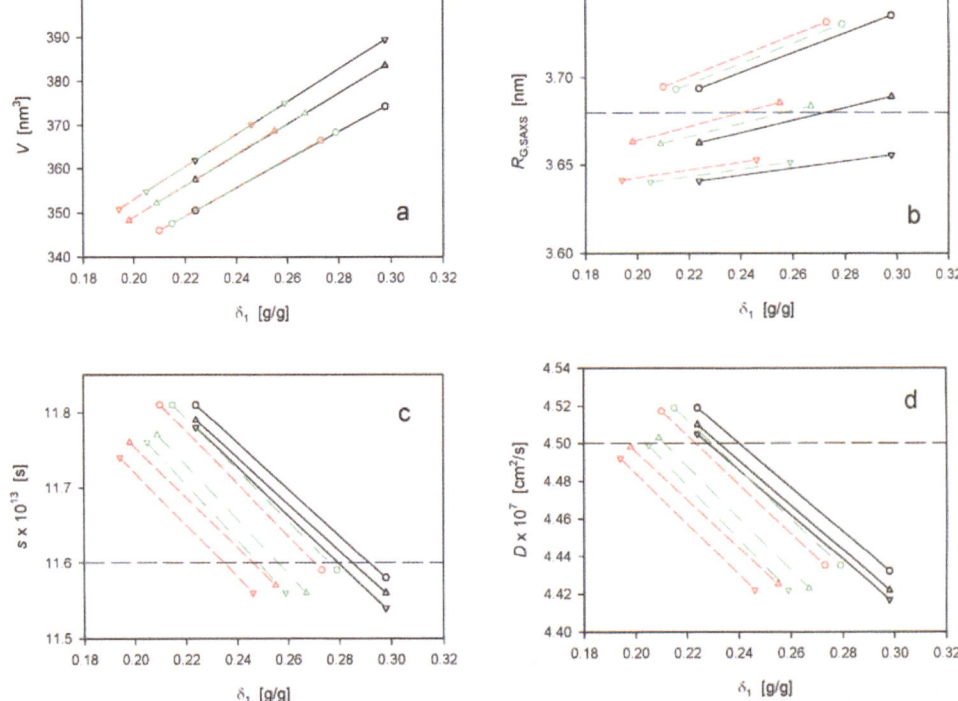

As may be taken from Fig. 5a and b, the calculated scattering curve $I(h)$ of the anhydrous model for citrate synthase runs above the experimental scattering curve. Among the hydrated models, $I(h)$ can be simulated best at low hydration ($\delta_1 \approx \delta_{1,1}$) by applying a water volume V_w of 0.0284 instead of 0.0245 nm^3. Similar conclusions can be drawn from the $p(r)$ functions shown in Fig. 5c and d. However, the tail ends of all calculated $p(r)$ functions run below the experimental function. This finding possibly indicates that the experimental function is influenced by the presence of small amounts of aggregates. As a consequence of this behavior, the experimental values for R_G and V would overestimate slightly the true quantities of the nonaggregated protein.

A critical comparison of the radii of gyration, calculated from the excess electrons, radii and coordinates of the beads ($R_{G,SAXS}$) or only from bead radii and coordinates ($R_{G,Hydro}$) indicates that the values obtained differ to some extent (Fig. 6). The differences between these two quantities increase with increasing V_w and δ_1. This follows from the regression lines, the slopes of which depend mainly on V_w, whereas the choice of the parameter set is much less important. The difference between $R_{G,SAXS}$ and $R_{G,Hydro}$ should be considered if the results from hydrodynamics are compared to those of solution scattering, especially when applying whole-body approaches (e.g., calculation of axial ratios from R_G and V).

Similar results as found for citrate synthase can be observed for catalase. Summarizing the findings in Fig. 7a–d, it can be stated that some of the experimental values (see Methods) are completely incompatible with our simulations ($V = 420.0$ nm^3, $R_G = 3.98$ nm, $s = 11.3 \times 10^{-13}$ s and $D = 4.1 \times 10^{-7}$ cm^2/s); however, simulations of the alternative values for R_G, s and D are successful at moderate hydration values ($\delta_1 \approx 0.2$–0.3 g/g).

In analogy to the results found for citrate synthase, the radii of gyration $R_{G,SAXS}$ and $R_{G,Hydro}$ of catalase diverge markedly, again depending on V_w and δ_1 (Fig. 8). The slopes of the regression lines exceed those observed for citrate synthase at the same V_w values (Fig. 6).

Conclusions

The combined use of advanced surface calculation and hydration algorithms, on the one hand, and efficient bead modeling approaches, on the other, turns out to be a powerful tool for the derivation of realistic models for hydrated proteins and predictions of structural and hydrodynamic parameters. A systematic variation of the parameters, which are required as inputs for surface and hydration programs, reveals their moderate influence on the hydrated models for citrate synthase and catalase and the molecular parameters derived therefrom. The largest differences occur between the results obtained for parameter sets A (only change of V_w in the hydration algorithm) and D (change of r_{probe} in the surface calculations, changes of V_w, r_w and d_w in the hydration program). With all parameter sets studied, however, realistic values for protein hydration ($\delta_1 \approx 0.2$–0.4 g/g) are obtained. Because of unavoidable deficiencies of experimental values, slightly different hydration models are required for fitting the parameters under analysis. Currently, no clear-cut decision can be made in favor of a definite density of bound water.

Application of different input parameters leads to differences in the positions of bound water molecules on the protein surface and the number of molecules assigned to certain AA residues. It makes no sense to overinterpret these differences, since in reality water molecules fluctuate permanently and very fast. Nevertheless the models of the hydrated proteins obviously suggest preferred places of residence of bound water molecules.

To make more detailed statements in this direction, further investigations are required, with respect to both the choice of proteins, on the one hand, and the calculation algorithms and input parameters to be applied, on the other.

Acknowledgements The authors are much obliged to Y.N. Vorobjev for making available the program SIMS and to J. García de la Torre for the program HYDRO.

References

1. Kumosinski TF, Pessen H (1985) Methods Enzymol 117:154–182
2. García de la Torre J (1989) In: Harding SE, Rowe AJ (eds) Dynamic properties of biomolecular assemblies. Royal Society of Chemistry, Cambridge, pp 3–31
3. Harding SE (1989) In: Harding SE, Rowe AJ (eds) Dynamic properties of biomolecular assemblies. Royal Society of Chemistry, Cambridge, pp 32–56
4. Harding SE (1995) Biophys Chem 55:69–93
5. Byron O (2000) Methods Enzymol 321:278–304
6. García de la Torre J, Huertas ML, Carrasco B (2000) Biophys J 78:719–730
7. Ashton AW, Boehm MK, Gallimore JR, Pepys MB, Perkins SJ (1997) J Mol Biol 272:408–422
8. Byron O (1997) Biophys J 72:408–415
9. (a) Spotorno B, Piccinini L, Tassara G, Ruggiero C, Nardini M, Molina F, Rocco M (1997) Eur Biophys J 25:373–384; (b) (1997) Eur Biophys J erratum 26:417
10. Zipper P, Durchschlag H (1997) Prog Colloid Polym Sci 107:58–71
11. Zipper P, Durchschlag H (1998) Biochem Soc Trans 26:726–731
12. Zipper P, Durchschlag H (2000) J Appl Crystallogr 33:788–792

13. Durchschlag H, Zipper P (2001) Prog Colloid Polym Sci: 119:121–130
14. Zipper P, Krebs A, Durchschlag H (2001) Prog Colloid Polym Sci: 119: 141–148
15. Cooper A, Harding SE (eds) (2001) Biophys Chem (in press)
16. Durchschlag H, Zipper P (2001) Biophys Chem (in press)
17. Bairoch A, Apweiler R (2000) Nucleic Acids Res 28:45–48
18. Kuntz ID (1971) J Am Chem Soc 93:514–516
19. Durchschlag H, Zipper P (1996) J Mol Struct 383:223–229
20. Durchschlag H, Zipper P (1997) J Appl Crystallogr 30:1112–1124
21. Durchschlag H, Zipper P (1998) Biochem Soc Trans 26:731–736
22. Durchschlag H, Zipper P (1999) Prog Colloid Polym Sci 113:87–105
23. Durchschlag H, Zipper P, Purr G, Jaenicke R (1996) Colloid Polym Sci 274:117–137
24. Chacón P, Díaz JF, Morán F, Andreu JM (2000) J Mol Biol 299:1289–1302
25. Kumosinski TF, Pessen H (1982) Arch Biochem Biophys 219:89–100
26. Durchschlag H, Zipper P (1997) Prog Colloid Polym Sci 107:43–57
27. Remington S, Wiegand G, Huber R (1982) J Mol Biol 158:111–152
28. Wiegand G, Remington S, Deisenhofer J, Huber R (1984) J Mol Biol 174:205–219
29. Sussman JL, Lin D, Jiang J, Manning NO, Prilusky J, Ritter O, Abola EE (1998) Acta Crystallogr Sect D 54:1078–1084
30. Durchschlag H, Zipper P, Wilfing R, Purr G (1991) J Appl Crystallogr 24:822–831
31. Wu J-Y, Yang JT (1970) J Biol Chem 245:212–218
32. Singh M, Brooks GC, Srere PA (1970) J Biol Chem 245:4636–4640
33. Fita I, Murthy MRN, Rossmann MG, Silva AM (1986) Acta Crystallogr Sect B 42:497–515
34. Malmon AG (1957) Biochim Biophys Acta 26:233–240
35. Lee JC, Timasheff SN (1974) Biochemistry 13:257–265
36. Sumner JB, Gralén N (1938) J Biol Chem 125:33–36
37. Tanford C, Lovrien R (1962) J Am Chem Soc 84:1892–1896
38. Sumner JB, Dounce AL, Frampton VL (1940) J Biol Chem 136:343–356
39. Connolly ML (1983) Science 221:709–713
40. Connolly ML (1985) J Am Chem Soc 107:1118–1124
41. Connolly ML (1993) J Mol Graphics 11:139–141
42. Vorobjev YN, Herman J (1997) Biophys J 73:722–732
43. Perkins SJ (1986) Eur J Biochem 157:169–180
44. Gerstein M, Chothia C (1996) Proc Natl Acad Sci USA 93:10167–10172
45. Svergun DI, Richard S, Koch MHJ, Sayers Z, Kuprin S, Zaccai G (1998) Proc Natl Acad Sci USA 95:2267–2272
46. Ebel C, Eisenberg H, Ghirlando R (2000) Biophys J 78:385–393
47. García de la Torre J, Navarro S, López Martínez MC, Díaz FG, López Cascales JJ (1994) Biophys J 67:530–531
48. Carrasco B, García de la Torre J, Zipper P (1999) Eur Biophys J 28:510–515
49. Zipper P, Durchschlag H (1999) Prog Colloid Polym Sci 113:106–113
50. Glatter O, Kratky O (eds) (1982) Small-angle X-ray scattering. Academic, London
51. Zipper P, Durchschlag H (2001) Physica A (in press)
52. Sayle RA, Milner-White EJ (1995) Trends Biochem Sci 20:374–376

Progr Colloid Polym Sci (2002) 191:141–148
© Springer-Verlag 2002

HYDRODYNAMICS

Peter Zipper
Angelika Krebs
Helmut Durchschlag

Prediction of hydrodynamic parameters of *Lumbricus terrestris* hemoglobin from small-angle X-ray and electron microscopic structures

P. Zipper (✉)
Physical Chemistry, Institute of Chemistry
University of Graz, Heinrichstrasse 28, 8010
Graz, Austria
e-mail: peter.zipper@uni-graz.at
Tel.: +43-316-3805415
Fax: +43-316-3809850

A. Krebs
Structural Biology, European Molecular
Biology Laboratory, Meyerhofstrasse 1
69117 Heidelberg, Germany

H. Durchschlag
Institute of Biophysics and Physical
Biochemistry, University of Regensburg
Universitätsstrasse 31, 93040 Regensburg
Germany

Abstract The 3.5-MDa extracellular hemoglobin from *Lumbricus terrestris* is a giant heteromultimeric hexagonal bilayer complex. The consensus model of the complex as derived from small-angle X-ray scattering (SAXS) and a 3D reconstruction obtained from cryoelectron microscopy (EM) were used as bases for the prediction of hydrodynamic parameters (sedimentation coefficient, s, diffusion coefficient, D, and intrinsic viscosity, $[\eta]$) by the bead modeling approach implemented in García de la Torre's program HYDRO. Since the number of beads in the initial EM and SAXS models was too high, appropriate data reduction had to be performed. For this purpose, the original hexagonal structures were mapped into a hexagonal grid to maintain symmetry details. Thereby the initial models, consisting of up to 23,000 beads of equal size and unequal density, were transformed to reduced models of about 500 beads of unequal size but equal density. A comparison of the results obtained for the various models reveals good agreement between predicted and experimental hydrodynamic parameters s and D, provided the models refer to hydrated volumes. In an alternative modeling approach, low-resolution models were generated directly by applying the genetic algorithm (GA) implemented in the program DALAI_GA2 by Chacón et al. to the experimental SAXS curve of *L. terrestris* hemoglobin and, for the purpose of testing, to the SAXS curve calculated from the initial hydrated EM model. While the resulting GA models are of peculiar shape and certainly without physical relevance, they nevertheless provide perfect fits to the experimental scattering functions and good predictions for s and D.

Key words Extracellular hemoglobin · Bead modeling · Hydrodynamic parameters · Electron microscopy · Small-angle X-ray scattering

Introduction

Solution scattering, electron microscopy (EM), and hydrodynamic methods typically yield low-resolution information, in contrast to high-resolution methods such as X-ray crystallography and NMR. The prediction of low-resolution properties of biopolymers from crystal structures can be a helpful tool for interpreting experimental low-resolution data. This argument also applies to the conversion of information on the low-resolution level itself, for example, the prediction of small-angle X-ray scattering (SAXS) behavior from EM data or the prediction of hydrodynamic properties from SAXS or EM structures. Such predictions require appropriate modeling of the biopolymer under consideration. For this purpose multibody approaches, for example, representing the particles by 3D assemblies of spheres (bead modeling), turned out to be very efficient [1, 2].

In previous articles we have reported various examples of predictions of low-resolution properties of proteins from crystal structures [3–6]. Some preliminary multibody predictions of hydrodynamic parameters from SAXS and EM structures have been presented recently [7]. This study continues our previous work and gives a detailed report on various approaches which we apply to predict hydrodynamic properties from SAXS and/or EM data.

The biopolymer used in this study was the giant extracellular hemoglobin (Hb) of the earthworm *Lumbricus terrestris*. Its properties and subunit structure can be explained by a "bracelet model", in which 12 dodecameric 200-kDa subunits, each composed of 12 heme-containing globin chains, and 36 heme-deficient linker chains form a hexagonal bilayer complex with a total mass of about 3.5 MDa [8–10]. An alternative subunit stoichiometry has been proposed by Zhu et al. [11].

The hexagonal bilayer (HBL) architecture of *L. terrestris* Hb is supported by 3D reconstructions at low and medium resolution obtained from cryoelectron microscopy [10, 12, 13]. Several models of slightly different structure have been proposed. In the model established by Lamy et al. [10] the upper half of the bilayer is rotated about 15° clockwise if compared to the lower half.

A SAXS study of the native HBL complex of *L. terrestris* was performed in the presence and absence of oxygen and yielded nearly identical results for the oxygenated and the deoxygenated state [14, 15]. Finally a 3D consensus model was derived from the SAXS data for the oxygenated state, in order to simulate the solution structure of the protein complex and its scattering behavior [14]. Later, this SAXS model was compared to a model based on 3D reconstructions of *L. terrestris* Hb from cryoelectron microscopic images [16]. The SAXS model as well as the EM model were recently used as bases for preliminary predictions of hydrodynamic parameters [7]. In this work we review the previous predictions briefly and complete them by including additional models and approaches.

Methods

Modeling

Different approaches were applied to obtain bead models which are appropriate for the prediction of hydrodynamic properties. Such bead models typically consist of several hundred but less than a thousand beads. Our previous approach to yield such models was to start from the SAXS consensus model or from the EM model of the protein, that means from models consisting of several thousand spherical elements, and to transform these initial models in such a way that the number of beads was reduced but the essential structural features were maintained [7]. A new alternative approach is to generate an appropriate hydrodynamic model directly from the experimental scattering curve by using a genetic algorithm (GA) [17, 18] for retrieving the low-resolution structural information. Both approaches are described with all necessary details in the following.

Initial SAXS model

The consensus model described in Ref. [16] was used as the initial SAXS model. The model consisted of 6,844 spheres of equal size (radius 0.66 nm) but unequal weight. It was obtained in a two-step procedure [14]. In the first step, about 600 different models, all biased to represent a HBL structure with eclipsed hexagons, were generated by trial and error and were tested for equivalence in scattering with the protein complex. In the second step, the 22 best-fitting models were superimposed and averaged, a procedure which resulted in spheres of different weights according to the different occupation densities of positions.

Initial EM models

The density map of a 3D reconstruction of native *L. terrestris* Hb [10], provided by F. de Haas and J.C. Taveau, served as the basis for two initial EM models. All voxels of the map, whose density exceeded a certain threshold, were selected and equal-sized spheres were placed at their positions. The weights of the spheres were derived from the density of the voxels, equating the upper limit of the arbitrary voxel density scale (0–255) with the difference between the calculated electron densities of the protein and water [16].

The total volume of the selected voxels (and of the corresponding spheres) is a measure of the molecular volume. Increasing or decreasing the density threshold for the selection of voxels is therefore equivalent to decreasing or increasing the volume. In the analysis and interpretation of 3D reconstructions, varying the threshold may provide better insights into the structure under investigation [13].

In our case, the density threshold was selected either to be 171, compatible with the expected anhydrous volume of the protein (a volume of $4,300 \pm 250$ nm^3 may be estimated from $M = 3,500 \pm 200$ kg/mol [15] and $\bar{v} = 0.733$ cm^3/g [19]), or 115, a value corresponding to a hydrated volume slightly higher than that derived from SAXS (Table 1). According to the voxel density distribution shown in Fig. 1, a threshold of 115 is at the lower limit of feasible values. The models resulting with the two thresholds consisted of 11,671 and 23,021 spheres, respectively. A final rescaling of the dimensions of the latter model was applied to improve the fit of calculated and experimental SAXS curves, $I(h)$, and distance distribution functions, $p(r)$ [16].

Reduced SAXS and EM models

In order to reduce the number of spherical elements efficiently without disturbing the symmetry relations, the given 3D structures of the SAXS and EM initial models, respectively, were first mapped into a hexagonal grid of cells of given edge length, and finally a sphere of appropriate radius was placed at the center of the mass distribution in each occupied cell [7]. Thereby, the initially large number of equal-sized spheres of unequal density was replaced with a much smaller number of spheres of equal density but unequal size in order to simulate the structure of the SAXS and EM models at a lower resolution. The edge length of the hexagonal cells was varied from 4.0 to 3.1 nm.

The reduction process itself was different for the SAXS model and the EM model. The volume of the initial SAXS consensus model is considerably higher than the volume found experimentally (Table 1). The higher value is due to the superposition of more than 20 different models, each of them having a volume close to the experimental one [14]. In this case, therefore, the reduction procedure had to yield a model whose volume is in accord with the experimental value. The volumes of the initial EM models, on the other hand, can be assumed to be the volumes of anhydrous and hydrated protein particles, respectively. In this case the volumes of the initial models had to be maintained in the reduction process.

Table 1 Comparison of experimental and calculated structural and hydrodynamic parameters of *Lumbricus terrestris* hemoglobin

	Beads	R_G (nm)	V (nm^3)	s (10^{-13} s)[a]	D (10^{-7} cm^2/s)	$[\eta]$ (cm^3/g)[a]
Experimental		10.71 ± 0.02[b]	$6{,}200 \pm 200$[b]	58.9 ± 0.2[c], 61.1[d]	1.61 ± 0.05[c]	
Initial small-angle X-ray scattering consensus model	6,844	10.68 ± 0.04[e]	8,279			
Reduced small-angle X-ray scattering models						
Cell edge = 4.0 nm	307	10.62	6,196	61.4	1.60	4.27
Cell edge = 3.1 nm	529	10.64	6,196	61.1	1.59	4.34
Genetic algorithm small-angle X-ray scattering models						
Bead diameter = 4.0 nm[f]	150	10.78	5,027	61.8	1.61	4.15
Bead diameter = 3.0 nm[f]	344	10.79	4,863	61.2	1.59	4.32
Rescaled[g]	344	10.81	6,483	60.5	1.58	4.47
Bead diameter = 2.6 nm[f]	482	10.78	4,436	60.6	1.58	4.42
Bead diameter = 2.2 nm[f]	749	10.84	4,176	59.5	1.55	4.66
Rescaled[g]	749	10.85	5,567	59.0	1.54	4.79
Initial electron microscopy model (anhydrous volume)	11,671	11.30	4,458			
Reduced electron microscopy models (anhydrous volume)						
Cell edge = 4.0 nm	263	11.28[h]/11.26[i]	4,458	63.6[h,i]	1.66[h,i]	3.88[h,i]
Cell edge = 3.1 nm	471	11.31[h]/11.28[i]	4,458	63.3[h]/63.2[i]	1.65[h,i]	3.93[h]/3.94[i]
Rescaled[j]	369	10.74[h]/10.72[i]	3,828	67.2[h]/67.1[i]	1.75[h,i]	3.30[h,i]
Initial electron microscopy model (hydrated volume, rescaled)	23,021	10.74	7,452			
Reduced electron microscopy models (hydrated volume, rescaled)						
Cell edge = 4.0 nm	269	10.70[h,i]	7,452	63.1[h]/62.6[i]	1.64[h]/1.63[i]	3.99[h]/4.08[i]
Cell edge = 3.1 nm	533	10.71[h]/10.72[i]	7,452	62.6[h]/62.2[i]	1.63[h]/1.62[i]	4.08[h]/4.17[i]
Rescaled[k]	533	10.70[h]/10.71[i]	6,200	62.9[h]/62.5[i]	1.64[h]/1.63[i]	4.01[h]/4.10[i]
Genetic algorithm electron microscopy models (hydrated volume)						
Bead diameter = 4.0 nm[f]	149	10.73	4,993	63.0	1.64	3.96
Bead diameter = 3.0 nm[f]	342	10.73	4,835	62.2	1.62	4.12
Rescaled[g]	342	10.74	6,446	61.5	1.60	4.25
Bead diameter = 4.0 nm[l]	153	10.73	5,127	63.0	1.64	3.95
Bead diameter = 3.0 nm[l]	335	10.73	4,736	62.4	1.63	4.07
Rescaled[g]	335	10.74	6,314	61.7	1.61	4.21

[a] Predicted values are based on $M = 3{,}500 \pm 200$ kg/mol [15]
[b] Ref. [15]
[c] Ref. [19]
[d] Ref. [24]
[e] Ref. [16]
[f] Initial search space: sphere
[g] Expansion of the beads in the program HYDRO to account for the packing factor of the hexagonal lattice [18]

[h] The densities of the initial model are taken into account
[i] The densities of the initial model are neglected
[j] Shrinkage of the beads and coordinates during data reduction to improve the fit of the radius of gyration R_G
[k] Shrinkage of the beads in the program HYDRO to improve the fit of the hybrated volume V
[l] Initial search space: oblate cylinder

These different constraints are taken into account by calculating the radius of a bead in the reduced model, r_{bead}, according to the general expression

$$r_{bead} = r_{ini} \sqrt[3]{\frac{\sum_{i=1}^{n} \rho_i}{\rho_{norm}}}, \quad (1)$$

where r_{ini} is the sphere radius in the initial model and ρ_i are the individual densities of the n spheres contributing to the new bead. According to the aforementioned constraints, ρ_{norm} was set equal to the maximum density of the spheres in the initial SAXS model ($\rho_{norm} = 1$) and to the average density of all spheres in the initial EM models ($\rho_{norm} = \sum \rho_i / n$), respectively.

It was not clear, however, whether for the purpose of hydrodynamic modeling the voxel densities had to be taken into account explicitly in the reduction process for the EM models or whether the densities could be neglected at all. For this reason, both variants were applied.

The reduced models are "space-filling" owing to a moderate overlap of the spheres. To check the influence of data reduction on the structural characteristics, scattering curves, $I(h)$, of initial and reduced models were calculated by means of the Debye formula and distance distribution functions, $p(r)$, were obtained therefrom by Fourier transformation [20].

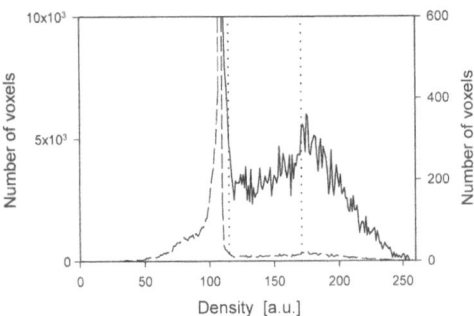

Fig. 1 Voxel density distribution of a 3D reconstruction of *Lumbricus terrestris* hemoglobin (Hb) as obtained from cryoelectron microscopy. The frequency of densities is characterized by a major water peak and a minor peak belonging to the protein (*dashed line*, left axis). Additionally, the minor peak is magnified for better visualization (*solid line*, right axis). The *dotted lines* indicate the thresholds 115 and 171, corresponding to the hydrated and the anhydrous volume, respectively

Models created by a genetic algorithm

In this alternative approach, SAXS-based models containing only a few hundred beads were created by applying the genetic algorithm [17] implemented in the program DALAI_GA2 (version DAL.LI-NUX) [18] to the experimentally obtained SAXS curve of *L. terrestris* Hb. Calculations were done in several consecutive steps, thereby decreasing the diameter of the beads gradually from 4 to eventually 2.2 nm. As the initial search space a roughly spherical shape was chosen. The models obtained from the GA consist of equal-sized, nonoverlapping spheres of equal density.

In order to test the power and efficiency of the GA itself, in two separate runs the algorithm was applied to the scattering curve that had been calculated for the EM model of the protein. The first run was performed with the spherical search space. In the second run, the caps of the spherical search space were cut off in order to yield an oblate search space.

Parameter predictions

Predictions of the hydrodynamic parameters s, D, and $[\eta]$ were performed by means of the program HYDRO [21], which had been modified by an ad hoc expression for the interaction tensor in order to improve the treatment of overlapping unequal spheres [5, 6, 22, 23].

Results

The experimental SAXS and hydrodynamic parameters (radius of gyration, R_G, hydrated volume, V, sedimentation coefficient, s, and diffusion coefficient, D) of *L. terrestris* Hb are given in Table 1, together with calculated and predicted values for structural and hydrodynamic parameters of various models, including also the intrinsic viscosity, $[\eta]$.

Initial and reduced SAXS models

Top and side views of the initial SAXS consensus model [16] are shown in Fig. 2, together with the analogous views of the reduced SAXS model corresponding to a cell

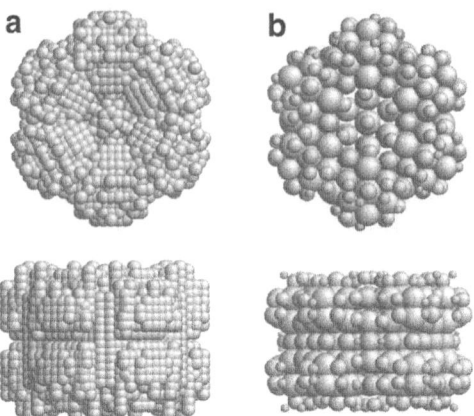

Fig. 2 Top and side views of **a** the initial small angle X-ray scattering (SAXS) consensus model of *L. terrestris* Hb and **b** the reduced model (cell edge = 3.1 nm) derived therefrom. The graphics were made with Rasmol (R. Sayle)

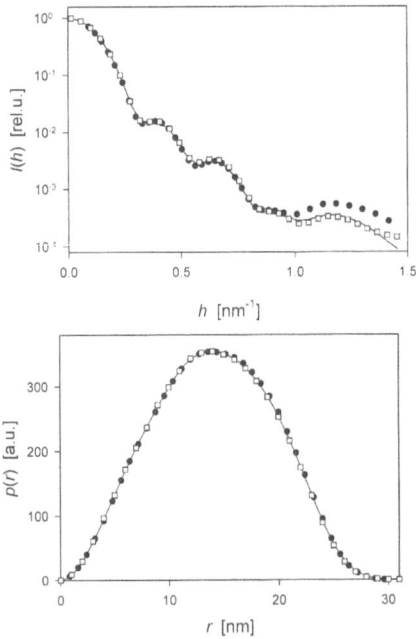

Fig. 3 Small-angle X-ray scattering curve, $I(h)$, and distance distribution function, $p(r)$, of *L. terrestris* Hb: experimental functions (*circles*); initial SAXS model (*lines*); reduced SAXS model (*squares*)

size of 3.1 nm. The experimentally obtained SAXS curve, $I(h)$, and pair-distance distribution function, $p(r)$, are compared in Fig. 3 with the scattering curves and $p(r)$ functions of the initial and reduced SAXS models, respectively. As follows from the figure, the calculated $I(h)$ and $p(r)$ curves of the models provide a very good, though not perfect fit to the experimental functions. For the scattering curves, however, this is only true in the range $h \leq 1.0$ nm^{-1}. A closer inspection of the tabulated parameters reveals that the reduced SAXS model slightly underestimates the radius of gyration of the protein, whereas its simulation of the experimental volume and hydrodynamic quantities s and D is nearly perfect, irrespective of the size of the cell edge used.

Initial and reduced EM models

As described in the Methods section, two lines of EM models were generated from the same 3D reconstruction obtained from cryoelectron microscopy. The voxel density distribution of the reconstruction is shown in Fig. 1. The main peak at position 108 on the arbitrary density scale is due to a great number of water molecules surrounding the protein particle in a shell of vitreous ice. The flat peak on the right of the main peak is due to the protein; the enlarged drawing of this peak reveals that its maximum is situated between 170 and 180 on the density scale. According to the use of different thresholds for the selection of voxels, 115 and 171, we derived models that correspond to the hydrated and the anhydrous volume of Hb, respectively.

The initial and reduced EM models related to the hydrated volume are illustrated in Fig. 4, the EM models corresponding to the anhydrous volume are shown in Fig. 5. The calculated $I(h)$ and $p(r)$ curves of the two initial [16] and the two reduced EM models are compared to the experimental $I(h)$ and $p(r)$ functions in Fig. 6. From the comparison it is obvious that the curves calculated for the initial and reduced models of the hydrated particle fit the experimental curves very well, whereas the curves for the models of the anhydrous particle exhibit large deviations.

The observed differences in the $I(h)$ and $p(r)$ curves of the EM models are responsible for the differences in the structural and hydrodynamic parameters of the models in Table 1. The R_G values of the reduced anhydrous model are considerably larger than the experimental radius of gyration, irrespective of whether the densities of the initial EM models were taken into account or neglected. A rescaling of the anhydrous model, performed in the data reduction process, is able to improve the fit of the R_G values, but diminishes at the same time the particle volume to an unreasonably low value and produces too high values for s und D.

The parameters of the hydrated EM model, on the other hand, are in fair accord with the experimental data, except the too high value for the particle volume. This discrepancy has already been caused by using the

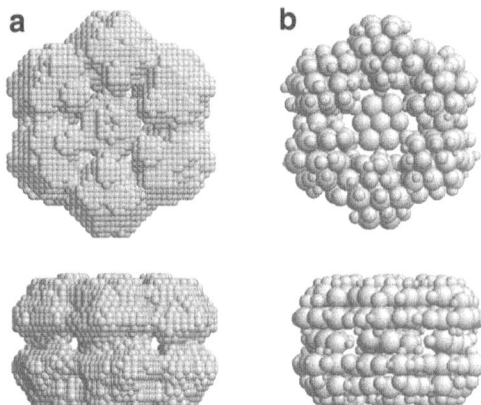

Fig. 5 Top and side views of **a** the initial anhydrous EM model of *L. terrestris* Hb and **b** the reduced model (cell edge = 3.1 nm, neglect of densities) derived therefrom

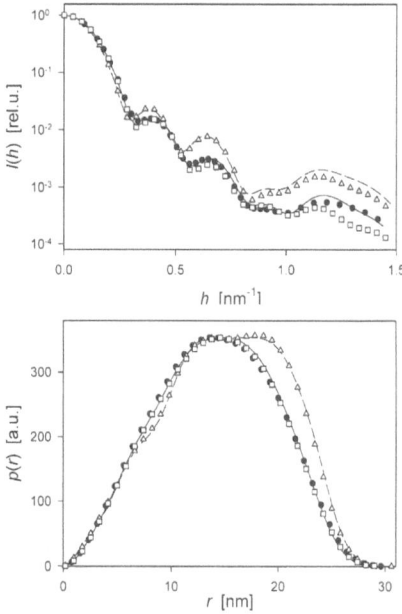

Fig. 6 SAXS curve, $I(h)$, and distance distribution function, $p(r)$, of *L. terrestris* Hb: experimental functions (*circles*); initial hydrated EM model (*solid lines*); reduced hydrated EM model (*squares*); initial anhydrous EM model (*dashed lines*); reduced anhydrous EM model (*triangles*)

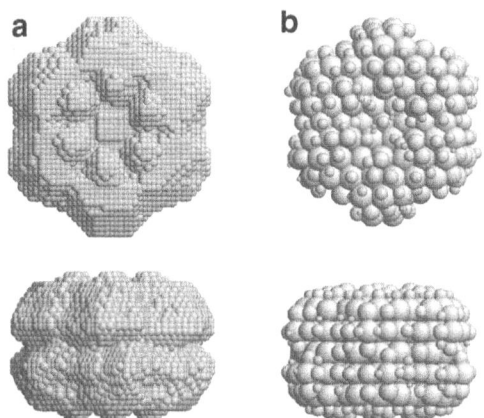

Fig. 4 Top and side views of **a** the initial hydrated electron microscopy (EM) model of *L. terrestris* Hb and **b** the reduced model (cell edge = 3.1 nm, neglect of densities) derived therefrom

threshold of 115 in creating the initial hydrated model. When the beads are simply shrunk in order to reduce V to the value derived from SAXS, the other data are altered only slightly. Differences of about the same size also result from the different treatments (use or neglect) of the densities of the initial model. On the other hand, the tempting alternative approach for solving the volume problem, namely a reduction in the volume of the initial model by increasing the threshold, turned out to be less successful: while models based on a threshold of about 138 indeed have volumes which are close to the value of V derived from SAXS, their $I(h)$ and $p(r)$ curves give poorer fits to the experimental SAXS curves, and the predicted hydrodynamic parameters of $s \geq 64$ and $D \geq 1.67$ are obviously too high.

Models generated by the GA

Application of the GA to the experimental SAXS curve of the protein, using a roughly spherical search space, resulted in a series of models of increasing resolution. Their scattering curves are nearly identical to the experimental curve up to the value of h corresponding

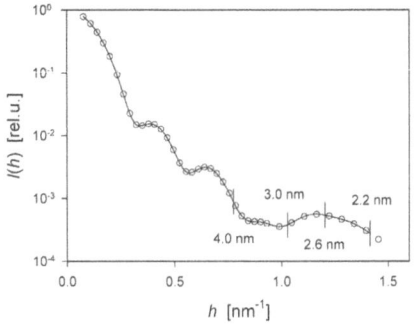

Fig. 7 Experimental SAXS curve, $I(h)$, of *L. terrestris* Hb (*circles*) and the scattering curve of the genetic algorithm (*GA*) model (bead diameter = 2.2 nm) derived therefrom (*line*). The *vertical bars* indicate the upper limits of the h ranges corresponding to the different bead diameters given in the figure

to the respective resolution (Fig. 7). Views of selected models are shown in Fig. 8. It is evident that these models possess a shape that is very different from the shapes of the EM and the previous SAXS models. A common feature of these GA models are cavities, which may lead to the appearance of clefts if the models are rotated in space. Obviously the cavities are necessary in order to achieve the almost perfect fit to the experimental SAXS curve. An inspection of the tabulated values for R_G and V of the models (Table 1), however, reveals the existence of deviations from the experimental data. Especially the volume of the models is definitely lower than the experimentally obtained V. This discrepancy is easily explained by the fact that the program DALAI_GA2 models the protein by spheres which are just touching their neighbors but do not overlap. The models are therefore not space-filling. In order to get the proper particle volume, the values obtained for V have to be divided by 0.75, the packing factor of the spheres in the hexagonal lattice [18]. It is tempting to simply increase the size of the beads in the GA models. This kind of rescaling causes only a slight increase in the R_G values, but leads to further alterations in the tail end of the scattering curves (not shown) and, therefore, counteracts the intentions of the GA modeling. Perhaps, using overlapping spheres in the GA modeling procedure might be a better solution to the problem.

As a consequence of the nearly perfect fit of scattering curves, the $p(r)$ functions of the GA models coincide with the experimental $p(r)$ function of Hb (comparison not shown); therefore, in spite of their different appearance, the GA models are equivalent in scattering to the protein molecules within the respective resolution, with the restrictions imposed by the volume problem discussed previously.

The equivalence in scattering may be considered as a structural equivalence. Indeed, the hydrodynamic parameters predicted for the GA SAXS models of *L. terrestris* Hb are very close to the experimentally found values (Table 1).

Fig. 8 Different views of GA SAXS models of *L. terrestris* Hb. The models correspond to bead diameters of **a** 4.0 nm, **b** 3.0 nm, **c** 2.6 nm, and **d** 2.2 nm

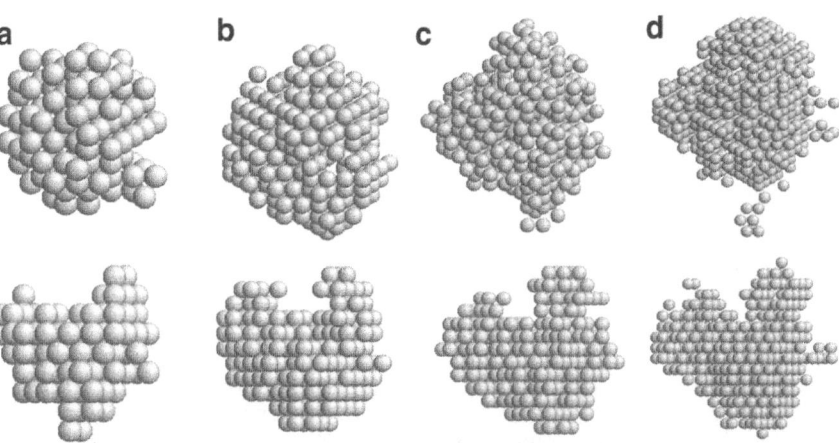

The tests of the GA, which were performed by applying the GA to the scattering curve of the initial hydrated EM model, led to models of similar shape and appearance as the GA models derived from the experimental SAXS curve (cf. Fig. 9 with Fig. 8), to similar excellent fits of the $I(h)$ and $p(r)$ curves (not shown) and to similar good predictions of the hydrodynamic parameters (Table 1). The choice of the type of search space had no obvious influence on the resulting models and predictions. The results obtained for the GA EM models exclude that the peculiar GA SAXS models are artifacts due to insufficiencies in the experimental scattering curve of *L. terrestris* Hb.

Conclusions

A critical inspection of the values listed in Table 1 reveals that good predictions of the hydrodynamic parameters s and D are possible when applying the reduced SAXS models, the reduced hydrated EM models (especially the models neglecting the densities), the GA SAXS models (except the models obtained for the highest resolution), and the GA EM models (provided a rescaling of the volumes is performed). By contrast, the reduced anhydrous EM models lead to erroneous anticipations of structural and/or hydrodynamic quantities.

It is not surprising that the previously mentioned SAXS and hydrated EM models mimic the hydrodynamic behavior appropriately, because they are based on well-founded information concerning the shape and structure of the HBL complexes of extracellular Hbs. The capability of the GA models to predict the hydrodynamic properties of *L. terrestris* Hb adequately, however, is somewhat unexpected since the GA models are rather peculiar and certainly do not represent any physical relevance. Some of the GA models, particularly those applying a too high resolution, exhibit some isolated spheres which are completely unrealistic from the biochemical point of view. In this context, however, it

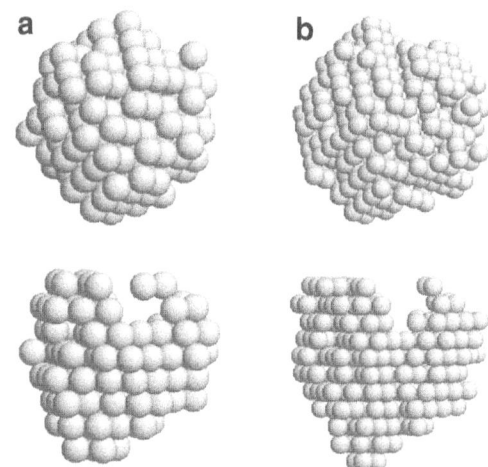

Fig. 9 Different views of GA EM models of *L. terrestris* Hb. The models correspond to bead diameters of **a** 4.0 nm and **b** 3.0 nm

has to be admitted that the GA models are ab initio models which do not exploit any existing information concerning symmetry, etc.

In spite of the shape deficiencies, the GA models can be utilized for hydrodynamic predictions. Obviously their equivalence in scattering with the Hb molecules is sufficient for the occurrence of hydrodynamic equivalence. In the past, we have introduced and successfully used $p(r)$ functions for testing multibody models with respect to their applicability for hydrodynamic predictions [5–7]. Whether the similarity of $p(r)$ functions of models equivalent in scattering is indeed a sufficient prerequisite for obtaining correct hydrodynamic predictions has to be tested in future investigations on a variety of molecules of different shape.

Acknowledgements We are grateful to Felix de Haas and Jean-Christophe Taveau for kindly providing the 3D reconstruction of *L. terrestris* Hb, to Fernando Díaz for making the program DALAI_GA2 available to us, and to J. García de la Torre for the program HYDRO.

References

1. Garcia de la Torre J, Huertas ML, Carrasco B (2000) Biophys J 78:719–730
2. Byron O (2000) Methods Enzymol 321:278–304
3. Durchschlag H, Zipper P, Wilfing R, Purr G (1991) J Appl Crystallogr 24:822–831
4. Durchschlag H, Zipper P (1996) J Mol Struct 383:223–229
5. Zipper P, Durchschlag H (1997) Prog Colloid Polym Sci 107:58–71
6. Zipper P, Durchschlag H (1998) Biochem Soc Trans 26:726–731
7. Zipper P, Durchschlag H (2000) J Appl Crystallogr 33:788–792
8. Vinogradov SN, Lugo SD, Mainwaring MG, Kapp OH, Crewe AV (1986) Proc Natl Acad Sci USA 83:8034–8038
9. Martin PD, Kuchumov AR, Green BN, Oliver RWA, Braswell EH, Wall JS, Vinogradov SN (1996) J Mol Biol 255:154–169
10. Lamy JN, Green BN, Toulmond A, Wall JS, Weber RE, Vinogradov SN (1996) Chem Rev 96:3113–3124
11. Zhu H, Ownby DW, Riggs CK, Nolasco NJ, Stoops JK, Riggs AF (1996) J Biol Chem 271:30007–30021
12. Schatz M, Orlova EV, Dube P, Jäger J, van Heel M (1995) J Struct Biol 114:28–40
13. Taveau JC, Boisset N, Vinogradov SN, Lamy JN (1999) J Mol Biol 289:1343–1359
14. Krebs A (1996) Thesis, University of Graz, Austria
15. Krebs A, Zipper P, Vinogradov SN (1996) Biochim Biophys Acta 1297:115–118

16. Krebs A, Lamy J, Vinogradov SN, Zipper P (1998) Biopolymers 45:289–298
17. Chacón P, Morán F, Díaz JF, Pantos E, Andreu JM (1998) Biophys J 74:2760–2775
18. Chacón P, Díaz JF, Morán F, Andreu JM (2000) J Mol Biol 299:1289–1302
19. Shlom JM, Vinogradov SN (1973) J Biol Chem 248:7904–7912
20. Glatter O (1982) In: Glatter O, Kratky O (eds) Small angle X-ray scattering. Academic, London, pp 119–165
21. García de la Torre J, Navarro S, López Martínez MC, Díaz FG, López Cascales JJ (1994) Biophys J 67:530–531
22. Carrasco B, García de la Torre J, Zipper P (1999) Eur Biophys J 28:510–515
23. Zipper P, Durchschlag H (1999) Prog Colloid Polym Sci 113:106–113
24. David MM, Daniel E (1974) J Mol Biol 87:89–101

Progr Colloid Polym Sci (2002) 119:149–158
© Springer-Verlag 2002

G. M. Pavlov

Evaluation of draining and volume effects in the interpretation of hydrodynamic data for linear macromolecules

G. M. Pavlov
Institute of Physics
St. Petersburg State University
Ulianovskaya str. 1, 198504
St. Petersburg, Russia

G. M. Pavlov
Laboratoire de Biophysique Moleculaire
Institut de Biologie Structurale
41 rue Yules Horowitz
38027, Grenoble, France

Abstract The equations describing the molecular-weight dependences of hydrodynamic values are compared for the cases of the excluded-volume effect and the draining effect. It is shown that they may be presented in a symmetric form. The equation obtained on the basis of the Gray–Bloomfield–Hearst theory of translational friction is applied to several polymer–solvent systems. The cases of a flexible chain with an excluded-volume effect and a semirigid chain with both excluded-volume and draining effects are considered. The intrinsic viscosity data are considered in a similar way.

Key words Molecular hydrodynamics · Excluded volume · Draining effects · Linear macromolecules

Introduction

Polymer investigations by the methods of molecular hydrodynamics are one of the bases of polymer science and molecular biophysics [1–5]. This information is essential for understanding the structure and conformation of isolated macromolecules. The methods of molecular hydrodynamics are being constantly improved, both experimentally and methodologically [6–11]. The fundamental basis is the understanding of the fact that their principal characteristics, such as translational diffusion, velocity sedimentation coefficients, and intrinsic viscosity depend differently on the molecular weight (MW), the size, and the shape of the macromolecules (particles).

However, there are problems concerning the basis of molecular hydrodynamics which have not been solved [12–15]. Some of these problems are of a general character caused by the absence of a general theory of a liquid and other theories concerning the polymer aspects.

For instance, it is known that molecular hydrodynamics is based on the Stokes equation: $f = 6\pi\eta_0 r$, obtained in the approximation of the solvent "sticking" (or bound) to the moving particle, where f is the translational friction coefficient, η_0 is the solvent viscos-ity, and r is sphere radius. If the solvent molecules "slip down" the surface of the moving sphere the resolution of the hydrodynamic equations leads to $f = 4\pi\eta_0 r$ [14]. The decrease in the coefficient 6π will evidently lead to a shift in the hydrodynamic parameters and, as a result, to a change in the hydrodynamic dimensions. This means that the same translational friction coefficient gives a hydrodynamic dimension 1.5 times higher or the volume of the particle more than 3 times higher. Moreover, the problem of the degree of "sticking" is related to that of determining the mass of the unit length of a linear polymer molecule in solution. This problem may be formulated as follows: should the molecular mass of the "sticking" solvent be included in the mass of the molecule in calculating M_L? In investigating polymers strongly interacting with the solvent (e.g., polyelectrolytes or cellulose in complex chelate solvents), it is probably necessary to take into account the mass of solvent bound to the polymer. The effect of preaveraging the Oseen hydrodynamic interaction tensor on hydrodynamic parameters (consequently, on the hydrodynamic chain size and also on the Flory–Mandelkern–Tsvetkov–Klenin hydrodynamic invariant) has often been discussed in the literature [16–18]. One of the unsolved problems is the separation of contributions of intramolecular draining and of excluded-volume effects on the chain size. Some of these problems are discussed here in considering the processing of experimental hydrodynamic data.

The methods of molecular hydrodynamics remain the basic methods in the study of properties of individual macromolecules. The main characteristics are the translational diffusion coefficient (D_0), the velocity sedimentation coefficient (s_0), the Gralen concentration coefficient (k_s), as well as the intrinsic viscosity ($[\eta]$). The molecular parameters of linear polymers obtained from the interpretation of experimental hydrodynamic values are the persistence length (a) [Kuhn segment length (A), $A = 2a$], the hydrodynamic chain diameter (d), and/or the parameter of thermodynamic polymer/solvent interaction (B or ε). To obtain these characteristics, it is necessary to know the mass per unit length of the polymer chain characterizing the linear density of the polymer substance ($M_L = M/L = M_0/\lambda$, where M_0 is the MW of the monomer unit, λ is the projection of the unit in the chain direction, $L = P\lambda$ is the contour length, and P is the degree of polymerization). Yamakawa and Fujii [19] called the value of M_L a shift parameter. It may also be called the structural parameter of the polymer chain. This parameter is commonly determined from the X-ray diffraction data or calculated using only the structural formula of the repeating unit. In rare cases of extremely stiff macromolecules [20, 21], this value can be estimated from the hydrodynamic data using the theories that describe the frictional behavior of the model of a weakly bent rod (cylinder) in the region of small relative contour lengths ($L/A < 2.3$, $L \gg d$) [22].

It is known that in recent years in addition to the persistence model a helical wormlike chain model has been developed [23, 24]. It is a hybrid of three simple models: a random coil, a rod, and a helix. This model is considered to be more universal for describing the hydrodynamic behavior of real macromolecules. However, it requires the introduction of some parameters which are difficult to determine and sometimes seem to be tentative (or approximate). Nevertheless, the analysis of experimental data shows that the use of only two parameters, the mass per unit length of the polymer chain (M_L) and the segment Kuhn length (A) can reduce the entire spectrum of hydrodynamic data for virtually the entire range of segment Kuhn lengths (e.g., the intrinsic viscosity data) to a dependence close to a certain curve (master curve) [25]. In this curve the values which have less influence on experimental values are the thermodynamic solvent quality (characterized by parameter ε, see later) and the hydrodynamic diameter (d) (determining chain draining). The set of these four parameters (A, M_L, d, and ε) probably describes more or less completely the hydrodynamic behavior of linear chain molecules.

Symmetrical form of excluded-volume equations and draining equations

The interpretation of the MW dependences of hydrodynamic characteristics requires the separation of contributions of intrachain draining and volume interactions to transport characteristics of macromolecules. Rigorous theories describing MW dependences of transport characteristics consider the case when volume interactions for models of a wormlike necklace or a wormlike cylinder are absent [19, 26–29]. Data on translational friction are considered when $L/A > 2.3$ in a system of coordinates

$$[s]N_A P_0 = M[D]P_0/k$$
$$= (M_L/A)^{1/2}M^{1/2} + (P_0 M_L/3\pi)[\ln A/d - \varphi(0)]$$
$$= f_1(M^{1/2}) \ , \qquad (1)$$

where N_A is Avogadro's number, k is Boltzmann's constant, $[s] \equiv s_0\eta_0(1-\upsilon\rho_0)^{-1}$, $[D] \equiv D_0\eta_0 T^{-1}$, η_0 is the solvent viscosity, $(1-\upsilon\rho_0)$ is the buoyancy factor of the polymer–solvent system, T the is absolute temperature, $P_0 = 5.11$ is Flory's hydrodynamic parameter, and $\varphi(0) = 1.431$ [25] (Hearst–Stockmayer) or $\varphi(0) = 1.056$ [26] Yamakawa–Fujii).

Viscometric data starting from results obtained for macroscopic Kuhn–Kuhn models are considered for relatively long chains [27–29] in a system of coordinates

$$M\Phi_0/[\eta] = (M_L/A)^{3/2}M^{1/2} + 0.89(M_L^2/A)[\ln A/d - \varphi(0)]$$
$$= f_2(M^{1/2}) \ . \qquad (2)$$

In this case, another Flory hydrodynamic parameter, Φ_0, in almost all theories with preaveraging of the Oseen tensor is $\Phi_0 = 2.87 \times 10^{23}$ [22] and $\varphi(0)$ in Ref. [26] is $\varphi(0) = 1.431$. In both cases the equilibrium chain rigidity is determined from the slope of the linear dependences and the hydrodynamic diameter is found from the intercept.

The procedures taking into account the influence of volume effects on transport characteristics were obtained on the basis of a linear approximation to the dependence of the swelling coefficient of flexible macromolecules (α) on the excluded-volume parameter (z) [30],

$$\alpha \equiv \langle h^2\rangle^{1/2}/\langle h^2\rangle_0^{1/2} = 1 + CZ \ ,$$

where $\langle h^2\rangle_0$ is the mean-square end-to-end distance of a Gaussian chain unperturbed by excluded-volume effects. For short chains the coefficient C depends on the reduced contour length L/A: $C = C(L/A)$ [31].

In the case of low excluded-volume effects, the data on translational friction are considered in a system of coordinates [32]

$$M^{1/2}/[s]N_A = k(M^{1/2}[D])^{-1} = K_{0f} + 0.2BM^{1/2}$$

$$= P_0[(A/M_L)^{1/2} + 0.2BM^{1/2}]$$

$$= f_3(M^{1/2}) \qquad (3)$$

(Cowie–Bywater plot) and the data on intrinsic viscosity in a system of coordinates [33, 34]

$$[\eta]/M^{1/2} = K_{0\eta} + 0.51BM^{1/2}$$

$$= \Phi_0[(A/M_L)^{3/2} + 0.51BM^{1/2}] = f_4(M^{1/2}) \qquad (4)$$

(Burchard–Stockmayer–Fixman plot). K_{0f} and $K_{0\eta}$ are the so-called macromolecular unperturbed dimension parameters, which are defined as

$$K_{0f} = P_0(\langle h^2 \rangle_0/M)^{1/2} = P_0(A/M_L)^{1/2} \qquad (5)$$

and

$$K_{0\eta} = \Phi_0(\langle h^2 \rangle_0/M)^{3/2} = \Phi_0(A/M_L)^{3/2} \quad . \qquad (6)$$

In these cases the slope determines the parameter of the polymer–solvent thermodynamic interaction (B) and the intercept determines the equilibrium chain rigidity (chain dimension unperturbed by thermodynamic interactions). These dependences are linear only in the range of moderate MW (low volume effects).

Both types of relationships (without or with volume interactions) can be represented in a symmetric form. Moreover, the relationships describing translational friction can, for instance, be extended at the expense of relationships describing intrinsic viscosity and vice versa. The transition can be carried out by using the equation

$$k/P_0[D] = M/[s]N_A P_0 = \langle h^2 \rangle_f^{1/2} \sim \langle h^2 \rangle_\eta^{1/2}$$

$$= ([\eta]M/\Phi_0)^{1/3} \qquad (7)$$

or

$$[s]P_0 N_A = (M^2\Phi_0/[\eta])^{1/3} \qquad (8)$$

following from the assumption that the size of macromolecules is equivalent (or similar) in the phenomena of translational ($\langle h^2 \rangle_f$) and rotational ($\langle h^2 \rangle_\eta$) frictions. In other words, a complete set of functional dependences can be formed. It describes the MW dependences of translational friction and viscosity for linear polymers.

$$[s]N_A P_0 = (M^2\Phi_0/[\eta])^{1/3} = f_1(M^{1/2}; A/d) \qquad (9)$$

$$M\Phi_0/[\eta] = [s]^3 P_0^3 N_A^3/M = f_2(M^{1/2}; A/d) \qquad (10)$$

The left-hand side of Eq. (9) corresponds to Eq. (1). Its right-hand side was obtained by the replacement of variables according to Eqs. (7) and (8) and was first developed in Ref. [35]. The left-hand side of Eq. (10) corresponds to Eq. (2). Its right-hand side was analogously obtained by replacing the variables. The following equations (Eqs. 11, 12) correspond to Eqs. (3) and (4)

and are presented in the form symmetric to Eqs. (9) and (10), respectively:

$$([s]N_A P_0)^{-1} M^{1/2} = (M^2\Phi_0/[\eta])^{-1/3} M^{1/2} = f_3(M^{1/2}; B) \ ,$$

$$\qquad (11)$$

$$(M\Phi_0/[\eta])^{-1} M^{1/2} = ([s]^3 P_0^3 N_A^3/M)^{-1} M^{1/2} = f_4(M^{1/2}; B)$$

$$\qquad (12)$$

The left-hand side of Eq. (11) corresponds to the Cowie–Bywater plot, and the left-hand side of Eq. (12) corresponds to the Burchard–Stockmayer–Fixman plot. This means that if one equation is known, it is possible to presume three additional dependencies on its basis.

Comparison of equilibrium chain rigidity values obtained by different equations

In some works the simultaneous influence of draining and thermodynamic swelling on the translational friction coefficient has been discussed [36–39]. Volume effects for a persistent necklace were taken into account for translational friction in Ref. [38] and those for intrinsic viscosity in Ref. [39]. The calculation was carried out by using the parameter ε characterizing the deviation of the statistical chain size from the Gaussian properties caused by volume effects. In this case for closely placed segment pairs when volume interactions have little effect, Porod statistics are applied. For distant segment pairs, volume effects are evaluated from the equation $\langle h^2 \rangle = N^{1-\varepsilon}A^2$, where ε is the parameter which characterizes the thermodynamic quality of the polymer–solvent system. As a result, the following equation may be obtained for translational friction coefficients at high $L/A > 2.3$ on the basis of Eq. (10) in Ref. [38]

$$[s]N_A P_0 = [3/(1-\varepsilon)(3-\varepsilon)]M_L^{(1+\varepsilon)/2}A^{-(1-\varepsilon)/2}M^{(1-\varepsilon)/2}$$

$$+ (M_L P_0/3\pi)[ln(A/d) - (A/3)(A/d)^{-1} - \varphi(\varepsilon)],$$

$$\qquad (13)$$

where $\varphi(\varepsilon)$ is the function tabulated in Ref. [38] and presented in Fig. 1. In a similar form Eq. (13) was first presented in Ref. [40].

The following Eq. (14) may be obtained for the nondraining limit from Eq. (13)

$$[s]N_A P_0 = [3/(1-\varepsilon)(3-\varepsilon)]M_L^{(1+\varepsilon)/2}A^{-(1-\varepsilon)/2}M^{(1-\varepsilon)/2} \ . \quad (14)$$

Equation (14) has the form of so-called equations of Kuhn–Mark–Houwink–Sakurada (KMHS):

$$s_0 = K_s M^{b_s} \ ,$$

$$D_0 = K_D M^{-b_D} \ ,$$

$$[\eta] = K_\eta M^{b_\eta} \ .$$

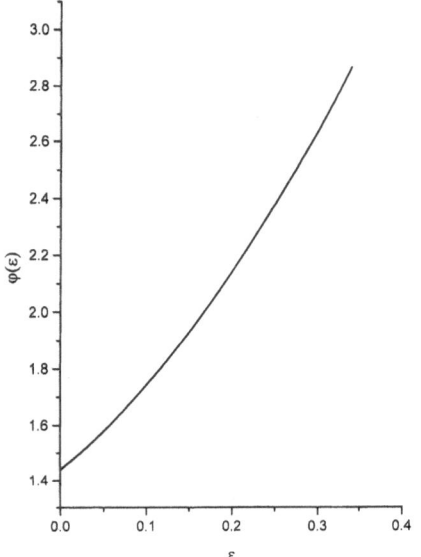

Fig. 1 Plot of the function $\varphi(\varepsilon)$ versus ε calculated according to Ref. [38]. This line may be represented as $\varphi(\varepsilon) = 1.439 + 2.545\varepsilon + 4.196\varepsilon^2$

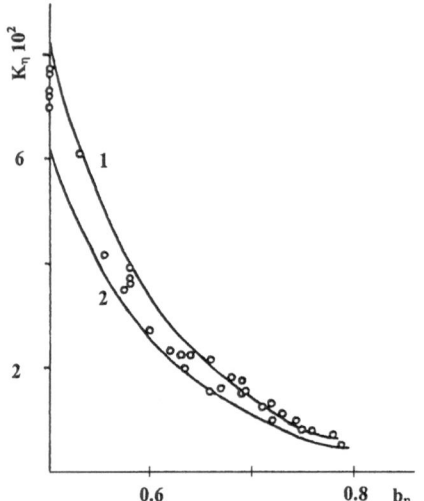

Fig. 2 Dependence of the Kuhn–Mark–Houwink–Sakurada parameter (K_η) on the exponent (b_η) for polystyrene. The curves obey Eq. (16) for $A = 18 \times 10^{-8}$ cm (*1*) and $A = 15 \times 10^{-8}$ cm (*2*). The *circles* represent experimental data from Ref. [41]

Since $(1-\varepsilon)/2 \equiv b_s$ [30], Eq. (14) is the basis of the corresponding scaling relationships connecting the MW and the velocity sedimentation coefficient. The relationship of K_s and b_s becomes evident from Eq. (14). For flexible-chain polymers at relatively high MW the following correlation $K_s = f(b_s)$ exists:

$$K_s \equiv (1 - v\rho_0)(N_A P_0 \eta_0)^{-1}[3/4b_s(1 + b_s)]M_L^{1-b_s}A^{-b_s} .$$

Viscometric data may also be interpreted on the basis of the theory of $[\eta]$ of a wormlike necklace taking into account the excluded-volume effects [39]. This theory allows us to obtain only the limiting expression in the range of high M:

$$[\eta] = \Phi(\varepsilon)A^{(3-3\varepsilon)/2}M_L^{-(3+3\varepsilon)/2}[1 + (5/6)\varepsilon + (1/6)\varepsilon^2]^{-3/2} \times M^{(1+3\varepsilon)/2} , \qquad (15)$$

where $\Phi(\varepsilon) = 2.87 \times 10^{23} (1-2.63\varepsilon + 2.87\varepsilon^2)$ [30].

Equation (15) also has the form of the KMHS equation. Since $\varepsilon = (2b_\eta - 1)$, Eq. (15) is also the basis of the corresponding scaling relationship (KMHS equation) at high M for flexible-chain polymers and describes the nondraining limit chain behavior. The relation $K_\eta = f(b_\eta)$ follows from Eq. (15):

$$K_\eta = \Phi_0(2.725 - 4.460\, b_\eta + 2.019\, b_\eta^2)A^{2-b_\eta}M_L^{-(1+b_\eta)} . \qquad (16)$$

The correlation between K_η and b_η for polystyrene molecules in different solvents is shown in Fig. 2; the original data are taken from Ref. [41].

The type of dependence on $[\eta]$ over the entire MW range can be obtained using Eqs. (8) and (13) as was first done in Ref. [42]. As a result we obtain

$$(M^2\Phi_0/[\eta])^{1/3}$$
$$= [3/(1 - \varepsilon)(3 - \varepsilon)]M_L^{(1+\varepsilon)/2}A^{-(1-\varepsilon)/2}M^{(1-\varepsilon)/2}$$
$$+ (M_L P_0/3\pi)[\ln(A/d) - (1/3)(A/d)^{-1} - \varphi(\varepsilon)] . \qquad (17)$$

We will first consider the data on flexible-chain polymers. Let us compare the values of equilibrium rigidity obtained from Cowie–Bywater and Burchard–Stockmayer–Fixman plots with those that may be obtained by using the Gray–Bloomfield–Hearst theory (Eq. 13) and by using Eq. (16). We will analyze as examples the data on translational friction for poly(methyl methacrylate) (PMMA) in acetone [43] (Fig. 3a), poly(1-vinyl-2-pyrrolidone) in 0.1 M sodium acetate [42] (Fig. 3b), and polyisobutylene in *n*-heptane [44] (Fig. 3c). As examples of viscometric data the results obtained for polystyrene in toluene [45] (Fig. 4a), poly(4-acetoxystyrene) in dioxane [46] (Fig. 4b), and PMMA in chloroform [47] (Fig. 4c) are used.

In these cases it is assumed that all the deviations of the b values from 0.5 are due to the excluded-volume effect and the parameter ε was calculated from $\varepsilon = (2b_\eta - 1)$ or $\varepsilon = (2|b_D|-1) = (1-2b_s)$ according to Ref. [30], where b_η, b_D, and b_s are the scaling indices in the corresponding KMHS plots.

Figures 3 and 4 illustrate the fact that the plots from Eqs. (3) and (4) are linear only in a limited MW range. The deviations from the linear dependence are observed at high and at very low MW; therefore, the choice of the MW range necessary for extrapolating the values to

Fig. 3 Cowie–Bywater plots (*1*) and the plots according to Eq. (13) of Gray–Bloomfied–Hearst theory (*2*) for **a** poly(methyl methacrylate) in acetone [43], for plot 2, $\varepsilon = 0.14$, $\varphi(\varepsilon) = 1.89$; **b** poly(1-vinyl-2-pyrrolidone) in 0.1 M sodium acetate [42], for plot 2, $\varepsilon = 0.16$, $\varphi(\varepsilon) = 1.97$; **c** polyisobutylene in n-heptane [44], for plot 2, $\varepsilon = 0.116$, $\varphi(\varepsilon) = 1.80$

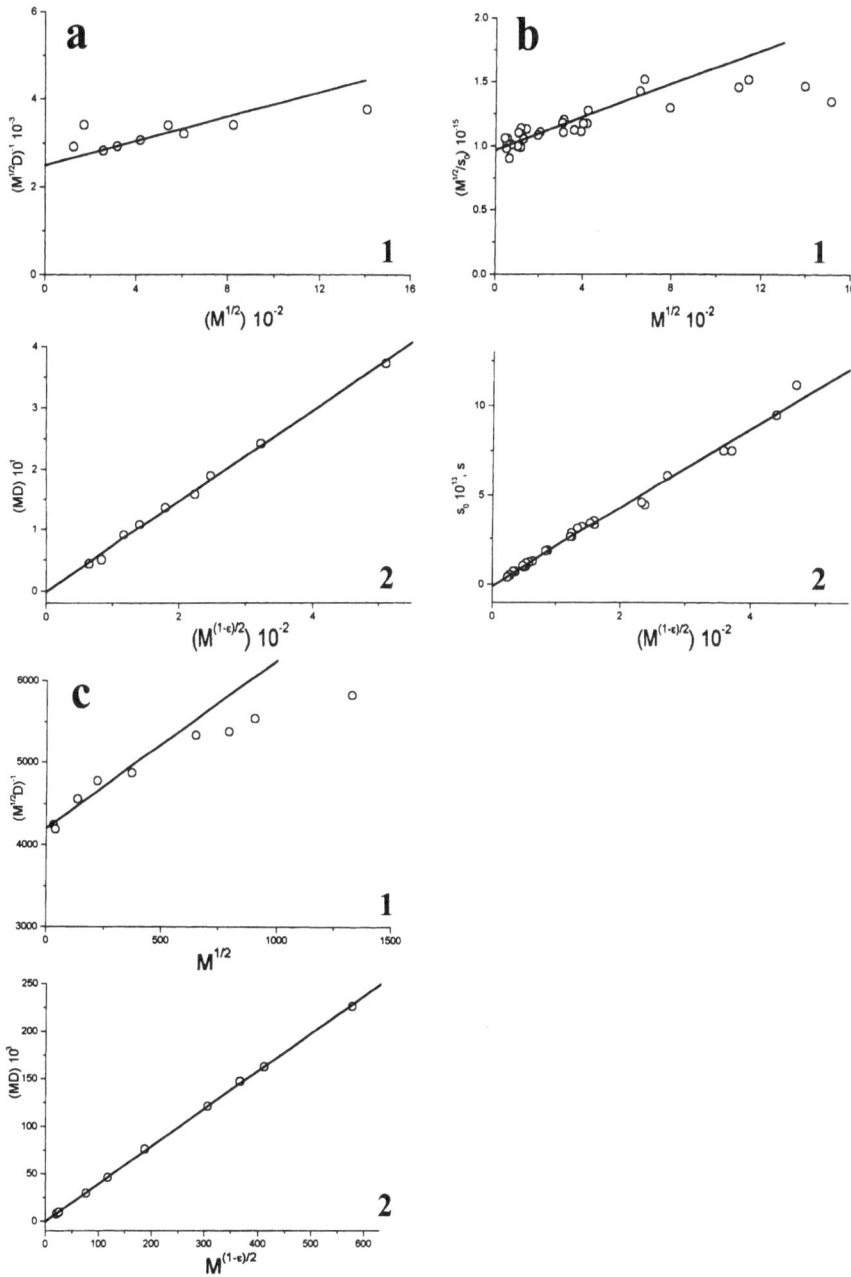

the value corresponding to $M \to 0$ is more or less arbitrary. Hence, the determination of unperturbed chain sizes is also arbitrary. However, the dependences obtained on the basis of Eqs. (13) and (16) by using the same experimental data are linear over the entire MW range with a limitingly high linear correlation coefficient ($r > 0.995$). As follows from Table 1 the values of the equilibrium rigidity obtained by the two methods are in good agreement. In this case the same values of the Flory hydrodynamic parameters were used: $P_0 = 5.11$ and $\Phi_0 = 2.87 \times 10^{23}$. The values of the hydrodynamic diameter are more uncertain in all cases. Moreover, it should be emphasized that it is necessary

to develop a complete theory describing the dependence of intrinsic viscosity over the entire MW range taking into account both the excluded-volume and the draining effects.

Semirigid polymer systems

The most complex case is presented by the systems of semi-rigid-chain molecules for which both the effects of intramolecular draining and those of excluded volume should be taken into account. Properly speaking, the Gray–Bloomfield–Hearst theory was intended for this

Fig. 4 Burchard-Stockmayer-Fixman plot (*1*) and the plot according to Eq. (17) for
a polystyrene in toluene [45], for plot 2, $\varepsilon = 0.19$, $\varphi(\varepsilon) = 2.10$;
b poly(4-acetoxystyrene) in dioxane [46], for plot 2, $\varepsilon = 0.10$, $\varphi(\varepsilon) = 1.74$;
c poly(methyl methacrylate) in chloroform [47], for plot 2, $\varepsilon = 0.173$, $\varphi(\varepsilon) = 2.03$

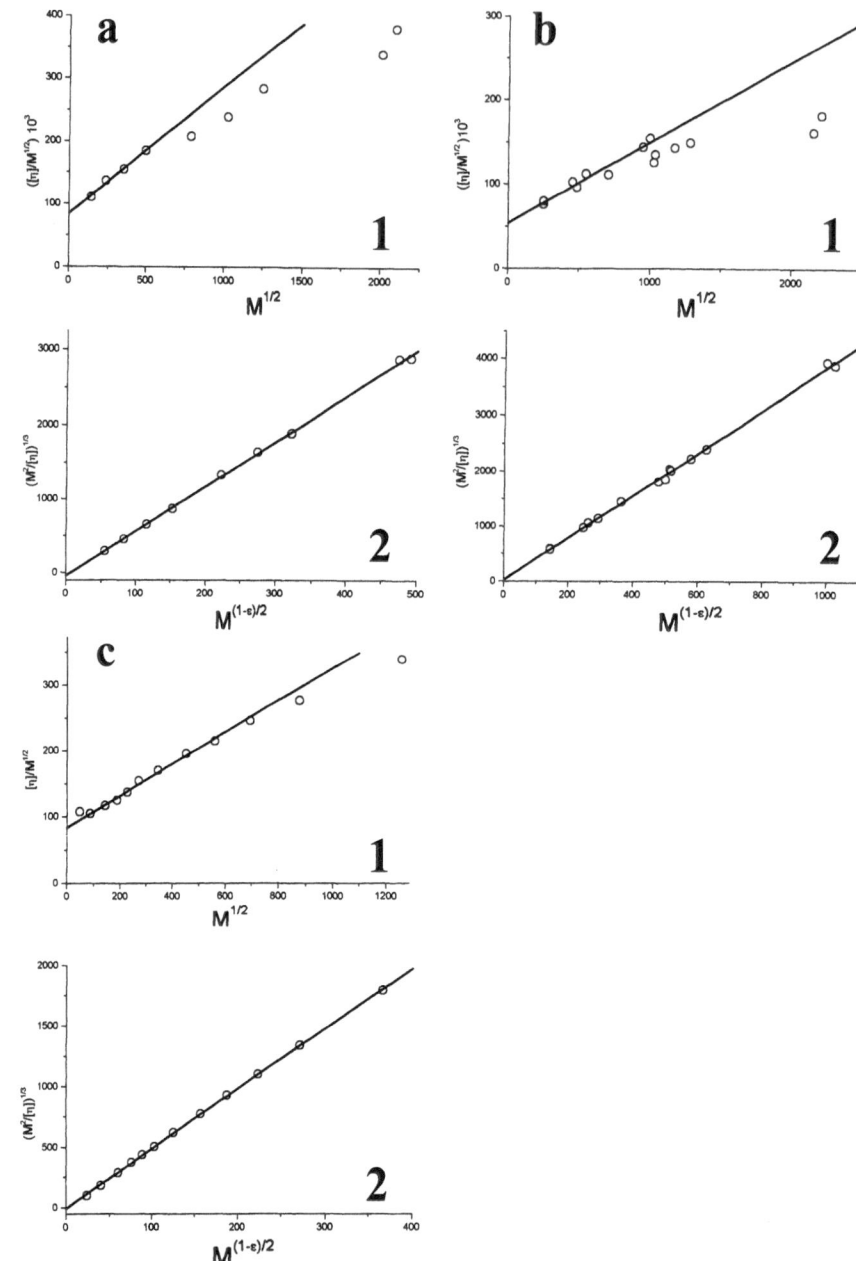

Table 1 Comparison of the Kuhn segment length estimations from the Cowie–Bywater and Burchard–Stockmayer–Fixman plots and from the equation of Gray–Bloomfield–Hearst theory

Methods and systems	$(A \pm \Delta A) \times 10^{-8}$ cm	r	$(A \pm \Delta A) \times 10^{-8}$ cm	$(d \pm \Delta d) \times 10^{-8}$ cm	r	Refs.
Translational friction	From Eq. (3)		From Eq. (13)			
Poly(methyl methacrylate)/acetone	16.7 ± 2.4	0.8964	16.1 ± 0.8	4 ± 2.7	0.9983	[43]
Poly(1-vinyl-2-pyrrolidone)/0.1 M sodium acetate	25.0 ± 1.1	0.8927	26.8 ± 1.2	7.5 ± 3.5	0.9950	[42]
Polyisobutylene/*n*-heptane	16.5 ± 0.7	0.9461	16.2 ± 0.3	3.0 ± 0.9	0.9999	[44]
Intrinsic viscosity	From Eq. (4)		From Eq. (17)			
Polystyrene/toluene	18.3 ± 1.4	0.9958	15.2 ± 0.4	7 ± 3	0.9996	[45]
Poly(4-acetoxystyrene)/dioxane	26 ± 5	0.9799	32.9 ± 0.9	4 ± 4	0.9991	[46]
Poly(methyl methacrylate)/chloroform	17.6 ± 0.9	0.9961	19.1 ± 0.2	3.2 ± 0.4	0.9999	[47]

case. Nevertheless, in the framework of this theory the problem of the separation of contributions to chain size provided by intramolecular draining effects from those due to excluded-volume effects is not solved. The possibility of using Eqs. (13) and (16) for the determination of the persistence length (Kuhn segment length) was illustrated for the case of flexible-chain molecules. In this case it may be assumed with considerable certainty that excluded-volume effects predominate. The intermediate cases will now be considered, and an attempt to take into account both contributions (draining and excluded-volume effects) will be made. The key problem in application of Eqs. (13) and (16) is the evaluation of the parameter ε characterizing volume chain swelling, where $\langle h^2 \rangle : M^{1+\varepsilon}$. In the case of flexible polymer chains it is assumed that the deviation of b_η is due solely to the excluded-volume effect. Here b_η is the scaling index in the relationship $[\eta] : M^{b_\eta}$. ε was calculated, for instance, from $\varepsilon = (2/3)\Delta b_\eta$; $\Delta b_\eta \equiv b_\eta - 0.5$. In the general case $\Delta b_\eta = \Delta b_{v\eta} + \Delta b_{d\eta}$, where $\Delta b_{v\eta}$ is the contribution of volume interactions to the value of scaling index and $\Delta b_{d\eta}$ $\Delta_{d\eta}$ is the contribution of the effects of intramolecular draining. The problem is to separate the deviation of the scaling indexes into two components, i.e., to estimate the value of ε in the case of partial draining when $\varepsilon = (2/3)\Delta b_{v\eta}$ and $\Delta b_{v\eta} \leq \Delta b_\eta$. The application of the Gray–Bloomfield–Hearst theory for such polymer systems by using both the data on translational friction and viscometric data is illustrated. We now discuss the results of hydrodynamic studies of one sulphatized aromatic polyamide based on isophthalic acid (PASAI). The polymer has the structural formula of the repeating unit represented in Fig. 5a. The experiments were carried out in an aqueous 0.1 M NaCl solution; the solvent density was 0.999 g/cm^3, the viscosity, $\eta_0 = 0.897$ mPas, and the buoyancy factor $(1 - v\rho_0) = 0.448$ at 25 °C. The conditions

of the polymer synthesis and the hydrodynamic studies have been described previously [48, 49]. In pure water, the polymer exhibits polyelectrolyte behavior [48], which can be inhibited to some extent by introducing a low-MW salt into the solution as well. In 0.1 M NaCl solvent, the primary polyelectrolyte effects do not perturb the molecular transport in the course of the hydrodynamic experiment, and the polyelectolyte nature of the polymer is manifested only as variation of short-range and long-range interactions in PASAI chains. In fact the following KMHS relationships were obtained: $[\eta] = 1.37 \times 10^{-3} \, M^{0.98 \, \pm \, 0.035}$, $r = 0.9962$; $D_0 = 2.87 \times 10^{-4} \, M^{-(0.61 \, \pm \, 0.024)}$, $r = -0.9955$; $s_0 = 5.88 \times 10^{-15} \, M^{0.39 \, \pm \, 0.024}$, $r = 0.9889$.

The parameters were calculated for the MW range $10 < M \times 10^{-3} < 145$.

Rotation in PASAI chains is mostly due to the presence of phenyl rings in the meta position. For a chain with completely free rotation about bonds (about which rotation is possible), the Kuhn segment length A_f can be calculated by methods of statistical mechanics. This approach has been successfully applied in studies of stiff-chain polymers of different types [22, 50, 51]. The repeating unit of PASAI and the corresponding system of virtual bonds are shown in Fig. 5b. To calculate A_f for a chain with a given structure of the repeating unit, it is possible to apply the Benoit result for a cellulose chain [52] with the same system of virtual bonds in the repeating unit:

$$A_f \lambda = \delta^2 + \Delta^2 (1 + \cos \alpha)/(1 - \cos \alpha) \ ,$$

$$\lambda = \delta \sin(\alpha/2) + \Delta \cos(\alpha/2) \ .$$

In the calculation, all bond lengths were to be $l = 1.4 \times 10^{-8}$ cm and all bond angles to be 120°. We obtained the following values: $\Delta = 16.8 \times 10^{-8}$ cm, $\delta = 1.1 \times 10^{-8}$ cm, $\alpha = 60°$, $\lambda = 15.8 \times 10^{-8}$ cm, and $A_f = 53.8 \times 10^{-8}$ cm. Here, λ is the projection of the repeating unit to the chain direction and A_f is the Kuhn segment length in the assumption of free rotation. The value obtained for A_f is the lower limit for the equilibrium stiffness of a neutral chain and an estimate for the structural contribution to the equilibrium stiffness of PASAI chain. In the polyelectrolyte theory [53, 54] the Debye–Huckel screening radius k^{-1} is used as a characteristic length unit. This is the radius of action of an electric field of a single charge placed in a medium containing other charges:

$$k^{-1} = (8\pi \lambda_B \mu)^{-1} \ ,$$

where $\lambda_B = (e^2/4\pi\varepsilon_0 \varepsilon k T)$ is the Bjerrum radius which characterizes the screening action of the solvent; e is the elementary charge, ε_0 is the absolute dielectric constant, ε is the dielectric constant of the medium, $\mu = (1/2)\Sigma c_i z_i^2$ is the ionic strength of the solution, cn_i is the molar

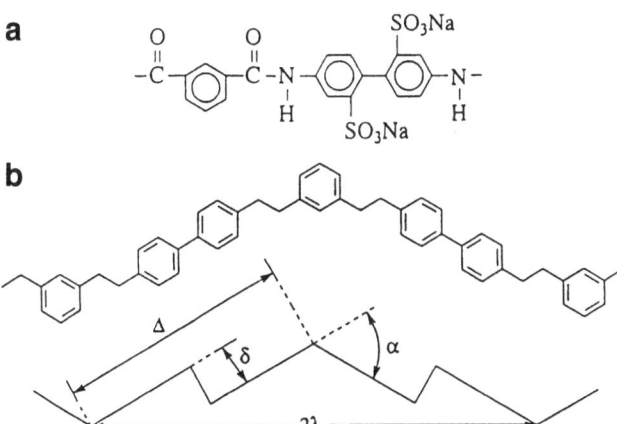

Fig. 5 a Structural formula of the repeating unit of sulfatized aromatic polyamide based on isophthalic acid and **b** the corresponding two repeating units and the corresponding system of virtual bonds. The structure in **b** is shown without the substituent groups

concentration of any mobile ion, and Z_i is the charge of the ith ion in e units.

The charges separated by the distance $l < k^{-1}$ along the chain are responsible for additional electrostatic short-range interaction, which results in increasing chain stiffness. The electrostatic contribution to the persistence length can be calculated using the Manning concept of counterion condensation [55] according to the formula

$$a_e = k^{-2}/4\lambda_B \ .$$

We accept the concept of stiffness additivity [56], according to which the total chain stiffness is the sum of the structural (A_{str}) and electrostatic (A_e) components:

$$A = A_{str} + A_e \ .$$

(Note that the assumption of stiffness additivity is probably valid only for polyelectrolytes containing charges of the same sign in the chain. This assumption is not valid for polyampholytes, in which both electrostatic repulsion and electrostatic attraction are possible. Note also that the stiffness additivity concept for polyelectrolyte chains containing charges of the same sign is opposite to the flexibility additivity concept [22] applied to neutral polymers, where the stiffness is governed by rotation freedom in chains.)

In our case we have $\lambda_B = 7.1 \times 10^{-8}$ cm, $k^{-1} = 9.6 \times 10^{-8}$ cm, and $A_e = 2a_e = 6.5 \times 10^{-8}$ cm. Charges separated by the distance $l > k^{-1}$ along the chain are responsible for the long-range electrostatic interaction. In our case, the contour length of a molecule is $L > k^{-1}$; therefore, it is necessary to consider also the electrostatic volume effects on the chain size.

An analysis shows that the lower limit of the equilibrium stiffness of PASAI chains allows us to classify them as semirigid molecules, i.e., the molecules that are usually capable of intramolecular draining. Comparison of the contour length of macromolecules with the Debye–Huckel screening radius also shows that electrostatic swelling must be considered in addition to draining effects. Because in this case both effects leading to deviation of the scaling indices from 0.5 are possible, let us vary the value of ε and for each ε value use Eq. (13). The lower limit ($\varepsilon_{min} = 0$) corresponds to the case when the volume effects are absent; the upper limit ($\varepsilon_{max} = 0.32$) corresponds to the absence of draining. Variation of ε in this range and use of Eq. (13) plotted in Fig. 6 gives us different values of A and d.

To interpret the hydrodynamic data, one needs to know the mass per unit length of a polymer molecule, M_L. One might reasonably suppose that polyelectrolyte molecules placed in solvent strongly bind some solvent molecules [1, 57] by strong electrostatic interaction. This binding of solvent molecules can be due to hydration of ions [58] and microclusterization of water by amide groups [59]. In this case, we calculate M_L using the relation

$$M_L^* = (M_0 + nH_2O)/\lambda \ ,$$

where M_0 is the molecular mass of a "dry" polymer chain and n is the number of water molecules strongly bound to a repeating unit of the macromolecule. On the basis of data taken from the literature [57–59], we accept $n = 12$. Thus, we obtain $M_L^* = 4.65 \times 10^9$ cm^{-1} (at $n = 0$ we have $M_L = 3.3 \times 10^9$ cm^{-1}). The value of M_L^* was used in the estimation of the equilibrium stiffness of a polymer chain. We must take into account that, first, A cannot be smaller than the analytical value $A_f = 60.3 \times 10^{-8}$ cm and, second, the estimates of the hydrodynamic diameter must make sense. The resulting dependences of A and d on ε are presented in Fig. 7. One can see that ε can vary in the range $0 \leq \varepsilon \leq 0.19$; therefore, $\Delta b_{v\eta} \leq 0.29$ and $\Delta b_{d\eta} \leq 0.19$. The contribution of volume effects to the deviation of scaling indices from 0.5 accounts for more than half of the total deviation.

A similar analysis can be performed on the basis of viscometric data. Let us consider the data on poly(n-alkyl

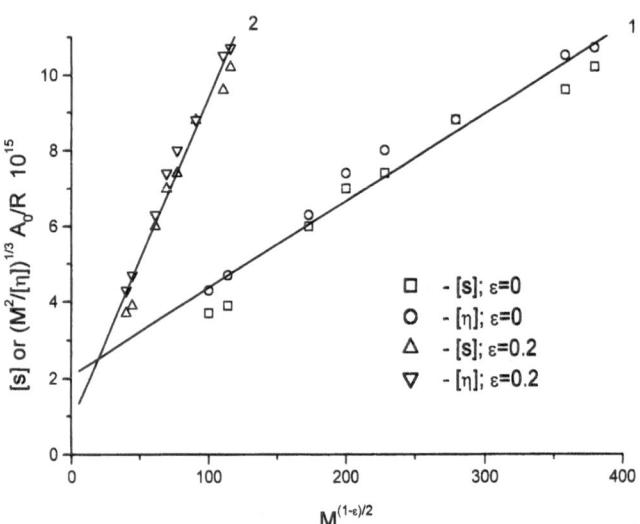

Fig. 6 Plots of $[s]$ and $(M^2/[\eta])^{1/3} A_0/R$ versus $M^{(1-\varepsilon)/2}$ M at $\varepsilon = 0$ (*1*) and 0.2 (*2*) for sulphatized aromatic polyamide based on isophthalic acid (*PASAI*) molecules. The plot corresponding to the intrinsic viscosity data was obtained using the theoretical value of the hydrodynamic invariant $A_0 = kP_0^{\iota} \Phi_0^{1/3} = 17.81 \times 10^{-10}$

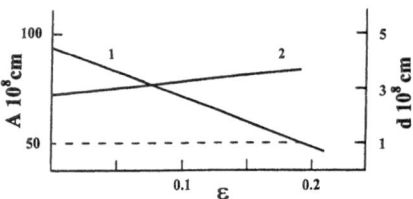

Fig. 7 Dependence of the Kuhn segment length (A) (*1*) and the hydrodynamic diameter (d) (*2*) of PASAI molecules on ε obtained from the plots in Fig. 11. The *broken line* corresponds to $A_{theor} = A_{str} + A_e$

methacrylates) [60]. It is known [22, 61, 62], that both the equilibrium rigidity of the main chains of poly(n-alkyl (methacrylates) and their transverse dimensions depend on side-chain length. Both of them increase when the side-chain length increases. The equilibrium rigidity values for these polymers (when the number of carbon atoms is greater than 10) are several times higher than those for flexible chains [2, 61]. Therefore, in this case partial draining is probable and must be taken into account. The Burchard–Stockmayer–Fixman plot in the modified coordinates is presented in Fig. 8. In this case the intercepts represent the chain rigidity. It is interesting in this case also to estimate the diameter of the chains. The plots of the data obtained in Ref. [60] for poly(tridecyl methacrylate) (C13) and poly(docosyl methacrylate) (C22) corresponding to Eq. (15) with the limiting value of ε are shown in Fig. 9. The dependences of the A and d values on ε are presented in Fig. 10 and reflect the reasonable tendencies in the variations of both values. The inequalities $A_{22} > A_{13}$ and $d_{22} > d_{13}$ are in the range of the ε values.

Conclusion

The initial Gray–Bloomfield–Hearst theory of translational friction describes the MW dependences of the sedimentation (translational diffusion) coefficient of flexible macromolecules in thermodynamically good solvents over a broad MW range from the oligomeric region ($M = 10^3$) to a very high MW ($M = 10^7$).

The estimation of equilibrium rigidity following the Gray–Bloomfield–Hearst theory virtually coincides with that obtained from a Cowie–Bywater plot. In the former

case the ranges of the volume effects and the MW are limited. In the intermediate cases of the coexistence of excluded-volume and draining effects, the variation of ε

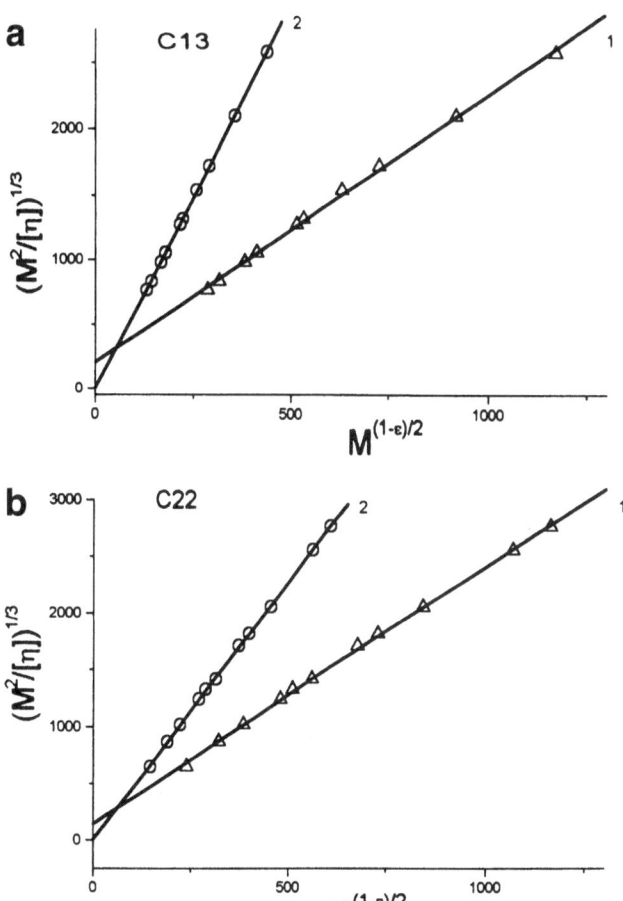

Fig. 9 Plot according to Eq. (17) for **a** poly(tridecyl methacrylate) at $\varepsilon = 0$ (*1*) and $\varepsilon = 0.14$ (*2*) and **b** poly(docosyl methacrylate) at $\varepsilon = 0$ (*1*) and $\varepsilon = 0.092$ (*2*)

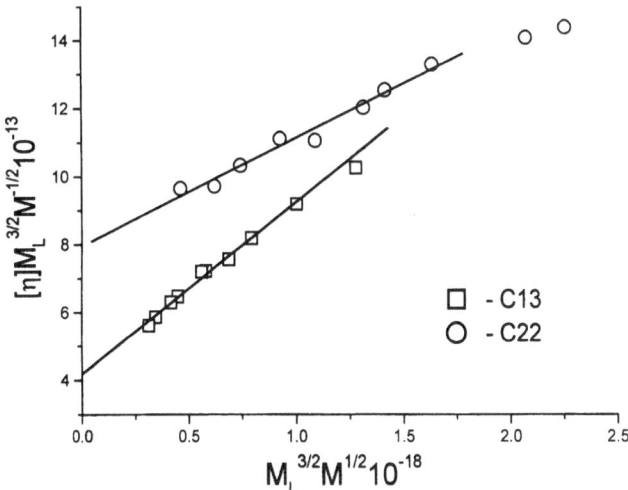

Fig. 8 Dependences of $[\eta]M_L^{3/2}M^{-1/2}$ on $M_L^{3/2}M^{-1/2}$ corresponding to a Burchard–Stockmayer–Fixman plot for poly(tridecyl methacrylate) (C13) and poly(docosyl methacrylate) (C22) fractions in tetrahydrofuran [60]

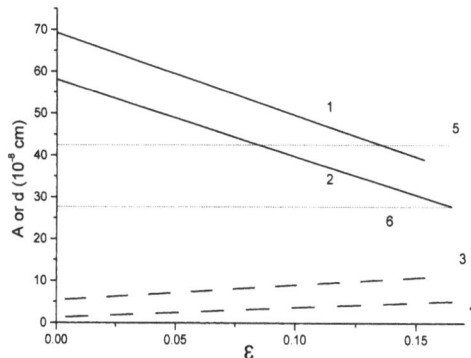

Fig. 10 Dependence A (*1, 2*) and d (*3, 4*) on ε obtained from plots in Fig. 9. The broken line (*5, 6*) corresponds to the estimations of A from the Burchard–Stockmayer–Fixman plot (Fig. 8). Lines *1, 3,* and *5* correspond to poly(docosyl methacrylate) data and *2, 4,* and *6* correspond to poly(tridecyl methacrylate) data [60]

gives us the possibility to find the limits of the equilibrium rigidity variation. In some cases the equilibrium rigidity value can be made more precise by taking into account the "reasonability" of the hydrodynamic diameter and/or the theoretical estimation of the A. The intrinsic viscosity data may be treated in a system of coordinates $(M^2\Phi_0/[\eta])^{1/3} = f_2(M^{(1-\varepsilon)/2})$ in a similar way. The A values obtained for flexible macromolecules virtually coincide with those obtained from Burchard–Stockmayer–Fixman plots. In the case of the intrinsic viscosity, the estimations are more or less qualitative because the theoretical numerical coefficients are not obtained reliably for the rotational movement of macromolecules.

References

1. Tanford C (1963) Physical chemistry of macromolecules. Wiley, New York
2. Tsvetkov V, Eskin V, Frenkel S (1970) Structure of macromolecules in solution. Butterworths, London
3. Cantor C, Schimmel P (1980) Biophysical chemistry. Freeman, San Francisco
4. Fujita H (1990) Polymer solutions. Elsevier, Amsterdam
5. Munk P (1989) Introduction to macromolecular Science. Wiley, New York
6. Harding S, Horton J, Rowe A (eds) (1992) Analytical ultracentrifugation in biochemistry and polymer science. Royal Society of Chemistry, Cambridge
7. Schuster T and Laue T (eds) (1994) Modern analytical ultracentrifugation. Birkhauser, Boston
8. Y Behlke (ed.) (1995) Analytical ultracentrifugation III. Prog Colloid Polym Sci 99
9. S. Harding and O. Byron (eds) (1997) Eur Biophys J 25
10. R. Yaenicke and H. Durchschlag (eds) (1997) Analytical Ultracentrifugation IV. Prog Colloid Polym Sci 107
11. H Coelfen (ed.) (1999) Analytical ultracentrifugation V. Prog Colloid Polym Sci 113
12. Wang S, Douglas J, Freed K (1985) Macromolecules 18:2464
13. Fujita H (1988) Macromolecules 21:179
14. Douglas J, Freed K (1994) Macromolecules 27:6088
15. Freed K, Douglas J (1995) Macromolecules 28:8460
16. Zimm B (1980) Macromolecules 13:592
17. Oono Y (1985) Adv Chem Phys 61:301
18. Garcia Bernal J, Tirado M, Freire J, Garcia de la Torre J (1991) Macromolecules 24:593
19. Yamakawa H, Fujii M (1974) Macromolecules 7:128
20. Yanaki T, Norisuye T, Fujita H (1980) Macromolecules 13:1462
21. Murakami H, Norisuye T, Fujita H (1980) Macromolecules13:345
22. Tsvetkov V (1989) Rigid-chain polymers. Consultants Bureau, New York
23. Yamakawa H, Fujii M (1976) J Chem Phys 64:5222
24. Yamakawa H (1997) Helical wormlike chains in polymer solutions Springer, Berlin Heidelberg New York
25. Pavlov G, Harding S, Rowe A (1999) Prog Colloid Polym Sci 113:76
26. Hearst J, Stockmayer W (1962) J Chem Phys 37:1425
27. Yamakawa H, Fujii M (1973) Macromolecules 6:407
28. Hearst J, Tagami Y (1965) J Chem Phys 42:4149
29. Yamakawa H, Yoshizaki T (1980) Macromolecules 13:633
30. Ptitsyn O, Eizner Yu (1959) Zh Tekh Fiz 29:1105
31. Yamakawa H, Stockmayer W (1972) J Chem Phys 57:2843
32. Cowie J, Bywater S (1969) Polymer 6:197
33. Burchard W (1961) Makromol Chem 50:20
34. Stockmayer W, Fixman M (1963) J Polym Sci C 1:137
35. Bushin S, Tsvetkov V, Lysenko L, Emelianov V (1981) Vysokomol Soedin Ser A 23:2494
36. Ptitsyn O, Eizner Yu (1958) Zh Fiz Khim 32:2464
37. Tschoegl N (1964) J Chem Phys 40:473
38. Gray G, Bloomfield V, Hearst J (1967) J Chem Phys 46:1493
39. Sharp P, Bloomfield V (1968) J Chem Phys 48:2149
40. Bushin S, Astapenko E (1986) Vysokomol Soedin Ser A 28:1499
41. Immergut I, Brandrup J (eds) (1989) Polymer handbook. Wiley-Interscience, New York
42. Pavlov G, Panarin E, Korneeva E, Kurochkin K, Baikov V, Uchakova V (1990) Makromol Chem 191:2889
43. ter Meer H, Burchard W, Wunderlich W (1980) Colloid Polym Sci 258:675
44. Osa M, Abe F, Yoshizaki T, Einaga Y, Yamakawa H (1996) Macromolecules 29:2302
45. Berry G (1967) J Chem Phys 46:1338
46. Arichi S, Sakamoto N, Himuro S, Miki M, Yoshida M (1985) Polymer 26:1175
47. Abe F, Einaga Y, Yamakawa H (1994) Macromolecules 27:725
48. Pavlov G, Korneeva E, Fedotov Yu, Polushina G, Andreeva I (1996) J Appl Chem 69:824
49. Pavlov G, Korneeva E, Fedotov Yu (1997) Vysokomol Soedin 39:1979
50. Tsvetkov V, Korshak V, Shtennikova I, Raubach H, Krongauz E, Pavlov G, Kolbina G (1979) Macromolecules 12:645
51. Pavlov G, Kozlov A, Yakopson S, Usova S, Efros L (1985) Vysokomol Soedin 27:30
52. Benoit H (1948) J Polym Sci 3:376
53. Dautzenberg H, Jaeger W, Kotz J, Phillip B, Seidel C, Stscherbina D (1994) Polyelectrolytes. Hanser, Munich
54. Barrat J, Joanny J (1996) Adv Chem Physics 94:1
55. Manning G (1969) J Chem Phys 51:924
56. Odijk T, Houwaert A (1978) J Polym Sci Part B Polym Phys 16:627
57. Rowland S (ed) (1980) Water in polymer. ACS symposium series 127. American Chemical Society, Washington, DC
58. Gordon J (1975) The organic chemistry of electrolyte solutions. Wiley, New York
59. Le Huy M, Rault J (1994) Polymer 35:136
60. Xu Z, Hadjichristidis N, Fetters L (1984) Macromolecules 17:2303
61. Andreeva L, Gorbunov A, DidenkoC, Korneeva E, Lavrenko P, Plate N, Shibaev V (1973) Vysokomol Soedin Ser B 15:209
62. Tsvetkov V, Andreeva L (1981) Adv Polym Sci 39:96

Progr Colloid Polym Sci (2002) 119:159
© Springer-Verlag 2002

Progr Colloid Polym Sci (2002) 119:160
© Springer-Verlag 2002

KEY WORD INDEX